Energy Consumption, Conversion, Storage, and Efficiency

Edited by Jiajun Xu and Bao Yang

Published in London, United Kingdom

Energy Consumption, Conversion, Storage, and Efficiency
http://dx.doi.org/10.5772/intechopen.100772
Edited by Jiajun Xu and Bao Yang

Contributors
Alberto Barragan-Garcia, Alberto Vidal-Sanchez, Ali Al-Qahtani, Andile Mkhohlakali, Bao Yang, Christophe Coquelet, Debjyoti Banerjee, Efren Diez-Jimenez, Faridreza Attarzadeh, Iuliana Matei, Jiajun Xu, Jiangtao Wu, Jian Lu, Jiyuan Tu, Jun Qiu, Lebohang Hlatshwayo, Mengqi Wu, Miguel Fernandez-Muñoz, Nan Gui, Nian Tang, Nonhlahla Ramashala, Patricia Scholczova, Sanele Nyembe, Shengyao Jiang, Sivuyisiwe Mapukata, Sunil Kumar, Takashi Hibiki, Tanjee Afreen, Thuy Tien Phan, Van-Phu Vu, Xibo He, Xingtuan Yang, Xu Liu, Zeeshan Farooq, Zhiqiang Yang

First published in London, United Kingdom, 2024 by IntechOpen
IntechOpen is the global imprint of INTECHOPEN LIMITED, registered in England and Wales, registration number: 11086078, 167-169 Great Portland Street, London, W1W 5PF, United Kingdom

British Library Cataloguing-in-Publication Data
A catalogue record for this book is available from the British Library

Additional hard and PDF copies can be obtained from orders@intechopen.com

Energy Consumption, Conversion, Storage, and Efficiency
Edited by Jiajun Xu and Bao Yang
p. cm.
Print ISBN 978-1-80355-744-1
Online ISBN 978-1-80355-745-8
eBook (PDF) ISBN 978-1-80355-746-5

Meet the editors

Dr. Jiajun Xu is a tenured professor in the Department of Mechanical Engineering, University of the District of Columbia (UDC), and a licensed Professional Engineer (PE) in Maryland, USA. He is the founding director of the NASA-funded Center for Advanced Manufacturing (CAM-STAR), DoD-funded Advanced Manufacturing Laboratory, and NSF-funded Multiscale Thermal Transport and Energy Conversion Laboratory. He joined UDC in 2013 as an assistant professor and received his tenure and promotion to associate professor in 2018. He is the recipient of the BEYA STEM Modern-day Technology Leader Award, the Excellence in Research Award, the Excellence in Student Mentoring Award, and the ONR summer faculty research fellowship. His research has been sponsored by the National Aeronautics and Space Administration, the National Science Foundation, the Department of Defense, the Department of Energy, and the US Department of Agriculture.

Dr. Bao Yang is a professor at the Department of Mechanical Engineering, University of Maryland, College Park, USA. He received a Ph.D. in Physics from the University of Science and Technology of China in 1998, and another Ph.D. in Mechanical Engineering from the University of California Los Angeles, USA in 2003. He is a recognized expert in advanced materials for thermal management, heat transfer, and energy conversion systems. He served as editor-in-chief for the *International Journal of Mechanical Engineering and Robotics Research*, associate editor for *Heat Transfer Engineering*, and an editorial board member for *Scientific Reports* and *Journal of Applied Mechanical Engineering*. He is a member of the American Society of Mechanical Engineers, Materials Research Society, American Physical Society, and International Thermoelectrics Society.

Contents

Preface

Welcome to *Energy Consumption, Conversion, Storage, and Efficiency*. In the contemporary landscape of global challenges, energy plays a pivotal role in shaping the future of our world. As we stand at the intersection of technological innovation, environmental stewardship, and economic development, understanding the intricacies of energy dynamics is more critical than ever.

This book serves as a comprehensive exploration into the multifaceted realm of energy, covering key aspects such as consumption patterns, conversion technologies, storage solutions, and the relentless pursuit of efficiency. Our journey through these topics is designed to provide readers with a holistic perspective on the complex and interconnected issues surrounding energy in our modern society.

In the pages that follow, we embark on a voyage that navigates the historical foundations of energy use, delves into the cutting-edge technologies propelling us forward, and scrutinizes the challenges and opportunities that lie on the horizon. The interdisciplinary nature of energy necessitates a nuanced approach, and this book endeavors to bridge the gap between scientific intricacies and accessible knowledge, making it suitable for students, researchers, professionals, and curious minds alike.

From the fundamentals of energy consumption patterns to the emerging frontiers of energy storage and the perpetual quest for efficiency gains, each chapter unfolds a piece of the overarching narrative that defines our energy landscape. We hope that this book will serve as a valuable resource, inspiring readers to contemplate the implications of their energy choices and encouraging a collective commitment to sustainable and responsible energy practices.

As we confront the complexities of the energy paradigm, let us embark on this intellectual journey together, armed with knowledge and a shared commitment to creating a future where energy is harnessed wisely, conserved judiciously, and employed efficiently for the benefit of all.

Jiajun Xu
Department of Mechanical Engineering,
Center for Advanced Manufacturing,
University of the District of Columbia,
Washington DC, DC, USA

Bao Yang
Department of Mechanical Engineering,
University of Maryland,
College Park, MD, USA

Section 1

Introduction

Chapter 1

Introductory Chapter: Energy Consumption, Conversion, Storage, and Efficiency

Bao Yang and Jiajun Xu

1. Introduction

Energy has emerged as the unspoken life force that sustains our world in the modern era. From powering our homes, propelling our vehicles, and fueling the engines of industry, energy underpins nearly every aspect of our lives. Energy influences our economies, geopolitics, and environment, and the importance of energy cannot be overstated. The global growth of energy demand is accompanied by rising prosperity and better life quality. Energy consumption, storage, conversion, and efficiency are interconnected components of the world energy system, each playing an important role in shaping our energy landscape. This chapter presents an introductory review of energy consumption, storage, conversion, and efficiency, inviting us on a journey into the intricate interplay of energy within our society.

2. Energy consumption

Energy consumption encompasses the utilization of energy in various forms for meeting human needs and fulfilling societal demands. This includes the utilization of electricity, natural gas, coal, oil, biomass and renewable solar, and wind, hydro, and geothermal energy as well. Energy consumption is directly linked to the prosperity of the world and is at the heart of our modern civilization. However, energy consumption is closely related to environmental problems. For example, carbon dioxide, the by-product of burning of fossil fuels, is the major cause of global warming or climate change [1].

The United States Department of Energy (DOE) publishes the nation's energy consumption and energy sources every year [2]. The US consumes 97.3 quadrillion BTU of energy in 2021, about 5% increase from 2020. A remarkable finding is that the renewable energy portion, including solar, wind, and biomass energy, increases significantly by more than 10%. This increase is consistent with the increasing public awareness of energy consumption and its environmental problems. Sustaining the growing energy consumption in a safe and environmentally responsible way remains a key challenge. Addressing this challenge requires a multifaceted approach that prioritizes the conversion of renewable energy into electricity, efficient energy storage, and sustainable and efficient energy consumption.

3. Energy conversion

Energy conversion involves transforming one form of energy into another to meet specific energy needs and applications. According to the first law of thermodynamics, energy can be transformed into different forms, but it cannot be created or destroyed [3]. Mechanical, chemical, thermal, optical, and electrical energy conversion processes take place around us all the time. Car engines transform chemical energy into mechanical energy. Solar panels convert optical energy directly into electricity. Space heaters convert electricity into thermal energy. LED bulbs convert electricity into optical or light energy.

Among these energy conversions, the conversion of various forms of energy into electricity is particularly interesting because electricity is as important as water and food to people in the modern era. Electricity is used for heating, cooling, and lighting and for operating cell phones and appliances in our daily lives. There are many different types of electric generators. The vast majority of electrical generators are based on Faraday's law of induction discovered in 1831 [4]. Other types of electrical generators include photovoltaic cells, fuel cells, and thermoelectric generators. Photovoltaic cells convert solar energy directly into electricity through the photovoltaic effect. Fuel cells produce electricity through an electrochemical process wherein a hydrocarbon fuel reacts with oxygen. A thermoelectric generator is composed of p- and n-type semiconductor legs that convert thermal energy directly into electricity through the Seebeck effect.

Renewable energy such as solar, wind, and geothermal energy has become one of the fastest growing sources of electricity generation around the world. Enhancing the efficiency and sustainability of energy conversion processes is critical for reducing greenhouse gas emissions, minimizing resource depletion, and mitigating environmental impacts.

4. Energy storage

The surplus energy is stored in different forms and can be used later as needed. There are many different types of energy storage, including batteries and thermal and mechanical systems. The first electrochemical battery, storing and releasing a charge through a chemical reaction, was invented in 1800 by Italian chemist Alessandro Volta [5]. The modern high-energy density lithium-ion batteries were commercialized in the 1970s [6]. Now, the batteries have been widely deployed in applications ranging from electric vehicles and consumer electronics to grid-scale energy storage projects.

Thermal energy storage systems store heat or cold. There are three main types of thermal energy systems: sensible heat, latent heat, and thermo-chemical heat storage. A water tank is often used to store sensible heat, while phase change materials are for latent heat storage. Thermo-chemical energy storage systems store heat or cold through reversible chemical reactions. The main types of mechanical energy storage systems are pumped hydro, flywheel, and compressed air. The pumped hydro storage uses two reservoirs with one at a higher elevation than the other and currently is the most widely used large-scale mechanical energy storage system [7].

Energy storage plays a pivotal role in bridging the gap between energy supply and demand, enabling the efficient utilization of various forms of energy. Energy storage technologies find applications across diverse sectors, including transportation, residential, commercial, industrial, and utility-scale energy systems. Many challenges

DOI: http://dx.doi.org/10.5772/intechopen.113887

to wide deployment of energy storage systems remain to be addressed. These include cost competitiveness, technological limitations, regulatory barriers, grid integration issues, and environmental considerations.

5. Energy efficiency

Energy efficiency refers to the use of less energy to achieve maximum output, thereby reducing costs, producing less waste, and mitigating environmental impacts. Technological advancements in energy-efficient appliances and industrial processes drive improvements in energy efficiency. For example, LED (light emitting diode) lighting consumes significantly less electricity compared to traditional incandescent lighting while providing the same illumination. Adding thermal insulation material in walls and roofs can minimize heat loss in buildings and then maintain the desired temperature more efficiently. Utilizing fuel-efficient vehicles, such as hybrid and electric vehicles, reduces fuel consumption and emissions. Combined heat and power systems generate both electricity and useful heat that may maximize energy efficiency compared to separate generation of heat and power.

The efficiency of electric generation systems is particularly interesting as electricity serves as the backbone of industries, infrastructure, transportation, and communication in the world. The second law of thermodynamics puts an upper limit on achievable efficiencies of heat engines and then heat engine-based electric generators such as steam turbine generators, gas turbine generators, and internal combustion engine generators [3]. The energy efficiency of electric generators depends on various factors including the type of generators, operating conditions, design, and technology. Continuous efforts to improve efficiency through advancements in technology and operational practices are crucial for reducing energy consumption and minimizing environmental impact.

Author details

Bao Yang[1] and Jiajun Xu[2]*

1 Department of Mechanical Engineering, University of Maryland, College Park, MD, USA

2 Department of Mechanical Engineering, University of the District of Columbia, DC, USA

*Address all correspondence to: jiajun.xu@udc.edu

References

[1] National Service Center for Environmental Publications (NSCEP). Published by United States Environmental Protection Agency

[2] U.S. energy facts explained [Internet]. 2024. Available from: https://www.eia.gov/energyexplained/us-energy-facts/ [Accessed: January 24, 2024]

[3] Moran MJ, Shapiro HN, Boettner DD, Bailey MB. Fundamentals of Engineering Thermodynamics. WileyPlus; Jan 2018. ISBN: 978-1-119-39138-8

[4] Faraday M. Faraday's Notebooks: Electromagnetic Induction. London: G. Bell & Sons, LTD, England for The Royal Institution of Great Britain; 1936

[5] Allerhand A. Who invented the earliest capacitor bank ("battery" of Leyden jars)? It's complicated. Proceedings of the IEEE. 2018;**106**(3):498-500. DOI: 10.1109/JPROC.2018.2795846

[6] Scrosati B. History of lithium batteries. Journal of Solid State Electrochemistry. 2011;**15**(7-8):1623-1630. DOI: 10.1007/s10008-011-1386-8. S2CID 98385210

[7] Rehman S, Al-Hadhrami L, Alam M. Pumped hydro energy storage system: A technological review. Renewable and Sustainable Energy Reviews. 2015;**44**:586-598

Section 2

Energy Consumption and Conversion

Chapter 2

The Environmental Kuznets Curve: Empirical Evidence from OECD Countries

Iuliana Matei

Abstract

The Environmental Kuznets Curve (EKC) hypothesis reflects the relationship between environmental quality and GDP per capita. The environmental quality decreases in early periods of GDP growth per capita until a certain point, and after that, it begins to increase. This paper investigates the presence of the EKC curve in OECD countries over the period 1997–2015 and identifies thresholds levels of GDP per capita leading to lower emissions per capita for these countries. Also, it points out the key role of energy from renewable sources. Based on nonlinear panel data methods, findings show that CO_2 emissions increase up to a certain level of income (10774$–44494$ per head) and then, they decrease. Policymakers are encouraged to consider economy, technology, and environment all together and handle the legal regulations they will implement, accordingly.

Keywords: carbon dioxide emissions, economic growth, renewable energy consumption, renewable electricity output, trade openness, urbanization, EKC, nonlinear panel models, OECD

1. Introduction

One of the biggest environmental challenges impacting the planet and humanity is climate change. To handle this challenge, various international initiatives aiming to rethink the energy-energy security-climate change nexus have been taken. Among them, the 2021 OECD International Program for Action on Climate (IPAC), the Earthshot Prize, and the 2020 European Green Deal (EGD) support countries progress toward net-zero greenhouse gas (GHG) emissions and a more resilient economy by 2050. They propose a mixt of channels and ambitious goals to mitigate climate change and provide more certainty to policymakers and investors regarding their future decisions on CO_2 emission levels. These emissions must be consistent with the EU's/OECD's goals of being climate-neutral by 2050 such as: reduce the GHG to at least 55% below 1990 levels by 2030; increase the share for renewable energy to more than 32%, improve energy efficiency at about 33%, sustain people and companies to find solutions to help fight climate change and so on.

But these challenges will not be easy to keep given that OECD or EU countries still rely on fossil fuels (such as coal, oil, gas, and petroleum products) for about

80% of their energy supply, notably for industry and transport that are the largest emitters of GHGs. New OECD data indicate that total fossil fuel support in 44 OECD and G20 economies rose by 10% in 2019 to 178 billion USD, ending a 5-year descending trend and undermining global efforts to mitigate climate change [1, 2].

In this context, the question that rises is whether economic growth leads inevitably to more pollution. Until the 1980s, the Club of Rome view (highlighting a negative impact of economic growth on the environment and natural resources) and the Dasgupta and Heal [3]'s view (pointing out a complementary relationship between economic growth and environmental improvement) have been the main approaches of the applied economic analysis. Since 1990s, the Environmental Kuznets Curve (EKC) penciled by Baloch and Wang [4] comes to the scene by suggesting an inverse U-shaped curve between GHG and income. Four main aspects explain it. *The first one* is income elasticity of environment demand. This means that as income increases, people are incited to consume more healthy products and pay more attention to the environmental quality while government is expected to put in place stricter rules on environment protection to improve it. *The second one* is the effect of economic scale, technology, and structure. The economic scale reflects the situation in which more economic growth means more energy consumption from nonrenewable sources and, as a result, more pollution [4]. The structural effect implies that structural changes may improve environment with economic development. Finally, the technology effect suggests that old pollution machines should be replaced by clean techniques of production, which are expected to positively affect the environmental quality. Overall, the EKC indicates that, at the beginning stage of development of a country, the negative scale prevails and the pollution increases (see, [5]). Once the income reaches a certain threshold, the structural and technological effect exceeds the scale effect leading to a fall of pollution levels. *The third aspect* refers to the international trade, which is a key factor of EKC validity. Trade has a dual, twin influence on the environment quality: a negative one through the scale effect or a positive one via the technological effect (see, [5]). *The last aspect* is that of FDI, which has two possible impacts on the quality of environment, too. On the one hand, the developing countries are tempted to lower their claim on the environment protection to attract more foreign investors, which aggravates environmental pollution; on the other hand, by getting technology by FDI from developed countries, most emerging countries help improve environment.

The empirical literature on the relationship between economic growth and environmental quality is considerable (e.g., [6–15]) and concentrates three main strands of studies. The first group of studies focuses on the growth-pollution nexus by testing the EKC validity with different time series and panel data models (e.g., [16–18]). The second group of studies focuses on the link between the economic growth and energy consumption from (non)renewable sources given that the GHG emissions are produced by fossil fuels (e.g., [19–23]). The last group of studies combines the two approaches by analyzing the link between energy consumption from renewable and nonrenewable sources, economic growth, and environmental quality [24]. Compared with the previous one, this wave of studies addresses concerns regarding biasness caused by omission of variables. By integrating variables on international trade, urbanization degree, foreign direct investment, and so on, one can assess the linear and nonlinear relationship between economic growth, pollution, and energy consumption with time-series and/or panel data. Overall, empirical studies conducted until now show diverging results on the growth-pollution nexus and explain this gap in findings by the use of various methods (with time series or panel data), length of periods, and/or the nature of country sample.

Regarding the OECD studies, some of them seek to empirically estimate the nonlinear relationships between income level and environmental degradation and dynamic causalities between environmental pollution and economic growth, along with other key explanatory macroeconomic indicators (e.g., [11, 25–27]). But the attempt to validate the EKC hypothesis still gives ambiguous outcomes. Frodyma [26] shows that the EKC models fail to explain the relationship between income and production-based emissions (or consumption-based emissions) when using an ARDL testing bound approach (due to the heterogeneity of the panel sample considered) over the 1970–2017 period. At the opposite side, Benjebli et al. [25] identifies an inverted U-shaped EKC curve for 25 OECD for the period 1980–2010 based on fully modified ordinary least squares (FMOLS) and dynamic ordinary least squares (DOLS) estimates. Differently, Madaleno and Moutinho [11] focuses on 15 EU countries (old members of OECD) and 12 Emerging EU countries over the 2008–2018 period. With dynamic fixed effects (DFE) and DOLS estimates, they identify a U-shaped EKC curve. Likewise, Churchill et al. [28] finds evidence that supports the EKC hypothesis in nine of the 20 OECD countries over the period 1870–2014 (a long-time horizon of 144 years). The EKC hypothesis is validated for Australia, Canada, Denmark, Finland, France, Japan, Spain, United Kingdom, and the United States, albeit with different turning points for income per capita. These findings are generally comparable to other empirical studies, which identified an EKC curve for OECD with panel data over shorter periods. For example, Dijkgraaf and Vollebergh [29] finds that 11 out of 22 OECD countries have a statistically significant turning point and verify an inverted U-shaped EKC pattern. Likewise, Apergis [30] and Shafiei and Salim [31] find evidence supporting an inverted EKC relationship for some OECD countries as well.

Another group of studies go further and compute the turning points in income per capita leading to lower emissions per capita. Galeotti [32] finds for OECD over the period 1960–1997 a turning point ranging between 15599.9$ and 21185.83$ (in 1990 US dollars). In the same vein, Churchill et al. [28] identifies greater turning points ranging from $19 978 to $84001, depending on whether or not financial development variable is included in the equation's model. Based on a panel smooth transition approach over the 2000–2018 period, for 75 countries (including some OCDE countries), Tatoglu and Polat [14] finds a turning point between $42053 and $42057. Also, Sun et al. [27] uses the Common Correlated Effect Mean Group (CCEMG) and Augmented Mean Group (AMG) methods for the period 1992–2015 and shows that growth output has significant positive effect on environmental pollution. The estimated turning points are greater ranging from 1479029.30 to 3277677.02. As we can observe, the EKC empirical evidence is still controversial, and there is no consensus on the income level at which environmental degradation begins diminishing.

This paper will contribute to the last strand of the empirical literature by providing evidence on the validity of the EKC curve for 34 OECD countries [33] and by identifying threshold levels of GDP per capita leading to lower emissions per capita for these countries. But, unlike previous studies, it will point out the key role of energy from renewable sources.

Therefore, the paper contributes to the existing empirical literature in the following ways. First, contrary to the most part of studies focusing on the role of energy consumption from fossil fuels in validating the EKC curve, the paper embodies indicators on energy from renewable sources, since OECD countries are considered as regions with a less energy-intensive economic structure and a more advanced development in renewable sources than emerging ones (e.g., [34]). Second, the paper includes a heterogeneous sample of countries with respect to their energy consumption mix over the period 1997–2015, which allow capturing disparities, if any, in the value of the threshold above and below which the effect of economic activity on environmental pollution may differ within this group of countries. The threshold is identified with a recent empirical model by Fouquau et al. [35] and González et al. [36] a panel smooth transition model (PSTR), which is the best model for tackling the heterogeneity in the effect of economic activity on the environmental degradation. Third, the paper includes two types of indicators for energy from non-fossil fuels: the renewable energy consumption and the electricity production from renewable sources per capita, excluding hydroelectric (kW h), the last being used for robustness purposes. Fourth, the findings provide further support to economic approaches investigating the renewable EKC curve (R-EKC curve). However, they go further by pointing out the non-monotonous relationship between environmental degradation and economic activity in combination (or not) with the renewable energy use, FDI, and urbanization degree. To my best knowledge, this aspect of the R-EKC curve has never been explored before with a PSTR approach. Finally, the paper provides useful insights for policymakers in supporting green economy to achieve the 2030 Agenda for Sustainable Development (via the IPAC and the EGD international initiatives).

The results show evidence for an inverted U-shaped relation between income and environmental pollution for OECD-34. However, the overall effect of output on the environmental degradation (through carbon dioxide emissions) is positive meaning that technology, preference, and environmental investments in clean energy are not sufficient developed in the energy consumption mix. The turning point or the income level in inflexion of R-EKC ranges between 10774.1 and 44967.7 dollars *per head* when the indicator of gas pollutant is the carbon dioxide emissions (such as in [17, 18, 27]).

The rest of the paper is structured as follows. Section 2 describes the data and presents the PSTR methodology. Section 3 provides the empirical results and discussion. The final section concludes.

2. Data and methodology

2.1 Data and descriptive statistics

The dataset contains 34 advanced countries of OECD and covers the yearly period 1997–2015. The following countries are included: Austria, Belgium, Finland, France, Germany, Greece, Ireland, Italy, Netherlands, Portugal, Slovak Rep., Spain, United Kingdom, Japan, United States, Australia, Canada, Denmark, Iceland, Luxembourg, New Zeeland, Norway, Switzerland, Turkey, Chile, Czech Rep., Estonia, Hungary, Israel, Korea Rep., Mexico, Poland, Slovenia, and Sweden.

In the models, the dependent variable is the natural logarithm of CO_2 emissions (metric ton per capita) for country i in year t, in line with prior literature (e.g., [31]). The vector of explanatory variables includes the following variables: real GDP per capita as a measure of output (GDP), ratio of exports and imports to GDP (TO) for openness to trade, the urban population (as % of total population-UR), foreign direct investment (as % of GDP-FDI), renewable energy consumption (% of total final energy consumption-RE), and the renewable electricity output (as % of total electricity output-REL), which is the share of electricity generated by renewable power plants in total electricity generated by all types of plants. The first three variables could have alternative impacts (they should be positively or negatively associated with CO2 emissions) while the last two variables are likely to be negatively linked to the environmental quality. All data come from the World Bank Development Indicators database and are in natural logarithms (i.e., each estimated coefficient is a constant elasticity of the dependent variable with respect to the independent variable).

The descriptive statistics on the selected variables for each country are displayed in **Table 1**. It can be observed that the lowest level of renewable energy consumption is 0.7% of total final energy consumption (United Kingdom) while the highest level is 77.8% of total final energy use (Island). The highest value of GDP per capita is up to 112418$ (Luxembourg) while the lowest level of GDP per capita corresponds to 6075$ (Poland). Regarding the CO2 emission per capita, the highest level is equal to 25.6% (Luxembourg) while the lower level of CO_2 emissions is for Chile, Mexico, and Turkey.

To address multicollinearity concerns, **Table 2** shows the matrix correlation among explanatory variables. The explanatory variables are not highly correlated so that they can be safely integrated into the model (except for the energy use from fossil fuels, which will be dropped from the final model).

2.2 PSTR model

The R-EKC curve is examined by using the empirical approach proposed by González et al. [36]. This recent method focuses on heterogeneous panels that allow estimated coefficients to vary both across countries and over time. The specification supposes the existence of an infinite number of intermediary regimes, and the coefficients depend upon these regimes. As it is considered a nonlinear panel model, it serves to capture a rise in the level of income that does not affect the income-pollution nexus linearly, but conditionally on the position in the income distribution.

Considering the level of income (GDP per capita) as a transition variable $q_{i,t}$, the PSTR model with two regimes and a single transition function can be written as follows:

$$CO2_{it} = \mu_i + \beta_0 GDP_{i,t} + \beta_1 GDP_{it} f\left(q_{i,t}; \gamma, c\right) + \zeta X_{i,t} + \varepsilon_{i,t} \tag{1}$$

where $CO2_{it}$ is the dependent variable (the carbon dioxide emissions per capita for country i at time t), μ_i the individual fixed-effects, $GDP_{i,t}$ is the GDP per capita of the country i, at time t, the $f\left(q_{i,t}, \gamma, c\right)$ is the transition function, and $\varepsilon_{i,t}$, the error term, which is i.i.d $(0, \sigma_\varepsilon^2)$. $X_{i,t}$ corresponds to the vector of the control variables including renewable energy use or renewable electricity output (RE_{it} or RE_{Lit}), urban population (UR_{it}), trade openness (TO_{it}), foreign direct investment (FDI_{it}). The transition function is continuous and integrable on the interval [0,1] and depends on three parameters: $q_{i,t}$, which is the transition variable, γ – the slope of the transition function, and c is the vector of location parameters such as $c = (c_1,\dots,c_m)'$, with m as the vector dimension.

There is no specific rule regarding the optimal number of thresholds in studying the income-pollution nexus. To identify the optimal number of thresholds, two tests are performed in the next section (the Lagrange Multiplier Wald test and the Lagrange Multiplier Fisher test), which indicate that m=2[1]. This means that there are two thresholds of "income" around which the effect of income on pollution is a

[1] This means that the final PSTR model will have two thresholds, and it can be written as follows:
$CO2_{it} = \mu_i + \beta_0 GDP_{i,t} + \beta_1 GDP_{it} f\left(q_{1i,t}; \gamma_1, c_1\right) + \beta_2 GDP_{it} f\left(q_{2i,t}; \gamma_2, c_2\right) + \zeta X_{i,t} + \varepsilon_{i,t}$.

nonlinear one. However, even in this case (m = 2), there are still a continuum of regimes that lie between the extremes (high pollution/low income and low pollution/high income). Therefore, as the transition variable $q_{i,t}$ increases, the effect of "income" progresses from β_0 in the first regime corresponding to $f(.) = 0$ to $\beta_0 + \beta_1$ in the second extreme regime corresponding to $f(.) = 1$ and so on. Between two extreme cases $f(.) = 0$ and $f(.) = 1$, the sensitivity/elasticity of pollution to income is computed by differentiating CO_2 emissions with respect to the level of income such as:

$$q_{i,t} = \frac{\partial CO2_{i,t}}{\partial GDP_{i,t}} = \beta_0 + \beta_1 * f\left(q_{i,t}; \gamma, c\right) \tag{2}$$

Furthermore, the sensitivity of CO_2 emissions to GDP per capita can vary under the two extreme regimes β_0 and β_0+β_1 and is a weighted average of parameters β_0 and β_1. Note that literature indicates that it is difficult to directly interpret the values of these parameters, but easier to interpret their sign as an increase or decrease in the elasticity, depending on the value of the transition variable, and the individual and time dimension given by the previous equation.

2.3 Pre-tests: test for linearity (homogeneity) hypothesis

Before estimating the PSTR model, I investigate homogeneity's model against the PSTR alternative. The Lagrange multiplier (LM) test of homogeneity based on the asymptotic $\chi 2$ distributions, their F-versions and the HAC versions are applied to each of the explanatory variables, they being potential "candidate" transition variables in the PSTR. The LM test looks at the null hypothesis of linearity (homogeneity) against the alternative logistic (m=1) or exponent (m = 2) PSTR model. The optimal number of transition functions is obtained by doing tests of no-remaining nonlinearity. The linearity supposes testing H0:γ = 0 or $H0{:}\alpha = \beta$, but in both cases the test is non-standard, since under *H0* the PSTR model has unidentified nuisance parameters. Usually, it is proposed to replace the transition function $f\left(q_{i,t}; \gamma, c\right)$ by its first-order Taylor expansion around γ = 0 and to test an equivalent hypothesis in an auxiliary regression such as:

$$c_{i,t} = \omega_i + \alpha CO2_{i,t} + \theta_1 CO2_{i,t}^2 + \varepsilon_{i,t} \tag{3}$$

In this first-order Taylor expansion, the parameter θ_1 is proportional to the slope parameter γ. Thus, testing the linearity against the PSTR model simply means testing: H0: $\theta_1 = 0$ in this linear panel model. This can be done by using standard tests such as the F-statistic.

2.4 Pre-tests: selecting the number of transition functions

The next step aims to identify the optimal number of thresholds (*m*) of the logistic transition function by using the LM homogeneity test. The procedure is like when testing the number of transition functions in the model. In the PSTR framework, it is assumed that the linearity hypothesis is rejected. Then the idea is to test whether there is one transition function (H0: m = 1) or whether there are at least two transition functions (H0: m = 2). Let now us consider the model with m = 2 regimes:

$$CO2_{it} = \mu_i + \beta_0{}'GDP_{i,t} + \beta_1{}'GDP_{it}f\left(q_{1i,t};\gamma_1, c_1\right) + \beta_2{}'GDP_{it}f\left(q_{2i,t};\gamma_2, c_2\right) + \zeta X_{i,t} + \varepsilon_{i,t} \tag{4}$$

If we replace the second transition function $\left(q_{2i,t};\gamma_2, c_2\right)$ by the first-order Taylor expansion around $\gamma_2 = 0$ and test linear constraints on the parameters, the model becomes:

$$CO2_{it} = \mu_i + \beta_0{}'GDP_{i,t} + \beta_1{}'GDPf\left(q_{1i,t};\gamma_1, c_1\right) + \theta GDP_{it} * q_{i,t} + \varepsilon'_{i,t} \tag{5}$$

The test of no-remaining nonlinearity is defined by H0: $\theta = 0$; consider SSR_0 the panel sum of squared residuals under H_0, i.e., in a PSTR model with one transition function. SSR_1 is the sum of squared residuals of the transformed model in (5). The testing procedure shows that given a PSTR model with $m = m^*$, the null hypothesis will be tested such as: H_0: $m = m^*$ against H_1: $m = m^* + 1$. If H_0 is not rejected, then the procedure ends. Otherwise, the null hypothesis H_0: $m = m^* + 1$ is tested against H_1: $m = m^* + 2$. The testing procedure continues until the first acceptance of H_0.

3. Empirical results and discussion

3.1 Pre-tests discussion

Before estimating the PSTR model, I performed a homogeneity test in which the income per capita is a transition variable (q). The results of this test (**Table 3** of Appendix) indicate whether a PSTR model can assess the effect of income on environmental pollution. The three statistics (LR, LMF, and LRT – the Wald, Fisher, and Likelihood ratio test statistics) indicate that the hypothesis of a homogeneous effect of income on environmental pollution is rejected. Furthermore, LM, LMF, and LRT p-values are smaller for the exponential model meaning that it is favored over the logistic one.

Considering these results, income influences environmental degradation in different ways. The test of remaining regimes (**Table 4** of Appendix) concludes on the existence of a two-threshold level of income. The LM and LMF statistics of these two tests are statistically significant at 1% and 5% levels, which indicate a rejection of the null hypothesis and the acceptance of at least two transition functions. **Table 5** of the Appendix complements the previous finding by indicating that three extreme regimes ($m = 2$) are enough to capture the nonlinearity of the income-pollution nexus based on the modulating effect of the income level. The optimal number of location parameters is given by the Akaike and Schwartz criterion and their values are also shown.

Table 6 of the Appendix gives thresholds of income per capita for 34 OECD economies. The threshold of income remains quite similar among different specifications ranging from 9.29 to 10.71. Because data are in natural logarithms, the exact value of the threshold is computed by applying an exponential function to the constant C. This transformation gives us the level of income per head around which income exerts a negative or a positive effect on the CO_2 emissions. Hence, the threshold of income ranges between about 10 774\$ and 44968\$.

For the advanced economies, the threshold of 25245\$ means that below this level of income, each group of countries will have higher levels of pollution. Conversely, it is only above this high threshold that these countries can improve the environmental degradation. Therefore, less income means more environmental pollution for these economies. An examination of the possibility of each country having more income based on the estimated threshold value of PSTR 1 and PSTR 2 (25245$ and 26132$ in the first regime) found that about a third of the advanced-economies countries (13) can respond to this threshold. These countries are Turkey and the new EU members. The rest of OECD countries have income per head superior to 25000$.

3.2 PSTR discussion

3.2.1 The R-EKC curve discussion

In **Table 7**, the results show that the direct effect of income on CO_2 emissions, measured by the b_0 coefficient, is positive and significant in all the PSTR estimates. This means that in early stage of development, more income means more pollution (such as in [31]). In the second and third regimes, the coefficients become negative and statistically significant for the PSTR 1 and PSTR 2 (in the second regime) and PSTR 3 and PSTR 4 (in the third regime), respectively.

The next rows in **Table 6** provide interesting insights: here the impact of income on pollution is conditional on the level of GDP per capita. More specifically, the b_1 coefficient, associated with the nonlinear component of the model, is always negative and significant at the 1% level (PSTR 1 and PSTR 2 including renewable energy consumption), with values ranging between −0.343 and −0.307. According to the exponential function, this implies that the elasticity of CO_2 emissions with respect to income changes from 0.34 or 0.42 (as b_0 takes these values in PSTR 1 and PSTR 2) to b_1, i.e., as pollution goes from high to low values. The shift between these two extreme regimes occurs around the associated endogenous threshold parameter c (shown in rows 6 and 7), which equals 25243$–44494$ (as in [14]). In the low pollution regime (indices ≥ 25243), the effect of income on environmental degradation is negative and statistically significant at the 1% level; in the extreme case (when $g\ (s_{it};\ \gamma,\ c) = 1$), all other things being equal, a 1% increase in the income gives a 0.31–0.34% reduction in CO_2 emissions (the coefficient in the second regime being equal to $\beta_0 + \beta_1$). Given that there is a continuum of possible points between these two extreme regimes, the elasticity is a weighted average of the parameters β_0 and β_1. This result implies that without a higher income level, advanced economies cannot optimally benefit from a sound environmental quality, and that any policy would be not so much effective.

The paper finds also that the impact of income on CO_2 emissions is nonlinear. Based on the three statistics (LR, LMF, and LRT) reported in **Table 3** of Appendix, the hypothesis of a heterogeneous influence of income on pollution is accepted. Thus, income affects CO_2 emissions in various ways. The test of remaining regimes (**Table 4** of Appendix) complements this finding and concludes on the existence of two thresholds for GDP per capita. Another argument in support of the nonlinear hypothesis is that the slope of the transition function differs between different regimes and the four PSTR. The higher the γ, the sharper the change from one extreme regime to another. For the PSTR 2, for example, results show that any effort in terms of decreasing pollution by a country just below the threshold value of 26132$ is likely to result in a sharp decrease in the elasticity of pollution with respect to income (from 0.42 to −0.34). However, for a country that is far below this threshold, the same effort will have little effect on the elasticity of pollution. Conversely, in the PSTR 1 it is found a smooth transition. This means that, unlike with the sharp transition previously defined, any effort to combat pollution, even by a country far below the threshold value, will always be recompensated (by a gradual fall in the marginal effect of pollution). Similar features are found for PSTR 3 and PSTR 4 including renewable electricity output among control variables.

3.2.2 The explanatory variables impact

Next, consistent with previous empirical literature, control variables are integrated such as trade openness, FDI, urbanization, and renewable energy consumption or renewable electricity output. These explanatory variables were almost significant and had the expected signs (except for urbanization).

Trade openness exerts a negative and statistically significant effect on pollution, which resonates with the technological effect of trade on environmental quality (contrary to [31] where a positive significant influence of trade openness on carbon emissions is found, validating rather a scale effect of trade on environmental pollution). Foreign direct investments are essential for improving productivity and increasing the competitiveness of an economy. This coefficient is negative or positive but statistically insignificant in all PSTR models. A counterfactual result is found for the urbanization. Indeed, an increase in the urban population should increase the level of emissions, particularly, carbon dioxide emissions; thus, countries that have higher urban populations are expected to pollute the environment more than other countries.

The coefficients of renewable energy consumption or renewable energy output (used for robustness purposes) are always negative and statistically significant at 1% level. This means that clean sources of energy reduce greenhouse emissions and protect environment. Therefore, governments must support the development of this burgeoning energy sector as well as the implementation of carbon taxes to discourage the use of fossil fuels (as a conventional energy source) and to protect environment. Furthermore, such "new-generation" energy policies are expected to be key for a complete decarbonization of the energy sector to achieve sustainable development goals of the IPAC or EU Green Deal by 2030 (which is in line with findings of [22]).

4. Conclusions

Despite a growing empirical literature, the income-pollution nexus remains a controversial topic. The puzzling results of previous studies incite to investigate it in-depth to better assess this issue. I examine the EKC curve in the case of OECD countries and tested the nonlinearity of the income-pollution nexus. To this end, I used the PSTR model of González et al. [36] to capture the heterogeneous effect of income across 34 advanced economies by using yearly data from the World Bank between 1997 and 2015. The overall impact of income on environmental degradation is slightly positive for the 34 OECD countries. This means that technological and structural effects do not exceed the scale effect regarding this relationship. Based on these results, I suggest that policymakers should consider supporting the development of this new energy sector by improving the quality of the existing clean technologies, by financing R&D investments in other promising renewable technologies and related infrastructure network to make renewable energy sources more competitive than fossil fuels and, to reduce, in this way, CO_2 emissions and enhance the energy security. Promoting regional cooperation and developing clean energy technologies will improve the environmental quality of countries, too.

Appendix

Variable	OBS	Mean	Std. dev.	Min	Max
CO2	646	9,1	4,1	3,1	25,6
GDP	646	34621,5	20844,8	6076,0	112417,9
TO	646	88,6	51,0	18,1	351,1
UR	646	88,6	11,3	50,7	97,9
REL	646	76,4	27,2	0,0	100,0
RE	646	27,3	15,6	0,7	77,8
FDI	646	5,6	11,5	−27,7	138,2

Table 1.
Summary statistics for all variables in this study.

	GDP	NE	RE	UR	TO	FDI	REL
GDP	1						
NE	0,69	1					
RE	0,03	0,08	1				
UR	0,43	0,41	−0,04	1			
TO	0,08	0,13	−0,02	−0,19	1		
FDI	0,40	0,26	−0,15	0,17	0,33	1	
REL	0,26	0,15	0,68	−0,07	−0,07	−0,02	1

Note: GDP – real GDP per cap, NE – energy use per capita, RE - renewable energy consumption; UR–Urbanization, TO-trade openness, FDI- foreign direct investment, REL - renewable electricity output.

Table 2.
The matrix correlation between the explanatory variables for OECD-34.

DOI: http://dx.doi.org/10.5772/intechopen.108631

Results of the test of linearity

Stats	Thresholds					
	m = 1		m = 2		m = 3	
PSTR 1 : Co2 = f(GDP, GDP2, RE, UR)						
LM	34.068***	(0.000)	10.609***	(0.014)	30.185***	(0.000)
LMF	11.301***	(0.000)	3.356**	(0.019)	9.803***	(0.000)
LRT	34.999***	(0.000)	10.697***	(0.013)	30.913**	(0.000)
PSTR 2: Co2 = f(GDP, GDP2, **RE**, UR, TO, FDI)						
LM	40.503***	(0.000)	34.724***	(0.000)	38.901***	(0.000)
LMF	8.121***	(0.000)	6.783***	(0.000)	7.587***	(0.000)
LRT	41.128***	(0.000)	35.692***	(0.000)	40.121***	(0.000)
Robustness Checks						
PSTR 3: Co2 = f(GDP, GDP2, REL, UR)						
LM	117.434***	(0.000)	33.100***	(0.000)	18.071***	(0.000)
LMF	45.101***	(0.000)	10.855***	(0.000)	5.756***	(0.000)
LRT	129.608***	(0.000)	33.978***	(0.000)	18.328***	(0.000)
PSTR 4: Co2 = f(GDP, GDP2, REL, UR, TO, FDI)						
LM	125.056***	(0.000)	37.713***	(0.000)	24.099***	(0.000)
LMF	29.143***	(0.000)	7.403***	(0.000)	4.588***	(0.000)
LRT	138.991***	(0.000)	38.859***	(0.000)	24.560***	(0.000)

*** *statistical significance at 1%*
** *statistical significance at 5% level*

Table 3.
LM, LMF, and LRT tests of linearity (p-values) for OECD-34.

Results of test for the number of regimes

Institut. Var.	Hypotheses	Test	Stat	(p-value)
PSTR with Pub. Debt				
PSTR 1	H0 : m = 0 vs H1 : m = 1	LM	34.068***	(0.000)
		LMF	11.301***	(0.000)
	H0 : m = 1 vs H1 : m = 2	LM	10.609***	(0.014)
		LMF	3.356**	(0.019)
	H0 : m = 2 vs H1 : m = 3	LM	30.185***	(0.000)
		LMF	9.803***	(0.000)
PSTR 2	H0 : m = 0 vs H1 : m = 1	LM	40.503***	(0.000)
		LMF	8.121***	(0.000)
	H0 : m = 1 vs H1 : m = 2	LM	34.724***	(0.000)
	LMF	6.783***	(0.000)	
	H0 : m = 2 vs H1 : m = 3	LM	38.901***	(0.000)
		LMF	7.587***	(0.000)
Robustness Checks				
PSTR 3	H0 : m = 0 vs H1 : m = 1	LM	117.434***	(0.000)
		LMF	45.101***	(0.000)
	H0 : m = 1 vs H1 : m = 2	LM	33.100***	(0.000)
		LMF	10.855***	(0.000)
	H0 : m = 2 vs H1 : m = 3	LM	18.071***	(0.000)
		LMF	5.756***	(0.000)
PSTR 4	H0 : m = 0 vs H1 : m = 1	LM	125.056***	(0.000)
		LMF	29.143***	(0.000)
	H0 : m = 1 vs H1 : m = 2	LM	37.713***	(0.000)
		LMF	7.403***	(0.000)
	H0 : m = 2 vs H1 : m = 3	LM	24.099***	(0.000)
		LMF	4.588***	(0.000)

***statistical significance at 1%.
**statistical significance at 5% level.

Table 4.
Test for the number of regimes for OECD-34.

Determination of the number of location parameters

PSTR 1	Optimal no. of transition functions	2
	Residual Sum of Squares	2.632
	Number of Parameters	13
	AIC Criterion	−5.441
	BIC Criterion	−5.351
PSTR 2	Optimal no. of transition functions	2
	Residual Sum of Squares	2.430
	Number of Parameters	19
	AIC Criterion	−5.492
	BIC Criterion	−5.361
Robustness Checks		
PSTR 3	Optimal no. of transition functions	2
	Residual Sum of Squares	3.534
	Number of Parameters	13
	AIC Criterion	−5.146
	BIC Criterion	−5.056
PSTR 4	Optimal no. of transition functions	2
	Residual Sum of Squares	3.088
	Number of Parameters	19
	AIC Criterion	−5.253
	BIC Criterion	−5.122

Notes: (i) For each model, the optimal number thresholds, denoted r (m), are determined according to a sequential procedure based on the LMF statistics of the hypothesis of non-remaining nonlinearity.
(ii) The RSS value is given for each couple (m; r). (iii) The AIC and BIC criterions are shown for each PSTR.

Table 5.
Determination of the number of location parameters for OECD-34.

Models	Speed of transit. – $\gamma 1$	Speed of transit. - $\gamma 2$	Threshold param- c1	Threshold param- c2
PSTR 1	22.6835	25.3073	10.1363	10.7031
PSTR 2	1273.8	28.2	10.1709	10.7091
PSTR 3	17.7708	10.1812	10.7002	9.2849
PSTR 4	10.9579	7.9411	10.7137	9.2925

Table 6.
Results of threshold values of GDP per capita for OECD-34.

Variable	PSTR 1		PSTR 2		PSTR 3		PSTR 4	
	coeff.	**t-Stat**	**coeff.**	**t-Stat**	**coeff.**	**t-Stat**	**coeff.**	**t-Stat**
GDP per capita	0.3372^{***}	10.5552	0.4180^{***}	10.374	0.8670^{***}	8.5244	0.9405^{***}	7.8088
GDP per capita*g_1 (.)	-0.3071^{***}	−4.2881	-0.3426^{***}	−4.833	−0.0508	−1.0829	−0.0503	−0.7263
GDP per capita*g_2 (.)	−0.0148	−0.3491	−0.0412	−0.709	-0.5227^{***}	−6.6361	-0.4600^{***}	−4.1124
Transition parameters								
Speed of transition – γ_1	22.6835		1273.8		17.7708		10.9579	
Speed of transition – γ_2	10.7031		28.2		10.1812		7.9411	
Threshold parameter – c_1	10.1363 **25242.895**		10.1709 **26131.585**		10.7002 **44364.7272**		10.7137 **44967.712**	
Threshold parameter – c_2	10.7031	**44493.572**	10.7091	**44761.336**	9.2849	**10774.095**	9.2925	**10856.2909**
Control variables								
Renewable energy consumption	-0.2784^{***}	−18.237	-0.2510^{***}	−15.724	-0.0456^{***}	−3.6457	-0.0433^{***}	−3.2781
Urbanization	-0.7158^{***}	−5.672	-0.6543^{***}	−4.864	-2.0202^{***}	−8.1128	-1.5299^{***}	−5.546
Openness	-	-	-0.1465^{***}	−3.408	-	-	-0.1722^{**}	−2.236
FDI	-	-	0.0055	1.236	-	-	−0.0199	−1.559
No. Obs	646 × 5		646 × 7		646 × 5		646 × 7	
No. Countries	34		34		34		34	

Note: i) *** – p-value < 1% ; ** - p-value < 5%; * - p-value < 10%; ii) the PSTR 3 and PSTR 4 include the renewable electricity output as a robustness checks exercise; iii) the coefficients of the control variables are those of the first linear regime; iv) in the second regime, RE coefficient is positive and statistically significant. The threshold parameter reflects the threshold of income around to which the income effect on pollution changes (from a positive one to a negative one or vice-versa).

Table 7.
PSTR estimates with the real GDP per capita as the threshold variable: 1997–2015, OECD-34.

DOI: http://dx.doi.org/10.5772/intechopen.108631

JEL Classification

Q53, Q56, F64, C54

Author details

Iuliana Matei
CES-University Paris 1, Paris, France

*Address all correspondence to: iuliana.matei@univ-paris1.fr

References

[1] OECD. Environment at a Glance Indicators. OECD; 2021

[2] Wolf S, Jonas T, Jahel M, Jaeger C. The European green deal—More than climate neutrality. Intereconomics. 2021; **56**(2):99-107

[3] Dasgupta P, Heal G. Economic Theory and Exhaustible Resources. Cambridge: Cambridge University Press; 1979

[4] Baloch MA, Wang B. Analyzing the role of governance in CO2 emissions mitigation: The BRICS experience. Structural Change and Economic Dynamics. 2019;**51**:119-125

[5] Bo S. A literature survey on environmental Kuznets curve. Energy Procedia. 2011;**5**:1322-1325

[6] Al-Mulali U, Ozturk I. The investigation of environmental Kuznets curve hypothesis in the advanced economies: The role of energy prices. Renewable and Sustainable Energy Reviews. 2016;**54**:1622-1631

[7] Álvarez-Herránz A, Balsalobre D, Cantos JM. Energy innovations-GHG emissions nexus: fresh empirical evidence from OECD countries. Energy Policy. 2017;**101**:90-100

[8] Armeanu D, Vintila G, Andrei JV, Gherghina SC, Dragoi MC, Teodor C. Exploring the link between environmental pollution and economic growth in EU-28 countries: is there an environmental Kuznets Curve? PLoS One. 2018;**13**:1-28

[9] Du X. Occurrence of turning point on environmental Kuznets curve in the process of (de)industrialization. Structural Change and Economic Dynamics, Elsevier. 2020;**53**(C):359-369

[10] Kisswani KM, Harraf A, Kisswani AM. Revisiting the environmental kuznets curve hypothesis: evidence from the ASEAN-5 countries with structural breaks. Applied Economics. 2019;**51**(17): 1855-1868

[11] Madaleno M, Moutinho V. Analysis of the new kuznets relationship: considering emissions of carbon, methanol and nitrous oxide greenhouse gases evidence from eu countries. International Journal of Environmental Research and Public Health. 2021;**18**:1e24

[12] Sadorsky P. Renewable energy consumption and income in emerging economies. Energy Policy. 2009;**37**(10): 4021-4028

[13] Simionescu M, Strielkowski W, Gavurova B. Could quality of governance influence pollution? Evidence from the revised Environmental Kuznets Curve in Central and Eastern European countries. Energy Reports. 2022;**8**:809-819

[14] Tatoglu FY, Polat B. Occurrence of turning points on environmental Kuznets curve: Sharp breaks or smooth shifts? Journal of Cleaner Production. 2021;**317**(1):128333

[15] Zafar MW, Mirza FM, Zaidi SAH, Hou F. The nexus of renewable and nonrenewable energy consumption, trade openness, and CO2 emissions in the framework of EKC: Evidence from emerging economies. Environmental Science Pollution and Controlled Series. 2019;**26**:15162-15173

[16] Apergis N, Payne J. CO_2 emissions, energy usage, and output in Central America. Energy Policy. 2009;**37**:3282-3286

[17] Bolluk G, Mert M. The renewable energy, growth and environmental Kuznets curve in Turkey: An ARDL approach. Renewable and Sustainable Energy Reviews. 2015;**52**:587-595

[18] Selden T, Song D. Environmental quality and development: Is there a Kuznets Curve for air pollution emissions. Journal of Environmental Economics and Management. 1994;**27**: 147-162

[19] Apergis N, Payne J. Renewable energy consumption and economic growth: evidence from a panel of OECD countries. Energy Policy. 2010;**38**:656-660

[20] Coers R, Sanders M. The energy-GDP nexus: addressing an old question with new methods. Energy Economics. 2014;**36**:708-715

[21] Constantini V, Martini C. The causality between energy consumption and economic growth: A multi-sectoral analysis using non-stationary cointegrated panel data. *Energy Economics*. 2010;**32**:591-603

[22] Matei I. Is there a link between renewable energy consumption and economic growth? A dynamic panel investigation for the OECD countries. Re vue d'économie politique. 2017;**127**:985

[23] Salim RA, Hassan K, Shafiei S. Renewable and non-renewable energy consumption and economic activities: Further Evidence from OECD countries. Energy Economics. 2014;**44**:350-360

[24] Bölük G, Mert M. Fossil, and renewable energy consumption, GHGs and economic growth: evidence from a panel of European union (EU) countries. Energy. 2014;**74**:439-446

[25] Benjebli M, Youssef SB, Ozturk I. Testing environmental Kuznets curve hypothesis: The role of renewable and non-renewable energy consumption and trade in OECD countries. Ecological Indicators. 2016;**60**:824-831

[26] Frodyma K. Revisiting the environmental Kuznets curve in the European Union countries. Energy. 2022;**241**:122899

[27] Sun H, Samuel CA, Amissah JCK, Taghizadeh-Hesary F, Mensah IA. Non-linear nexus between CO2 emissions and economic growth: A comparison of OECD and B&R countries. Energy. 2020;**212**:118637

[28] Churchill SA, Inekwe J, Ivanovski K, Smyt R. The environmental Kuznets Curve in the OECD: 1870–2014. Energy Economics. 2018;**75**:389-399

[29] Dijkgraaf E, Vollebergh HR. A test for parameter homogeneity in CO2 panel EKC estimations. Environmental and Resource Economics. 2005;**32**(2):229-239

[30] Apergis N. Environmental Kuznets curves: new evidence on both panel and country-level CO_2 emissions. Energy Economy. 2016;**54**:263-271

[31] Shafiei S, Salim RA. Non-renewable and renewable energy consumption and CO2 emissions in OECD countries: A comparative analysis. Energy Policy. 2014;**66**(C):547-556

[32] Galeotti M. Reassessing the environmental Kuznets curve for CO2 emissions: A robustness exercise. Ecol. L'Economat. 2006;**57**(1):152-163

[33] Grossman G, Krueger A. Environmental impacts of a North American free trade agreement. NBER working paper; 1991: 3914.

[34] IEA 2015 Report: Recent energy trends in OECD excerpt from: Energy

Balances of OECD Countries, IEA 2015 Statistics.

[35] Fouquau J, Hurlin C, Rabaud I. The Feldstein-Horioka Puzzle: A Panel Smooth Transition Regression Approach Economic Modelling. 2008;**25**(2):284-299

[36] González A, Teräsvirta T, van Dijk D, Yang Y. Panel Smooth Transition Regression Models. CREATES Research Paper 2017-36. Department of Economics and Business Economics, Aarhus University; 2017

Chapter 3

Sustainable Energy Efficient Industrial Facility Design

Zeeshan Farooq and Ali Al-Qahtani

Abstract

Industrial facilities are rapidly changing over the years, various technological advances have led to break-through its performance. The transformation is fetched from inventions, development of sophisticated processes/technologies, digitalization, the list is incessantly growing. These technical progressions and stratagems to realize it on the life-cycle basis forces the industries to explore sustainability concepts within the scope. Also, the mission to chop back GHG emissions to protect environment is needed than ever before. Consequently, the revolution and evolution of products, processes, technologies, resources, and sustainability requirement to mitigate climatic anomaly beams on the industrial sector and ordained it to cater an approach toward optimizing all efforts. The challenge is to develop a method by accommodating all the technological advances and sustainability requirements to transform facilities into a much better performing asset. The answer is to construct and preserve industrial facilities to be energy efficient which enable them to be competitive and environmentally friendly. The chapter illustrates, through case study, the 'plant lifetime retrofitability concept in industrial facilities. Its objective is to develop the most energy efficient facility design via arranging each future modification to remain energy efficient throughout its lifecycle without deterrents and utilizing minimum capital. Finally, a way is proposed to develop inherently sustainable energy efficient design in order to cater the benefits throughout the industrial facility life-cycle.

Keywords: energy efficiency, sustainability, energy integration, pinch technology, process integration, retrofit, GHG emissions reduction, facility design, and oil & gas

1. Introduction

Industrial sector relies on fossil fuels as the primary source of energy that is in control of approximately one-third of world-wide energy usage. A major percentage of energy consumed is disbursed to provide utilities for oil & gas facilities in order that energy may well be engendered for other sectors. The adverse impact of fossil-fuel combustion has instigated efforts to attenuate carbon emissions, whilst environmental regulations have a considerable effect on the energy's cost. Furthermore, it's a competitive world with ever-increasing energy prices, industrial communities are convened to brainstorm over conservation of energy resources and profitability. Energy efficiency should be a foundational strategy to support plans to fulfill the climate policy goals moreover as for being lucrative. Energy efficiency with its integral role in energy intensive industrial processes is critically essential to realize desired energy-saving potential and reduction in Carbon Dioxide gas (CO_2) emissions.

The economics of industrial production, environmental conservation realities and global energy supply limitations are persistent concerns for the whole industrial sector. Regardless of where on turns, there's an unswerving appeal to preserve energy, reduce carbon emissions, and safeguard environment for future generations. These enablers made energy efficiency as an entrenched thanks to conduct the business. In fact, energy efficiency for any industrial facility (i.e. several processes/plants tied with a central utilities-plant) is extremely dynamic and wish to be improved throughout its lifetime, nonetheless of how energy-efficient facility design was achieved during design-phase. On the opposite hand, in its entire lifetime, a plant undergoes various retrofit subjected to high-profile operational changes/improvements/profits. The changes that include process disturbances, uncertain feedstock conditions, and products demand are short-term, requiring no major retrofit. However, long-term transformations comprise of must process more raw feed, feedstock changes, improve performance etc. warrants the debottlenecking of the whole facility and comprises key retrofits. In summary, may be stated that, the retrofitting of a plant may well be needed numerous times, in its lifespan, to upgrade energy efficiency and satisfy the required increase in production rates.

The main challenge is to develop the proposed facility design into a Sustainable Energy-Efficient Design, which implies a design that's capable of improving its energy efficiency supported by the energy-capital trade-off dynamics. Conceptually the problem will be formulated as a bi-level programming problem with two objectives, the minimum disruption within the energy utility supply (zero deficiency) and minimum energy utility consumption. In such cases the energy utility is minimized subject to least energy deficiency in supply. The outer problem of optimization of minimum energy consumption should be solved at the best-case scenario of the inner problem which is that the least possible deficiency in energy utility supply to the process plant.

On the macro level, there are three energy system components i.e. generation, distribution and utilization. Throughout the process facility design, the main target is to boost all three components of energy i.e. energy demand (process/user) minimization, efficiently fulfill energy demand (efficiently site-wide utility/supplier) and efficiently transport energy from supplier(s) to user(s). Objectives are to minimize: waste in energy, fresh resources and capital (de-bottlenecking) in these three components. This may be done via the continual upgrade of the efficiency of energy system components in generation, distribution and utilization. However, the utilization component has a unique feature, where its boundaries do not seem to be completely dictated by the process/users. Therefore, the space of improvement within this component is way wider than the others. The interaction among all three components specified waste from one component to be utilized by others and also modifications of one component supported the other components waste availability will suffice in developing energy efficient process design. Moreover, as highlighted earlier that over the last forty plus years several design standards/criteria aside from safety; health and environment besides capital and operating costs are considered, like switch-ability; flexibility; reliability; maintainability; availability; controllability; operability; acceptability and then on. Although, almost every industrial facility has to be retrofitted several times throughout its lifetime to satisfy its objectives which was coined as "retrofit with retrofit in mind approach". "Retrofit-ability" as a design requirement has not been addressed or may be coined within the chemical engineering design community until recently. It's worthwhile to say that the chapter emphasize to develop sustainable energy efficient design by reducing the process facility demand i.e. the utilization energy component of the process.

As an everlasting progression, sustainability may be a constantly embryonic field i.e. Every new facility that's developed has the potential to be more sustainable and energy efficient than the preceding as new technologies and methodologies are sprouting incessantly. There are numerous ways to conserve energy throughout the planning, construction, and operation of a facility — it's all about optimizing existing systems to further efficient settings. The expanse of energy ingesting in manufacturing has turn out to be a strategic factor in determining whether the business will remain competitive or not. There are certain rudiments of the manufacturing systems which are very difficult to alteration once the process facility is designed and meant functioning. Every manufacturing system or industrial facility design involves amassing together separate elements into a connected or a coherent whole i.e. Synthesize with the activities during which the varied process elements are combined together to generate the flowsheet of the system in order to meet certain objectives. Thus, each facility design may be a result of process synthesis which represents the configuration of the varied pieces of equipment and their interconnection represented on a flowsheet. It's almost impossible to induce a sustainable energy efficient facility design if the definition includes process synthesis as a part of the definition, humans are always in pursuit of excellence through knowledge and it has no practical limits. New discoveries will be made and new processes are going to be synthesized to attain the identical outcomes with meager energy and capital in an exceedingly more environment friendly manner. Because the facility could not be designed to cater inventions, consequently the inclusions of it within the sustainable energy efficient facility design definition is not defensible. Hence, the definition of sustainable energy efficient facility design does not embrace process synthesis but it lies within the response of the question i.e. the way to intensify facility design for a specified process which is capable of improving its energy efficiency grounded on the energy-capital trade-off throughout its life cycle. The foremost hurdle to achieve such a design is its acceptance to "retrofit", because the design has to be adapted several times during its life-cycle to be energy efficient. It will not be erroneous to conclude that the aspect of retrofit-ability could be a precondition for a sustainable energy efficient design [1, 2].

1.1 Retrofits

It is palpable that "Retrofits" are very significant to accomplish desired goal of developing sustainable energy efficient facility design. Retrofit/Revamp is defined as modification of a plant or facility to boost efficiency, reduce emissions and/or change capacity, and make it adequate for a replacement design or operating condition. Chemical engineering design community over the last forty plus years has considered several design criteria aside from safety; health and environment (SHE) besides capital and operating costs, like switch-ability; flexibility; reliability; maintainability; availability; controllability; operability; acceptability and then on. It's true that any industrial facility has to be retrofitted several times during its lifetime to meet its purposes. However, "Retrofit-ability" as a design obligation has not been addressed or maybe coined within the chemical engineering design community till some recent works, it necessitates attention of the community as emissions reduction and competitiveness are sought than ever before [3–5].

There might be various objectives for a retrofit, as an example, GHG emissions reduction, utilities savings, throughput increment, modifying appropriate network topology, upgrading heat transfer units, installing additional heat transfer area, re-piping streams and re-assigning heat recovery matches etc. Pinch technology imparts a systematic technique based on thermodynamics, design and economics heuristics for saving energy in processes and on entire sites. This procedure helps optimize thermal heat recovery, and immediately became popular as a theoretically elegant yet feasible approach to scheme Heat Exchanger Networks (HENs). Heat exchanger networks (HENs) are widely applied in industrial projects over the past decades because they offer significant energy and economic savings. Applications of HEN integration is divided into two categories i.e. grassroots and retrofit design. In oil refining, retrofit design are way more common than grassroots applications. The retrofit objective is to spot an economical HEN, subject to many design and operating constraints. However, implementing aforementioned retrofit strategies in practice could also be very difficult, because of constraints associated with the topology, safety and maintenance which regularly exist during a Heat Exchanger Networks (HEN) design. Besides, the capital cost is typically high due to considerable piping and civil works required for the retrofit and potential production losses during process modification. Nowadays and since late seventies of the last century, another important frequent change is going on because of the continual escalation in energy prices at a rate that's above plant equipment cost. It also warrants continual modification of the facility's HEN to reinforce the plant energy efficiency along its life time (sometimes reaching up to 30 years). In such cases, the HEN retrofit objective/task is to supply a practically implementable cost-effective HEN design modification that satisfies the new process objective and its new operating constraints. There are many possible modifications for an existing HEN to retrofit the first design to the new objective. It can include a mixture of all possible process operating and design condition modifications, existing HEN topological/structural modifications, and existing HEN unit design modifications and parametric modifications (such as heat transfer enhancement to boost U) furthermore. For example, fossil oil distillation plants grassroots design including pre-heat train (PHT) is well established, but without considering Retrofit-ability criterion during the plant design phase. The design and retrofit of the crude oil pre-heat train remains to this point the topic of the many research and development work, thanks to its importance in any crude oil refinery, since the petroleum distillation system is among the biggest energy consumers within the industrial community, and also the problem is not trivial with many decision variables and constraints owed to the high interaction between the pre-heat train and also the fractionation columns of the crude distillation plant. The retrofit of the crude distillation plant including PHT may be a task that can be conducted at least 4 to 5 times along the crude oil refinery lifetime not only because of the necessity for energy saving, GHG emissions reduction but also more importantly for throughput increase. Since the atmospheric and vacuum crude distillation towers designs are highly interlinked to the crude distillation plant pre-heat train (PHT), retrofit of any one system goes to severely impact the remaining systems. All of those objectives require heat duties within the PHT to be changed, surface areas to be altered, pressure drop in the PHT is varying, the requirement for adding new heat exchangers units, the option of changing units' sequence, the compulsion to split streams, the requisite even for new streams matching, the stipulation to change the atmospheric and/or vacuum towers internals, obligation to change crude pumps so on. Such situations will bring hard constraints to any plant owner to start out any retrofit on the idea of energy saving or energy-based

GHG emissions reduction particularly, unless it's absolutely necessary for unit de-bottlenecking via the furnace debottlenecking to permit throughput increase. In such situations usually, many good opportunities to save lots of energy consumption and reduce energy based-GHG emissions are overlooked. Although, retrofits are inevitable for nearly all facility during its lifetime but none of the facilities are designed to cater retrofits, the approach to develop facility design contains a hindrance to several improvements but especially to energy efficient modifications. Therefore, there's dire need of assimilating retrofit-ability as a design criterion within the facility design so as to stay energy efficient through its lifecycle [1, 2].

1.2 Sustainable energy efficient facility design

Retrofits are almost inevitable for a facility to be energy efficient throughout its life-time, but it emanates with shut-down, structural (process/HEN topology) and plot-plan limitations. Altogether, these constraints make retrofit infeasible and consequently facility energy inefficient at some point in their life-time. Thus, to achieve sustainable energy efficient design a much-sought criterion is that the design that may be successfully retrofitted during the course of its life time. Retrofit-ability could be a pre-requisite criterion for the sustainable energy efficient facility design and wishes to be entrenched within the design.

A healthy-aging design supported by retrofit-ability criterion i.e. no shut-down obligation, no proposed topology alteration or no-plot plan limitations for the execution of all energy efficiency projects will be energy efficient through-out its life time. Subsequently, "Sustainable Energy Efficient Facility Design" will be demarcated as a design for a given facility which shall be capable to acclimatize any essential modification for energy efficiency (i.e. trade-off between energy cost reduction and capital investment) at any point in its life-time with none restraint owed to its existing design. The sustainable energy efficient design is capable of adjusting to the dynamic trade-off between energy cost reduction and capital investment supported the principle of energy targeting without feasibility limitations because of present design throughout its life time. It shall be developed as a design supported energy-capital trade-off changing aspects from the beginning of its life-cycle till the termination. All essential modifications shall be planned and provision is formed within the initial design to include deviations with the identical or less capital as for a brand-new design with the identical energy values.

Minimum Energy Requirement (MER) of a facility may be a subject that ride the methodology pragmatic to focus on the demand, targets among the plants may be generated supported direct/indirect integration or the other technique. Although, Pinch approach which is widely utilized in industry to get energy targets utilizes indirect targeting methodology i.e. each process has energy targets (based on direct integration) and also the integration among the processes is indirect i.e. via buffer systems like through steam headers. Direct or a mix of direct and indirect integration also called hybrid integration among different processes/plants is capable to boost energy efficiency of the facility drastically and also provide opportunities to boost facility design through plot plan optimization but has several constraints and really limited usage within the industry. Any method which might generate energy targets supported energy and capital trade-off i.e. a design criterion like global approach temperature at the beginning and end of the facility lifecycle might be utilized to develop a sustainable design from energy efficiency perspective. This criterion will define the initial design and every amendment to reach in the ultimate design such current changes offers no hindrance to the subsequent modifications. Hence, the methodology employed to attain energy efficiency within the process facility design is one among the important factors for the development of sustainable design. The methodology employed to develop energy efficiency has to be systematic. As, development of sustainable design desires to foresee energy efficient facility design supported the energy-equipment cost trade-off values for this and further. Accordingly, methods supported on impromptu energy efficiency improvement are extremely difficult to develop current design and predict future modifications to reach the final design [3–5].

2. Methodology

Process design has evolved from inventions of recent design based upon experiments to today's novel process flow-sheet synthesis based-upon optimization. It responded to through a generation of process design based-upon the concept of stand-alone unit operations and therefor the integration between unit operations. The development of varied systematic design techniques supported on process integration has made it a promising tool in resource conservation activities. Improvement opportunities shall be identified either as process modification or design improvement as compared to the base case energy consumption regardless of the approach applied. As some approaches try to improve the energy performance of the facility design on extempore basis rather than developing it with an organized approach. Hence, it's important to define the base case energy consumption for measuring the performance enhancement so as to gauge the advance. The section is devoted to a number of approaches together with their advantages and dis-advantages which will use to boost the energy performance [6].

2.1 Primitive approach

In primitive facility design approach (**Figure 1**), only core process was designed and utility was provided for all energy consumption i.e. heating, cooling or shaft work with none concept of energy efficiency or perhaps utility targeting. Consequently, the two elements were designed separately with none emphasis on energy efficiency and also the process design is directly dictating the utility system design and there aren't any interactions among systems to boost the facility design. This approach was in practice when the efficient utilization of the resources wasn't priority in any aspect facility design. Nowadays, the approach is sort of extinct for the mass production in process industries, can be found in practice for very small-scale production or for batch processes or processes where the production mechanism is not fully known or could not be controlled with needed accuracy.

The approach is incredibly simple to implement and used generally for complex batch processes, because it is incredibly difficult to integrate them. As process and

Figure 1.
Primitive approach of facility design.

utilities systems are designed separately, only individual systems might be improved and advantages of integration cannot be realized utilizing this approach. The most advantages are simplified design and lower initial investment. There's no element of energy efficiency from demand reduction perspective, hence obtaining a sustainable energy efficient design utilizing the approach is sort of impossible.

2.2 Conventional approach

At this time, the efficient utilization of resources and particularly of energy has become a matter of priority in many aspects of industrial and social planning. The accepted reason is that the rapid increase within the cost of key resources which has made a reappraisal of the many adopted practices necessary. Additionally, there seems to be a case for efficient resource utilization whether it would not be economically justified yet: dwindling raw materials will tend to steer further price increases - and scarcities - within the future and conservation in time will help to arrange for such a course of events. Industry was beholding to enhance for energy conservation and to reduce negative environmental impacts but there was no systematic approach to attain it within the past. The improvements were only supported by observations and lessons learned from the previous experiences, sometimes energy optimization problem was very narrowly defined to realize improvements locally which could have adverse effect at a higher level. Accordingly, the facility design approach as shown in **Figure 2**, is improved but only on unpremeditated basis, still no interactions were studied to boost the facility design systematically. The sole system exists was that, usually the starting design of the new facility starts with the best available existing facility design but not always. This can improve the system design for one facility supported on the previous facility design but could not predict the long run design improvements for the same as no mechanism to predict future performance with the needed design modifications.

The conventional approach is predicated on the design of heat recovery system but on impromptu basis, the advantage of it's that some energy efficiency improvements may be subsequently identified. The energy efficiency improvements are supported by experience and without considering the interaction among the systems but supported by design and operation of comparable preceding system. The approach could end in facility with better energy efficiency and straight forward to design due to segregated system but lack benefits of integration. There's no provision within the approach to review interaction among the system to switch the system design. The approach is considered as an unstructured approach to boost energy performance without any feedback mechanism from each system. Moreover, targets supported by trade-off energy values and capital cost do not seem to be generated before the designs as a result future design performance/improvement are very tough to include into design. Consequently, improvised approach to enhance energy efficiency could not lay the inspiration of sustainable energy efficient facility design.

Figure 2.
Conventional approach of facility design.

2.3 Pinch approach

Energy Integration is that the corner stone of process integration science. While heat recovery was the initial focus of process integration, the scope has been expanded during the late 80's and 90's till the end of the last century to take account of most aspects of process synthesis and design. The single most important concept and therefore the one that gave birth to the sector of process integration is that the heat recovery pinch, discovered independently by Hohmann/Lockhart, Umeda et al. and Linnhoff et al. The Pinch concept is a systematic tool that provides critical information for a complete plant or perhaps site level. The concept is additionally generally applicable in areas aside from waste heat recovery. The main target of this approach is on basic energy recovery and utility targeting, utility selection, integration of heat pumps, heat engines, furnaces, distillation columns, etc., heat exchanger network synthesis and retrofit, combined heat and power integration and process optimization for energy integration [7, 8].

2.3.1 Problem formulation

Pinch technology is the technique that provides a systematic methodology for energy saving in processes and even total sites. The methodology is predicated upon thermodynamic principles which identify economically feasible energy efficiency improvements. Pinch Analysis was first developed within the late 1970s as a method for optimization of thermal heat recovery, and rapidly gained wide acceptance as a theoretically elegant as based on first and second law of thermodynamics yet practical approach to the design of Heat Exchanger Networks (HENs). Within a short period of time, it's evolved into a general methodology for optimization, based on the principles of process integration.

In Pinch approach, firstly, the core of every process is designed with fixed flow rates and temperatures yielding the heat and mass balance for the process. Then the design of a heat recovery system is completed for every process. Next, the remaining duties are satisfied by the use of the appropriate utilities. Process integration using pinch technology offers a completely unique approach to come up with targets for minimum energy consumption before heat recovery network design. Heat recovery and utility system constraints are then considered within the design of the core process. Interactions between the heat recovery and utility systems are also considered. The pinch design can reveal opportunities to adjust the core process to improve heat integration. The pinch approach is exclusive because it treats each process with multiple streams as a single integrated system. At site level, the waste heat is recovered using utility exchanger network i.e. utility or buffer systems are developed to recover heat from one process and utilize it for other process to attenuate net utility consumption at site level and opportunity to integrate cogeneration. It's been applied successfully not only to energy systems (heat recovery, pressure drop recovery, power generation), but also to fresh water conservation, wastewater minimization, production capacity de-bottlenecking, and management of chemical species in complex processes.

The main advantage of the Pinch approach (shown in **Figure 3**) is that the designed facility is going to be very energy efficient up to the amount that systems are modified for better integration based on the interaction among them. The elements which are missing is the design retrofit-ability (i.e. future design modification as per dynamic energy prices), direct heat integration among the processes, and

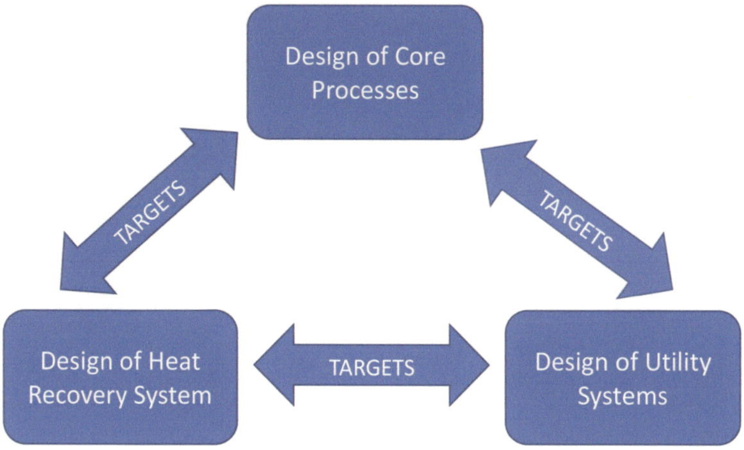

Figure 3.
Pinch approach for facility design.

waste energy recovery technologies. These elements are going to be incorporated into extensions of pinch approach in future but retrofit-ability can be added to pinch approach in current format or future extensions. It's worthwhile to mention that retrofit-ability is about provision of predicted future modifications into the current design and also the prediction might be better depending upon the approach. The elements direct heat integration and waste energy recovery technologies will affect level of energy efficiency while retrofit-ability is serving to succeed toward the extent whether high or low. Thus, retrofit-ability can be considered as a design criterion for any energy efficiency approach which is capable of predicting future designs from energy efficiency perspective. Pinch approach has all the eminence to qualify the prerequisite of an approach which might be adapted to develop sustainable design. As pinch approach may well be applied to predict the minimum energy consumption of a specified process for a given energy-capital trade-off. Henceforth, the longer-term energy consumption from the process may be predicted, supported by the energy-capital trade-off forecast. Therefore, the capital investment requirement to scale back the energy demand may well be extracted from the pinch approach and utilized to make the idea of developing sustainable energy efficient design.

2.3.2 Solution methodology

For each process within the site, Pinch analysis (Linnhoff et al., 1983) are often accustomed to set the minimum energy targets through thermo-economic analysis of potential recovery and process integration, and to spot heat exchanger network design and synthesis opportunities. All processes heat flow above and below the process pinch temperature is summed up for every temperature interval to get the site sink and source profiles respectively. Transfer of heat between the site source and sink streams takes place through the utility system, the quantity of heat recovery being estimated by the overlap between the two profiles. It also links the likelihood of cogeneration within the site supported by the design of the utility system. The methodology shown in **Figure 4** uses an iterative procedure to get targets for various utility system configurations and feedback among the systems. Many commercial software is accessible for heat integration within a process but only some for the site

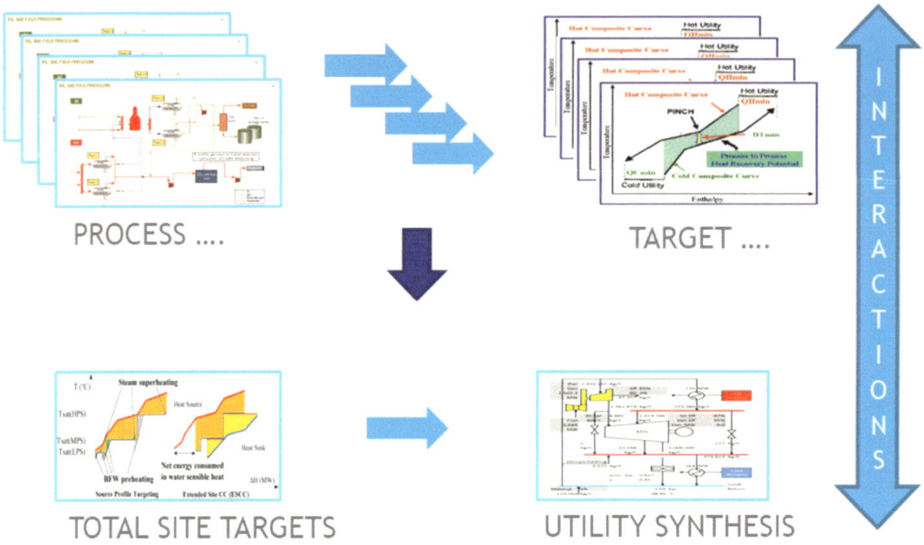

Figure 4.
Pinch methodology to achieve energy efficient facility design.

integration, none is out there for the utility synthesis. Even commercially available software has limitations in automatic design generation and would better utilized by the expert within the field [9–11].

2.3.3 Outcomes and limitations

The outcome from pinch approach may be a system that has benefits of energy recovery and in-process modifications for all processes and supplementing with energy utility system as shown in **Figure 5**. This approach is predicated on the economic trade-off between energy cost and equipment cost but only for a given set of values at some extent of time. The draw-back or limitation of such design is that, it does not consider the long run energy values escalation as compared to equipment cost and also future environmental regulations. These constraints are a hindrance to create the facility competitive from energy efficiency, environment or margins perspective throughout its life-cycle. All future retrofits will either require higher Capex for execution or infeasible as no provision were made within the original design for the mandatory alteration. Although, pinch design is predicated on energy values and equipment cost trade-off for the beginning of facility life-cycle but never considers the long run trade-off within the design.

One way to urge "Sustainable Energy Efficient Facility Design" is to urge sustainable design for every process within the facility which shall be modified for energy efficiency at any point in its life-time with none restraint due to its current design. A crucial feature of Pinch approach is that the ability to spot Performance Targets before the design planning is started. The ability to spot the target is the minimum requirement to appreciate sustainable design because it sets the footing of future designs and also the modifications required to attain it. The element of retrofit-ability is missing in Pinch approach but the developed design has enough flexibility (degree of freedom) to embrace this extra constraint. The approach must be extended beyond present Pinch Methodology to develop designs for all predicted trade-off values

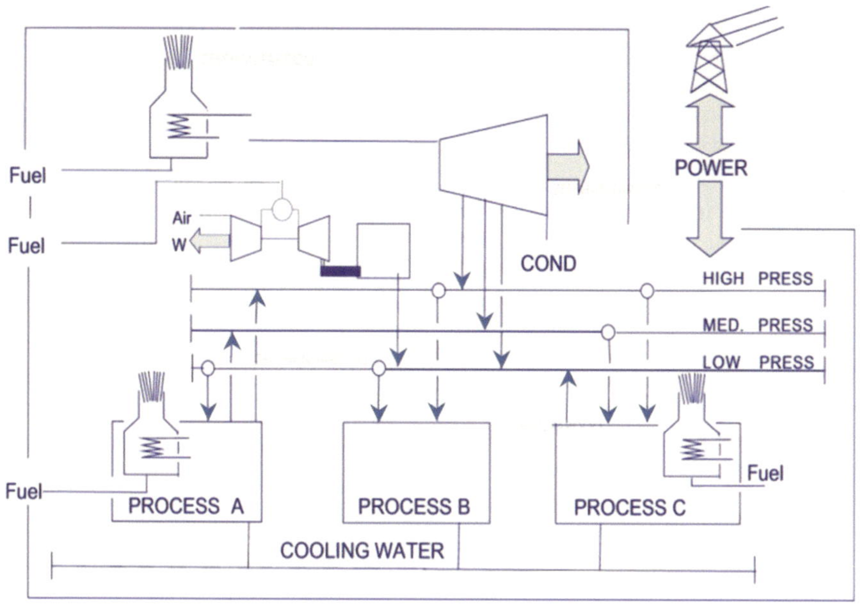

Figure 5.
Facility design based on pinch approach.

for the upcoming facility lifecycle. All the developed solutions for ever-increasing energy-capital trade-offs shall be compiled together to merge into one design which shall be modified within the future to urge subsequent designs. Finally, the availability of all identified future modifications shall be incorporated in the specified design and provisions are made for the subsequent modifications. The design developed might be modified with none hindrance at any point supported by the trade-off and could be considered as sustainable energy efficient process design. The sustainable process design approach may well be applied for each process of the facility to urge a "Sustainable Energy Efficient Facility Design". The facility design developed with these principles has all the elements of sustainability but lacks elements which are highlighted earlier as pinch approach limitations.

Although, Pinch Methodology is incapable of developing sustainable energy efficient design but the pinch framework may be utilized to develop it. Elements which are vague in Pinch approach is retrofit-ability, process to process direct unification, and waste energy utilization technologies needs scientific community attention. In future, a framework has to be developed to handle of these limitations and take away deficiencies to develop a well-integrated energy efficient retrofit-able design which may be "Sustainable Energy Efficient Facility Design".

3. ADU-VDU plant sustainable energy efficient design

Oil refineries are vital to the global economy and at the identical time major consumers of energy. Petroleum refineries are under increased pressure to reduce emissions of greenhouse gases, mainly CO_2 to accommodate with the upcoming stricter environmental regulations. Energy efficiency optimization may be a means solution to GHG emissions reduction because of its impact on energy consumption at the source. Heat exchangers play a significant role in crude oil refineries in energy saving, in general. Distillation is that the main consumer of energy in a crude oil refinery, and heat exchangers connected together in what's called a preheat train are essential to dramatically reduces the thermal duty of the atmospheric crude unit furnace. Crude distillation could be a primary processing operation in refineries throughout the planet and requires heat, steam and cooling to work. Although the crude distillation unit (CDU), that consists of both atmospheric distillation unit (ADU) and also the vacuum distillation unit (VDU), is not the foremost energy-intensive plant within the petroleum refinery, in terms of energy per barrel, every barrel of oil that is processed within the petroleum refinery passes through this unit/plant, making it the most important energy consumer, of the entire energy consumed, in fossil oil refineries.

Crude distillation process separates fossil oil into fractions in keeping with boiling point so that down-streams processing units/plants are often charged with feedstock that meets their particular specifications.

Crude oil separation process is accomplished by first fractionating crude petroleum at essentially atmospheric pressure and then feeding the high-boiling fraction, called topped crude or reduced crude, from the atmospheric distillation tower bottoms to a second fractionation tower that's operated under vacuum conditions.

The petroleum oil vacuum distillation unit is employed to avoid the high temperatures necessary to vaporize topped crude at atmospheric pressure. This unit reduces the hazard of thermal cracking, product discoloration and equipment fouling because of coke formation.

Before entering the atmospheric distillation tower flash zone, the petroleum oil charge is heated to the required desalting temperature, desalted, heated again to separate light fractions vapor during a pre-flash drum or pre-flash tower, heated again before the atmospheric unit furnace using product streams and column reflux streams, referred to as pump-arounds. The desalted and pre-flashed petroleum oil charge is heated up within the atmospheric distillation furnace(s) to about 375 C.

Topped crude from the atmospheric tower bottom sometimes called reduced crude is mixed with steam and pre-heated to about 390 C to 450 C before routed to the vacuum distillation tower. A system of vacuum pumps or steam ejectors is employed to form vacuum within the vacuum distillation column for the separation of high boiling temperatures cuts without its chemical degradation.

Crude oil distillation plants design including pre-heat train (PHT) is well established. However, the design and retrofit of the crude oil pre-heat train remains up to now the topic of many research and development because of its importance in any crude oil refinery, since the crude oil distillation system is among the largest energy consumers in industry, and therefore the problem is not trivial with many decision variables and constraints because of the high interaction between the pre-heat train and also the distillation columns of the crude distillation plant.

Nowadays, the retrofit of the crude distillation plant including PHT may be a task that may be conducted a minimum of 4 to 5 times along the fossil oil refinery lifetime not only because of the requirement for energy saving, GHG emissions reduction but also more importantly for throughput increase, change; for product mix/specification; (more gasoline than Diesel or vice versa) and permanent change in the API of the processed crude. Since the atmospheric and vacuum crude distillation towers designs are highly interlinked to the crude distillation plant pre-heat train (PHT), any retrofit of one system goes to severely impact the other. All of those objectives require heat duties within the PHT to be changed, surface areas to be changed, pressure drop in the PHT is changing, the necessity for adding new heat exchangers units, the requirement for changing units' sequence, the necessity to streams splits, the requirement even for brand new streams matching, the requirement to alteration the atmospheric and/or vacuum towers internals, the need to change crude pumps and then on. Such situations will bring hard constraints to any plant owner to start out any retrofit on the premise of energy saving or energy-based GHG emissions reduction particularly, unless it's absolutely necessary for unit de-bottlenecking via the furnace debottlenecking to permit throughput increase. In such situations usually, many good opportunities to avoid wasting energy consumption and reduce energy based-GHG emissions are going to be overlooked.

The feasibility of pre-heat train design modifications within the fossil oil distillation plant depends not only on the retrofit needs of the pre-heat train but also on the constraints associated with distillation towers. The interaction between the atmospheric and vacuum distillation towers, products and inter-coolers (top, middle and bottom pump-arounds) of both columns' conditions beside its hydraulic situations create a multifarious problem to the process owners that forces them to significantly re-consider on the premise of energy saving only and/or emissions reduction any design modification especially if the desired modifications need long downtime of the plant, to be implemented. Such constraints make the choice makers of any petroleum distillation plant avoid completely any try to change the pre-heat train design and only considering the modifications which mostly accept the initial pre-heat train design with minimal changes. The petroleum refineries owners have very valid point to behave this manner.

The moves of the heat exchangers units within the crude distillation plant pre-heat train from one location to a different for re-sequencing heat exchangers units, as an example, even between the same streams, will need a crane work inside the process area and definitely a crude distillation plant down time to implement it as a project. To maneuver one heat exchanger unit within the crude distillation plant pre-heat train to a new location to be matched with another stream will even be harder not only because it needs crane work and down time but another more involved engineering project to design the new pipework required and also the pipe rack capability to accommodate the new portion of the piping system including the pre-heat train re-piping required, civil work, instrumentation and control modifications material of construction selection, safety study/HAZOP. In many situations within the crude distillation plant's pre-heat train area, congestion will not allow such modifications in the slightest degree and if it allows it; the pipework modifications could be very expensive. Besides, in such situations the re-use of existing heat exchangers units, a minimum of from surface area point of view and/or materials of construction are going to be another unfeasible situation to contemplate for enhancing the pre-heat train energy efficiency.

Adding of new heat exchangers to reinforce the energy efficiency of the crude distillation plant pre-heat train via re-matching of streams between the crude oil cold stream and also the hot products for instance or the splitting of crude stream in one side and/or the towers products or pump around streams on the opposite side, whether or not the initial/original design may find it to be beneficial from energy saving point of view, most of the times the prevailing topology of the original design does not allow it because of the previous constraints and no way to proceed on the premise of energy saving merits alone. In many other situations the original/grassroots design of the pre-heat train does not have any merit in doing modifications to avoid wasting energy without revisiting completely the crude distillation plant pre-heat train grassroots design plot plan and re-do the design from scratch. Therefore, if we miss the "right" crude distillation plant pre-heat train grassroots design from the beginning, we are going to be constrained with the existing distillation towers and pre-heat train grassroots design plot and there'll be very limited opportunity to boost the pre-heat train energy performance. In conclusion, the initial design cannot be improved in the least along its lifetime. Therefore, it'll be beneficial to the petroleum refineries to form the pre-heat train design of the crude oil "right" from the beginning with inherent capability to capture waste energy with no topological modification of its original design. The inventions [US Patent 10,494,576 B2 (2019) & US Patent 10,822,551 B2 (2020)] renders energy efficient healthy aging design of crude oil refineries distillation units' Pre-Heat Trains. The invention is related to new energy efficient configuration of integrated crude oil atmospheric and vacuum distillation units' pre-heat train. The invention renders novel pre-heat sustainable design from energy consumption efficiency and fossil fuel-based GHG emissions' points of view along the fossil oil refinery lifetime solely; through the pre-heat train heat exchangers' surface areas manipulation (**Figure 6**). The novel pre-heat topology design is fixed and "right" from the star of the petroleum refinery commissioning up to the refinery end-of-service. It enables the cold crude oil stream of medium grade and mixed grade crude oils uses the identical topology with minimum energy consumption, compared with prior art, within the crude furnace before the atmospheric distillation column with none structural modifications along the petroleum refinery lifetime through some heat exchangers surface areas manipulation only.

This invention enables the crude oil refineries to know ahead of time their plot plan needs for future crude distillation units furnace debottlenecking and/or energy saving projects. The invention shown embodiment exhibits the small print of the pre-heat train's design for the Minimum Approach Temperatures Range of 30 C to 15 C and further the thermal duties of heat exchangers (Q) in MW and temperatures in degree C (the recommended values for these designs). The energy saving compared with state-of-art brand new refineries pre-heat train configuration is up to about 30 MW of fuel saving. This saving can increase even more by up to about 50% to save up to about 50 MW of fuel using the identical invention configuration with more surface area manipulation only as mentioned above while the prior art designs do not have this capability. Taking into consideration that oil refineries can live about 50 years, the missed opportunity in both fossil fuels saving and fuel-based GHG emissions reductions are huge. Taking also into consideration that every barrel of oil going to petroleum refineries worldwide goes through this pre-heat train, the worldwide missed opportunity will be significant and increasing with time.

The developed designed shall be further developed till the end of its life-time and shall be designed for every interval of Minimum Approach Temperatures which suits normal shutdown for instance 5 C i.e. Minimum Approach Temperatures of 30 C, 25C, 20C and 15 C for a life-cycle of 20 years with a shut-down cycle of each 5 years. Because the detailed design is completed and provisions were made during the initial design, during planned shut-down exchangers areas/shell shall be installed to induced the desired energy efficiency. The design is prepared to be implemented with an awfully short cycle and might be called "off the shelves design", as all requirements to be energy efficient supported on capital and energy values trade-off may be fulfilled with none hindrance. As ADU-VDU plant heat demand is almost constitutes of its all energy requirement, the propose design can be assumed to be sustainable energy

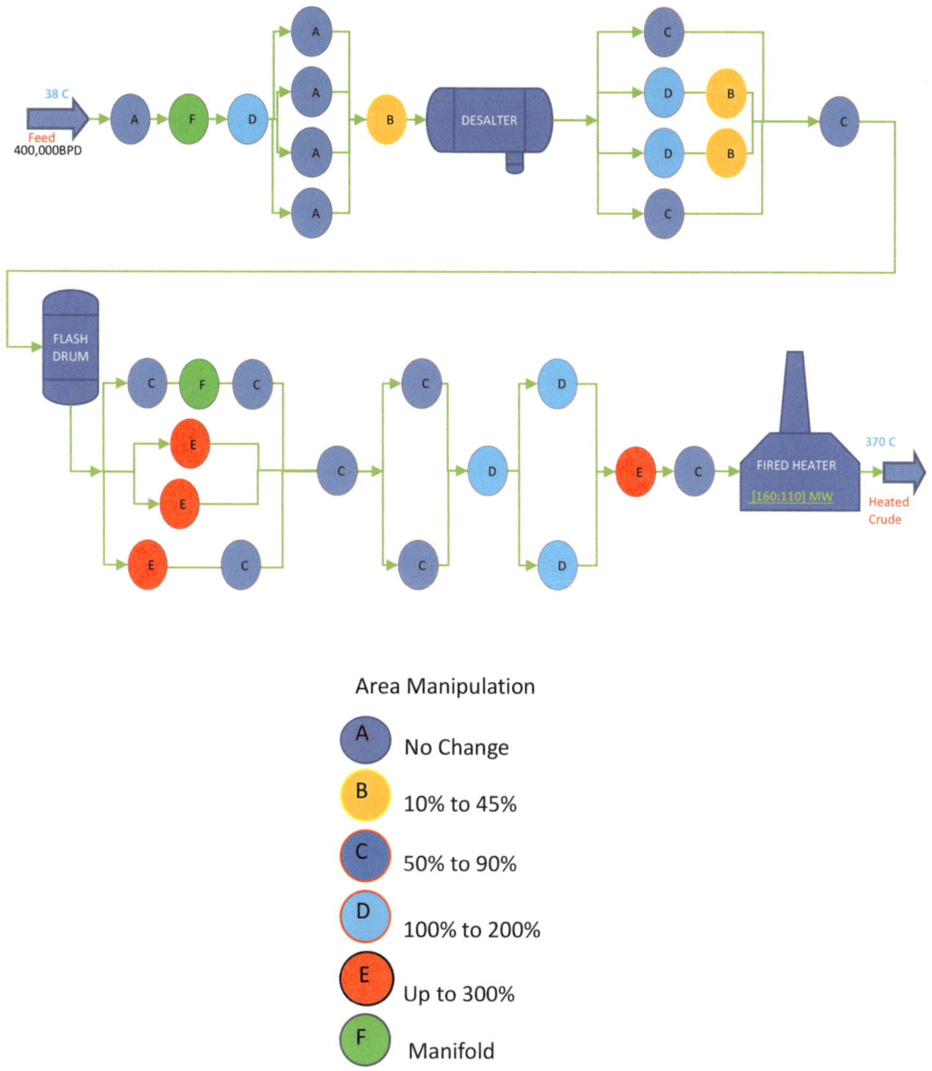

Figure 6.
ADU-VDU sustainable energy efficient design.

efficient design. For a whole sustainable design all energy elements i.e. heating, cooling, shaft power and waste energy recovery technologies has to be designed in the same manner [3–5].

4. Path forward

Nowadays integration is moving away from unit operations and focuses on phenomena; like reaction, heat and mass transfer that occur within a single piece of equipment. Future systematic techniques for process integration will not only explore for integration between units but also integration within units.

Process sites produce useful products in one or more process units and use a centralized utility system to satisfy the heat and power demands of the process units. Improvements in process technology, heat recovery philosophy and enhanced throughput are a number of the various reasons why the process changes must be implemented in an exceedingly site. Because of the interactions that exist between the various subsystems (e.g. processes, heat exchanger networks and utility systems), a change in one has a knock-on effect on the other subsystems. Many of the concepts and methods developed within the past for grassroots design, operational management, retrofit and debottlenecking scenarios, have focused on utility systems and process units independently. A clear strategy and systematic approach therefore must be developed to integrate all processes and utility systems and determine the synergies among them. Developing an efficient design of energy systems (utility system) requires an honest understanding of every subsystem in compass different processes. An effective design method is one that integrates the various subsystems within the site and captures the advantages instituted on interactions between them. Another element which needs attention is that the energy available finally after all integration i.e. energy which could not be employed in the present form and will be utilized through energy recovery technologies to the form energy which is required within the site. Physical insights and graphical methods of research is accustomed to study these interactions between the various parameters within the site and estimate the impact of process changes on site targets. However, as they do not consider the prevailing constraints within the site, they are less rigorous compared to mathematical programming techniques.

The approach (**Figure** 7) are going to be very effective in achieving energy efficient facility design because it will unify all processes, utilities, heat recovery and waste energy utilization to come up with targets and lay the inspiration to capture direct/indirect/hybrid integration among the plants. The resultant designs are going to be optimized for the full facility as even the plot plan is going to be optimized from direct and indirect integration perspective. Moreover, the retrofit-ability aspects of the design are going to be also covered within the approach, the resultant design will not only energy efficient today but easily modified as per trade-off to stay energy

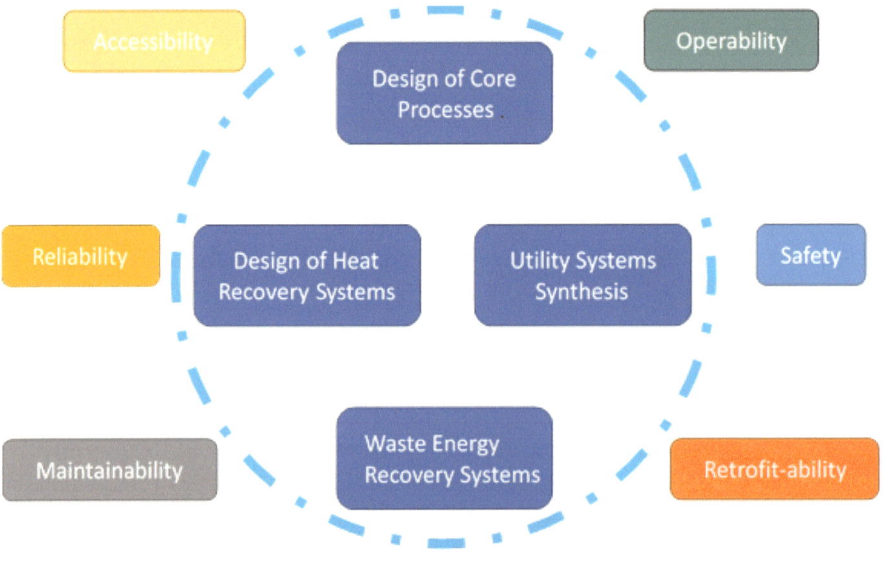

Figure 7.
Approach to facility design via integrated process-recovery-utility systems.

efficient throughout its life cycle. Because it is extension to the Pinch approach it has all its benefits and further improvement by unifying all facility design, low-grade energy recovery and utility synthesis. It's very difficult to magically establish such a global solution to figure from scratch. First, there should be the fundamental ingredients in situ, namely the desire of site-wide plants to participate and second proper tools and insights to conduct the analysis. These basic ingredients can then be enhanced and improved upon, with correct support structure in situ. The key to developing a successful sustainable energy efficient design is to determine the best integration among different plants and its utility system. Specifically, it's the material and energy flows relationship among the various plants which allow establishing optimal linkage to method a fruitful inter-dynamic structure. The other element is that it also looks to the application of energy technologies to the facility which starts with the quantification of the available waste energy that will not be utilized for integration during its life-time.

The method should be consisting of the holistic approach for total site targeting supported on the pinch technology accompanied by mathematical programming techniques, which is that the most generally used, where it allows waste heat from processes to be used as a source of heat in other processes. The waste heat sources are converted to steam, and then passed to processes that are in an exceedingly heat deficit condition via steam system infrastructure. To spot the external heating and cooling requirements of a group of individual plants from the central utility system, the temperature/enthalpy data from individual plants are first required to be extracted from the plant after the thermal integration of its hot streams to be cooled and cold streams to be heated using individual plant' grand composite curve. These grand composite curves define each plant' thermal heat deficiency and thermal heat surplus after intra-plant heat integration. The collection of grand composite curves of the whole site are then accustomed graphically add all thermal deficiencies to draw the entire site heating demand curve, and add all thermal surpluses to draw the entire site cooling demand curves. Such two curves are superimposed on one graph with the present and/or suggested steam generation levels and steam supplying levels to search out the minimum total site external energy utilities requirement and naturally best indirect inter-plants thermal integration, using the site steam system. During this method, intra-integration is completed first, which implies only waste heat of one plant is shared with other plants' members (which is not proactive form of cooperation). In this method the direct integration among adjacent plants, even using steam as a buffer, is not systematically addressed, where plant A, for example, with a surplus heat shall generate steam from its hot streams after its integration with its cold streams and send it first to the central utility and then the central utility sends it back to plant B leading to energy quality degradation and potential for mismatch between supply and demand. In this method, the mismatch within the number of steam levels required by the site users' generation and utilization leads to energy loss with very high possibility [12, 13].

The worldwide commercial software for transitioning industrial complexes to integrated-processes-utility-energy recovery facility does not exist and even the software for planning synthesis/design of the new ones for energy efficiency optimization do not seems to be much. The current state-of-art methods focus only on indirect integration using steam, after all for obvious reasons.

In brief, a mathematical programming formulation-based software is solved successively to enumerate all possible combinations of solutions ranked from best objective function value to least for every "user-selection" type of plants' integration

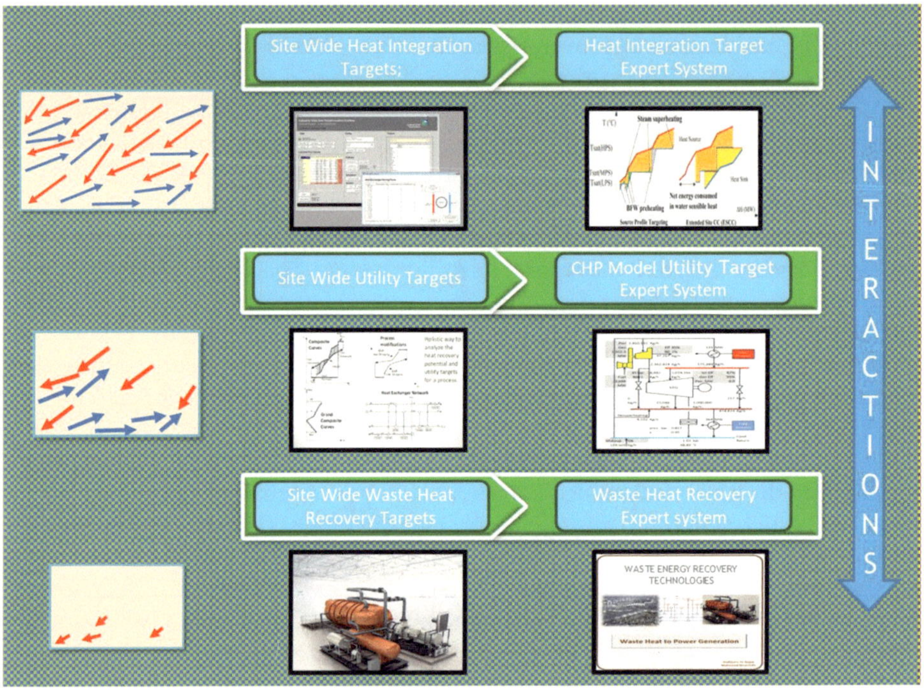

Figure 8.
Future methodology to achieve energy efficient facility design.

(membership in an alliance to avoid wasting energy) to spot the plants for inter-plants integration and also the ones for intra-only integration together with waste energy utilization technologies as shown in **Figure 8**. This may give an energy efficient design for a given economic trade-off and also has capability to optimize facility plot-plan for better integration opportunities but nevertheless it must be developed to be capable of life-time retrofit-ability. This needs solutions to be generated for all future economic trade-off supported on capital and energy costs forecasts, these solutions shall be compiled together to urge a design which can be modified from initial to final design with none hindrance. Provisions shall be provided for all the longer-term future modifications within the current design to form it a "Sustainable Energy Efficient Facility Design". The end result of the analysis is going to be highly integrated design as shown in **Figure 9** which is retrofit-able throughout its lifecycle and incorporated waste energy utilization technologies.

5. Overall summary and recommendation

5.1 Summary

The approach to attain energy efficiency in facility designed is incredibly important in achieving sustainable best-in-class energy efficient facility design. ADU-VDU case is presented to produce insight for developing plant level sustainable energy efficient design which may well be extended to the facility level. Since the facility operates as an integrated system of plants, units and streams, it must be understood in and of itself and it must be treated in and of itself. Research within the area of process systems

DOI: http://dx.doi.org/10.5772/intechopen.108829

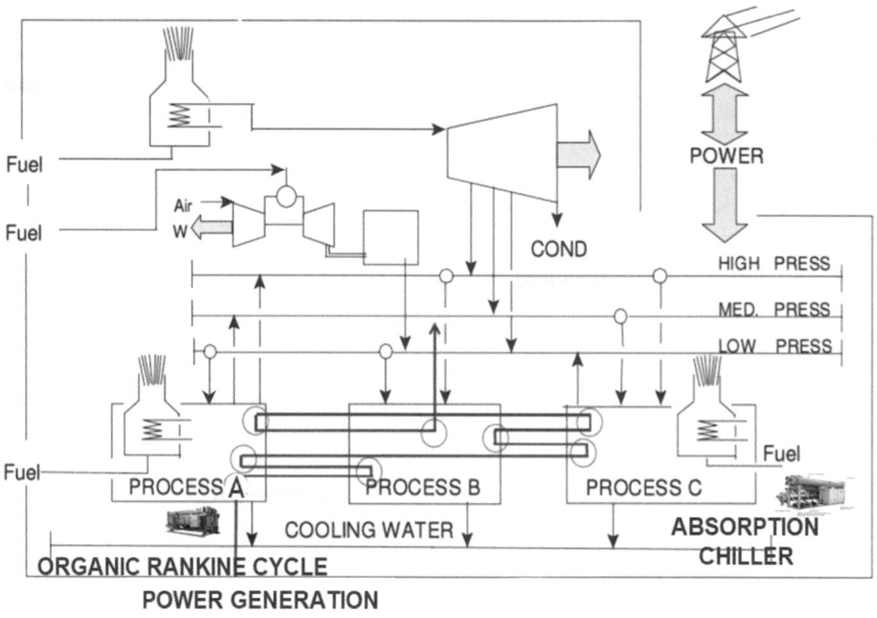

Figure 9.
Future energy efficient facility design.

engineering has led to a motivating breakthrough referred to as process integration. This breakthrough has expanded the influence of engineering design to methodically and insightfully address systems of multiple plants, units and streams employing a consistent basis of fundamental concepts of chemical and systems engineering. Process integration is defined as a holistic approach to design and operation that emphasizes the unity of the process. Thus, unifying all processes specified all design intents are met is that the way to achieve sustainable energy efficient design. All design aims are addressed well except Retrofit-ability, as an example Reliability, Accessibility and Maintainability is fulfilled by RAM Study while Safety and Operability requirements are accomplished by HAZOP Study. As, in its entire lifetime, an industrial facility undergoes various retrofit subjected to changes that features but not limited to process disturbances, uncertain feedstock conditions, product demand, throughput increase, feedstock changes, improve performance etc., it has to be considered as a criterion for the new design. In summary, it is stated that, the retrofitting of a plant may be needed numerous times, in its lifespan, the most effective way to approach it's through sustainable energy efficient facility design because it will fulfill most of retrofits objectives.

5.2 Challenges

There are several challenges which hefts to be considered so as to accomplish sustainable energy efficient facility design actually, these are highlighted to lure attention from the researchers within the field;

5.2.1 Retrofit-ability as design criterion

Inclusion of retrofit-ability as a design criterion ends up in a design which will be continuously valid for whole life span of a facility and at the same time reminiscent to a new energy efficient design of the upcoming facility. Retrofit-ability insertion as a design criterion at process level up to facility level is kind of challenging because the facility designs has to be developed for all energy-capital trade-offs furthermore as other scenarios. The developed designs must be compiled together to urge a design which can be subsequently modified without hindering the old design specified as the primary design being most inexpensive and least efficient (in terms of energy consumption not energy-capital trade-off) and shall be improved with minimal capital investment throughout its life-cycle. All designs must be further developed to form provision to incorporate all identified alterations within the facility from start till end. Developing a mathematical framework to capture all elements of the sustainable energy efficient design is difficult and lack of insights; preferably it can cause to integrate processes/stream which are never studied from the unification perspective and data/information is unavailable for unifying them. The foundation of integration starts with process, which approach ascends up to facility level, leading to unifying the entire facilities process with options of direct, indirect, hybrid integration options among the processes. Consequently, springing up with sustainable energy efficient facility design is extremely difficult, even the plant level design is difficult to realize. As shown even one sustainable process design generates lots of values to the facility by improving its performances, hence, it's advisable that in developing sustainable design evolutionary-approach may be adapted versus the more time-consuming revolutionary-approach.

5.2.2 Mathematical complexity

Analysis, like the problem definition phase, the targeting phase (approximations and heuristic rules that fail), likewise the design and optimization phase (multiple scenarios, topology traps, etc.). In theory, Mathematical Programming overcomes of these limitations however, a number of the corresponding models are extremely difficult to resolve. Moreover, the problem itself is incredibly difficult to formulate as economic trade-off predictions for the full life-cycle of a plant and develop feasible designs which might fulfill all trade-offs and provision for the inclusion of all needed modifications from start to end into a single achievable design are the parameters that cannot be easily obtained in mathematical equations. Even for a given process the insight from the prevailing design will be very difficult to formulate mathematically. If we utilize approach where all processes-utility-waste recovery are attempted to be unified which include parameters like plot plan modification's economical evaluation etc. the formulations are tremendously difficult to obtain. This is often another excuse to focus and develop sustainable design through evolutionary-approach instead of with the more time-consuming revolutionary-approach. Finally, it should be instructive to say that Mathematical Programming could provide a framework but very difficult to urge Automatic Design, which implies that time is needed to induce solutions with experience-based directions for high-level decisions [14–16].

5.2.3 Engineering design

To transform a design into a sustainable energy efficient design, it's important to develop conceptual designs into engineering designs which has all future modifications for implementation with none aggravations. Because the requirement is to develop facility life cycle healthy aging design, which implies it's embedded with many designs into a single design. The necessities are to develop each design for a period and therefor the subsequent modifications afterwards, the modification must be engineered in an exceedingly way that it shall be swiftly implemented. It needs extensive design experience to capture these elements into equipment design and develop it further to other details like piping layout, foundations, plot plan etc. It needs time to conduct analysis and also due to proposed design modifications, new scenarios for several engineering activities are going to be created, as an example HAZOP or RAM Study/assessment require more times to incorporate future retrofits, which is kind of challenging to realize. Finally, to appreciate such design the foremost important equipment is heat exchanger which must be designed with better understanding about future enhancements and well flattered by other designs like piping layout, foundations etc. Available tools like 3D models would definitely help in visualizing solutions and consequently within the design development.

5.2.4 Waste energy recovery technologies

One of the challenges of Sustainable Energy Efficient Facility Design is that the application of waste energy recovery technologies, as these technologies are evolving almost each year. It'd be difficult to define the energy efficiency of the facilities because of its application, as these technologies have very different efficiencies yet as different application will ends up in very changed facility efficiencies. It'd cause the requirement of developing an expert system which incorporates all commercially available waste energy recoveries technologies and a framework for their application which is quite challenging.

6. Conclusions and recommendations

To develop sustainable energy efficient facility design evolutionary-approach shall be adapted rather than the more time-consuming revolutionary-approach. The concept is to develop sustainable energy efficient process design of every individual process, which itself is challenging but might be realized. Subsequently, once all the processes are sustainable the facility is going to be sustainable in energy efficiency develop options for inter/intra/hybrid integration together with energy recovery technologies to form the facility "best in class" from energy perspective.

Definitely obtaining a sustainable energy efficient facility design is extremely challenging but the advantages from its enormous, even a design from only heat integration perspective are great. If a design, even at a process level from heat integration perspective is produced such it's fixed topology i.e. no new stream matches and energy efficiency improvements by area manipulation only will ends up in high energy as well as monetary savings. This is demonstrated through ADU-VDU sustainable design for a 400,000 BPD refinery, the advantages include energy savings of about \$100 million NPV (depending on energy vales) together with about 90 kilo tons of CO_2 emissions reduction annually. The modification to boost energy efficiency does not require extended shut-down for energy projects but may well be drained normal shut-down window. Moreover, lesser capital expenditure is required for the design modifications as compared to retrofitting the conventional design case.

Figure 10 summarizes the benefits of Sustainable Energy Efficient Design benefits, as the solutions are "off the shelves" design i.e. pre-engineered, it does not have shut-down, structural and/or plot plan limitations. Retrofits in normal design

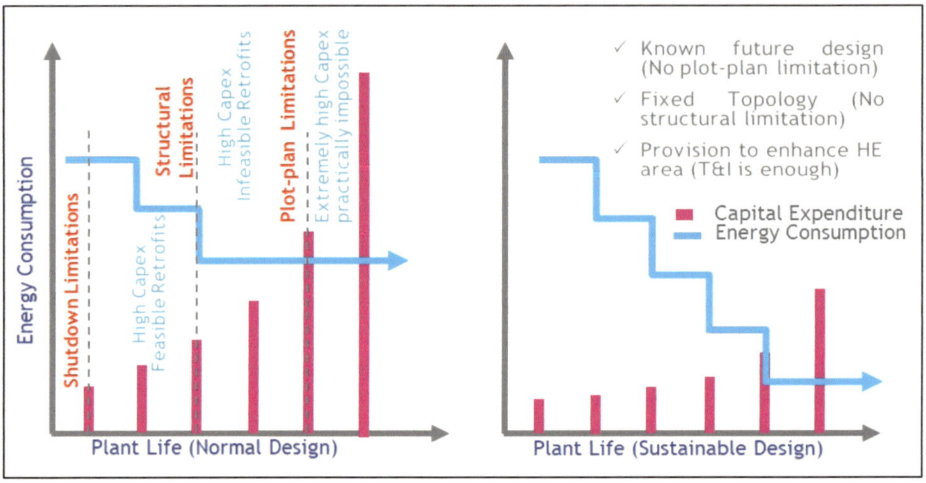

Figure 10.
Comparison between Normal versus sustainable energy efficient design.

are generally limited initially by shut-down opportunities as initially energy savings projects are feasible but later obstructed by structural limitations which implies the topology can either restricted or resulted in retrofits with very high capital and ends up with the area constrains may leads to impracticable retrofits. The constrains of shut-down, structural and plot-plan limits the facility to be energy efficient solely up to a particular level which will not able to be energy efficient supporting the economic trade-off for its life-time. On the other hand, the capital needed to do the modifications is much less in sustainable design as the provision for the modifications were made during initial design. Consequently, the initial as well as final facility design are very close to be energy efficient based on the economic trade-off. Hence, it is highly recommended for all chemical engineering community to come together for developing methods, techniques and frame-work to achieve "Sustainable Energy Efficient Plant Design" and extend it to achieve the ultimate goal of "Sustainable Energy Efficient Facility Design".

Nomenclature

ADU	Atmospheric Distillation Unit
BPD	Barrels Per Day
C	Degree Celsius
CDU	Crude Distillation Unit
CO_2	Carbon Dioxide
GHG	Green House Gas
HAZOP	Hazard and Operability
HE	Heat Exchanger
HEN	Heat Exchanger Network
MER	Minimum Energy Requirement
MW	Mega Watt
NPV	Net Present Value
PHT	Pre-Heat Train

Q	Heat Duty of Heat Exchanger
RAM	Reliability, Availability and Maintainability
SHE	Safety Health and Environment
T&I	Turndown and Inspection
U	Overall Heat Transfer Coefficient
VDU	Vacuum Distillation Unit

Author details

Zeeshan Farooq[1*] and Ali Al-Qahtani[2]

1 Energy Transition Engineering Department, Saudi Aramco, Dhahran, Saudi Arabia

2 Energy Systems Division, Energy Transition Engineering Department, Saudi Aramco, Dhahran, Saudi Arabia

*Address all correspondence to: zeeshan.farooq@aramco.com

References

[1] Farooq Z, Almutairi S. Low-cost retrofits to improve efficiency at Aramco refinery. Oil & Gas Journal. 2022;**120**(1):41-45

[2] Farooq Z, Almutairi S. Pinch analysis spots design retrofits to support future unit upgrade. Oil & Gas Journal. 2022;**120**(2):31-36

[3] Noureldin MB, Farooq Z. "Refinery Pre-Heat Train Systems and Methods". 2020. USP 10,494,576 (2019), USP 10,822,551 B2

[4] Noureldin MB. "Systems and Program Product for Heat Exchanger Network Energy Efficiency Assessment and Lifetime Retrofit". 2012. USP 8,150,559

[5] Noureldin MB. "Methods for Heat Exchanger Network Energy Efficiency Assessment and Lifetime Retrofit". 2012. USP 8,150,560

[6] Biegler LT, Grossmann IE, Westerberg AW. Systematic Methods of Chemical Process Design. New Jersey: Prentice Hall; 1997

[7] IPIECA. Pinch analysis. 2013. Available from: https://www.ipieca.org/resources/energy-efficiency-solutions/efficient-use-of-heat/pinch-analysis/

[8] Linnhoff B. Pinch analysis-state of the art overview. Transactions Institution Chemical Engineering Research and Design. 1993;**71**(A5):503-522

[9] Linnhoff B, Hindmarsh E. The pinch design method for heat exchanger networks. Chemical Engineering Science. 1983;**38**:745-763

[10] Gundersen T, Naess L. The synthesis of cost optimal heat exchanger networks: An industrial review of the state of the art. Computer and Chemical Engineering. 1988;**12**:503-530

[11] Ciric AR, Floudas CA. Heat exchanger network synthesis without decomposition. Computers and Chemical Engineering. 1991;**15**:385-396

[12] Micieta B, Markovic J, Binasova V. Advances in sustainable energy efficient manufacturing system. MM Science Journal. 2016;**06**:918-926

[13] YenLiew P, WaiLipTheo SRWA, JengShiunLim ZAM, JaromírKlemeš J, Varbanov PS. Total site heat integration planning and design for industrial, urban and renewable systems. Renewable and Sustainable Energy Reviews. 2017;**68**:964-985

[14] Floudas CA, Ciric AR, Grossmann IE. Automatic synthesis of optimum heat exchanger network configurations. AICHEJ. 1986;**32**:276-290

[15] Ven Reisen JLB, Polley GT, Verheijen PJT. Structural targeting for heat integration retrofit. Applied Thermal Engineering. 1998;**18**(5):283-294

[16] Asanta NDK, Zhu XX. An automated approach for heat exchanger network retrofit featuring minimal topology modification. Computer & Chemical Engineering. 1996;**20**:7-12

Chapter 4

A Review on Phase Change Materials for Sustainability Applications by Leveraging Machine Learning

Sunil Kumar and Debjyoti Banerjee

Abstract

Phase change materials (PCMs) have been envisioned for thermal energy storage (TES) and thermal management applications (TMAs), such as supplemental cooling for air-cooled condensers in power plants (to obviate water usage), electronics cooling (to reduce the environmental footprint of data centers), and buildings. In recent reports, machine learning (ML) techniques have been deployed to improve the sustainability, performance, resilience, robustness, and reliability of TES platforms that use PCMs by leveraging the Cold Finger Technique (CFT) to avoid supercooling (since supercooling can degrade the effectiveness and reliability of TES). Recent studies have shown that reliability of PCMs can be enhanced using additives, such as nucleators and gelling agents, including for organic (paraffin wax) and inorganic (e.g., salt hydrates and eutectics) PCMs. Additionally, material compatibility studies for PCMs with different metals and alloys have also garnered significant attention. Long-term studies for demonstrating the material stability and reliability of candidate PCMs will be summarized in this review book chapter.

Keywords: phase change materials (PCMs), thermal energy storage (TES), machine learning (ML), sustainability, material compatibility

1. Introduction

Phase change materials (PCMs) are pivotal in advancing sustainability across various industries, offering enhanced energy efficiency, reduced environmental impacts, and bolstered sustainability. Central to thermal energy storage (TES) and thermal management application (TMA), PCMs facilitate innovative solutions in cooling systems, electronics, and building designs. Their application ranges from conserving water in power plant condensers to minimizing the environmental footprint of data centers through efficient cooling strategies. Additionally, PCMs revolutionize building architecture by improving energy efficiency and lessening dependence on traditional heating, ventilation, and air conditioning (HVAC) systems, aligning with sustainability objectives. This chapter delves into the application of PCMs in

 IntechOpen

TES systems, exploring their use with additives like nucleators and gelling agents to improve performance in both organic and inorganic forms. It addresses material compatibility with various metals and alloys to ensure the long-term reliability of PCM-based solutions. Moreover, the chapter underscores the potential of machine learning (ML) in enhancing the sustainability, performance, and reliability of PCMs. Through the Cold Finger Technique (CFT) and ML, it tackles challenges such as supercooling, showcasing the significant role of PCMs in sustainability efforts and the innovative impact of ML on optimizing their application and environmental benefits.

1.1 Phase change material (PCM)

PCMs are commonly employed to enable efficient TES in compact configurations, leading to the categorization of these TES systems as Latent Heat Storage Units (LHSUs) [1]. TES platforms can be categorized as illustrated in **Figure 1**.

PCMs store energy during melting or solidification through both sensible and latent heat. Initially, a PCM begins in a subcooled solid state, warming up *via* sensible heating until it reaches its phase transition point. At this juncture, energy input causes a phase transformation, maintaining a constant temperature within a portion of the PCM. This phase change absorbs energy, crucial for the transformation process across its volume, eventually transitioning into a liquid phase. Further energy input then increases the temperature, facilitating sensible heat storage based on temperature change (ΔT) and the PCM's specific heat capacity. Although the storage of latent heat, determined by the PCM's enthalpy, is the primary method in PCM TES systems, a minimal amount of sensible heat storage also occurs in both phases, constituting a small portion of the total energy stored. Melting and solidification behaviors, based on the thermal energy stored in the PCM, are depicted in **Figure 2**.

1.2 A brief history of PCM research

Research on PCMs has evolved significantly since the 1970s and 1980s, initially focusing on enhancing building comfort and energy efficiency. This foundational

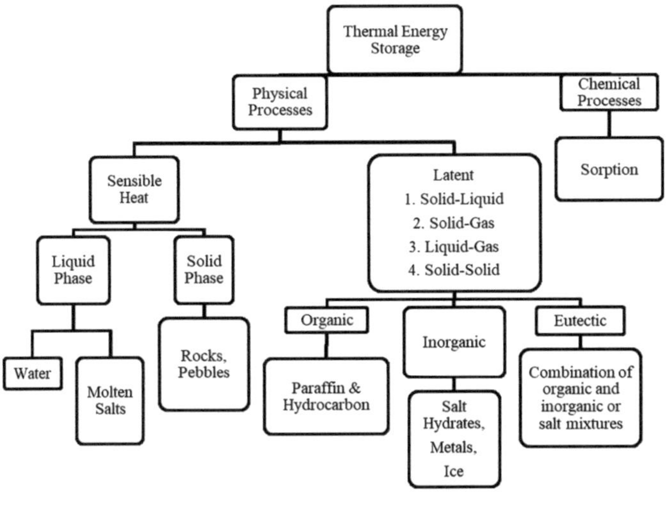

Figure 1.
Classification of thermal energy storage (TES) materials [2].

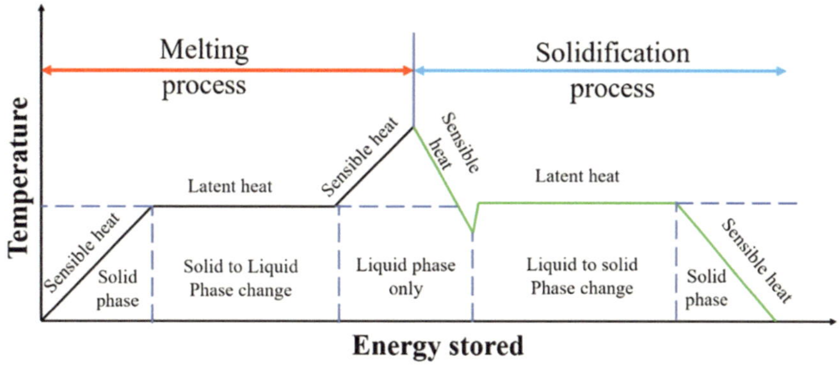

Figure 2.
Melting and solidification behavior of PCM.

work has led to the innovative use of PCMs in construction materials for sustainable building practices, demonstrating their importance in clean energy solutions, such as in large-scale solar power plants and domestic hot water systems [3, 4]. NASA's advancements in thermal management for space shuttles during the late 1970s further underscored PCMs' utility [5]. By the early 1990s, PCM research expanded to address electronic cooling challenges in high-performance computing, marking the beginning of its application in high heat flux scenarios [6]. The early 2000s saw PCM research diversify into both high- and low-temperature applications, including waste heat recovery and utilization, essential for sustainable energy practices [7]. More recently, the integration of ML techniques has marked a significant milestone, enhancing PCM selection, design, and utilization. This has led to accelerated discoveries of optimal PCMs, improved thermal energy storage systems, and advanced thermal management strategies, showcasing the potential for ML to revolutionize PCM research and applications in sustainability and advanced technology domains [8].

1.3 Literature survey

PCMs are integral to TES systems, celebrated for their ability to store heat energy efficiently in compact designs through the significant enthalpy changes during phase transitions. Ideal PCMs exhibit high latent and specific heat capacities, excellent thermal conductivity, and minimal volume change, and are nontoxic and noncorrosive, ensuring effective heat transfer [9]. TES, utilizing PCMs, is crucial for balancing thermal energy supply and demand, leading to the development of compact and efficient Latent Heat Thermal Energy Storage Systems (LHTESS) [1, 10]. These systems operate within a limited temperature range, focusing on the phase transition temperature [11]. PCMs have broad applications across various domains, including underfloor heating [12], building energy management [13], solar power [14], waste heat recovery [15], domestic water heating [16], electronics thermal management [17], passive cooling [18], and thermal insulation for unmanned underwater vehicles [19], showcasing their versatility and effectiveness in enhancing energy efficiency and management.

Nanofluids have enhanced TES by incorporating nanoparticles to enhance specific heat capacity [20]. PCMs are grouped into organics (e.g., ice, paraffin), inorganics (e.g., salt hydrates [$LiNO_3{\cdot}3(H_2O)$]), and eutectics [21]. Organic PCMs, for instance, paraffin wax, offer uniform freezing and reliability but have low storage capacity (i.e., needs larger quantities) and poor power ratings due to low thermal conductivity. Solutions like finned tubes and metal matrices have been explored to enhance power ratings [22, 23]. Inorganic PCMs like salt hydrates have high latent heat capacity but suffer from subcooling, phase segregation, degradation, poor nucleation, significant volume changes, and corrosiveness [24, 25]. Efforts to address these issues have been ineffective and costly. Eutectic Phase Change Materials (EPCMs), such as $Na_2CO_3{\bullet}10H_2O$-$Na_2HPO_4{\bullet}12H_2O$, offer stable PCM with high enthalpy and durability. EPCMs can achieve unique phase change temperatures, by adding stearic acid. However, EPCMs also face challenges like low thermal conductivity, phase separation, supercooling, and corrosion due to their diverse PCM blend [26, 27].

Subcooling or supercooling poses challenges to the power rating and reliability of thermal energy storage (TES) systems by delaying solidification in phase change materials (PCMs), crucial for time-sensitive applications. The phenomenon is quantified by the difference between the phase transition and nucleation temperatures, typically assessed using the T-History method [28]. Factors like nucleation types, surface finish, and experimental conditions significantly impact subcooling. To mitigate subcooling in salt hydrates, researchers emphasize the importance of nucleation methods. Seeding techniques, introducing nucleating agents with compatible lattice structures [29], and the Cold Finger Technique, which initiates nucleation by creating a cold spot [30], are explored alongside dynamic methods employing high-pressure and shock waves to prompt PCM solidification, enhancing the thermal performance of Latent Heat Thermal Energy Storage Systems (LHTESS) [31].

The CFT addresses subcooling by leaving part of the PCM unmelted for nucleation, with controlling the melt fraction posing a challenge. ML, particularly through an artificial neural network (ANN) employing radial basis functions, offers a robust solution. It predicts the PCM's melt fraction more accurately than traditional methods, improving thermal capacity utilization and mitigating subcooling. This approach is effective across various conditions, given sufficient data for model training [32].

2. PCM and their applications in TES

PCMs are cutting-edge tech with potential to boost efficiency, reduce environmental impact, and enhance sustainability in engineering applications. The area of PCM application has been discussed in the following sections.

2.1 Thermal management of electronics

Moore's Law, which forecasts the doubling of computing chips' density and processing power every 2 years, has significantly increased the need for advanced cooling solutions due to higher heat dissipation and reduced form factors. This trend poses challenges for thermal engineers in developing efficient chip-cooling technologies to prevent chip damage and maintain performance by keeping temperatures below critical levels. Portable electronic devices, designed for intermittent high computational activity, require cooling systems that can efficiently manage sudden heat spikes without draining the battery or accelerating aging. Phase change materials (PCMs) are ideal for this purpose, absorbing energy peaks and gradually releasing heat during standby, thus maintaining a stable operating temperature and effectively managing the thermal load. A study by Joshi and Pal demonstrated the efficacy of PCMs, like paraffin and eutectic alloys (**Figure 3**), in providing passive thermal management for mobile devices, showcasing their ability to control system temperature during variable power operations [33, 34].

A study by Amon and Vesligaj on the impact of transient loads on electronics, with a duty cycle involving a 10 W peak for 45 minutes followed by 30 minutes of standby, showed that Phase Change Materials (PCMs) effectively mitigate temperature fluctuations, limiting the temperature to 31°C compared to 37°C without PCMs [35]. This demonstrates that PCMs provide reliable passive cooling for small electronic packages, reducing the need for moving parts found in active cooling systems. Incorporating PCM-based heat spreaders addresses design challenges in modern applications by meeting geometric constraints and minimizing solidification time.

2.2 Advances in PCM technology for data center cooling

Data centers, crucial for digital infrastructure, comprise IT equipment, infrastructure support, communication links, and security devices, operating continuously. This operation generates significant heat, requiring management for reliability and sustainability. User load varies diurnally, with "peak hours" during the day and "off hours" at night, challenging cooling systems, especially as server upgrades alter thermal characteristics.

Data centers utilize various cooling strategies, including Computer Room Air Conditioning (CRAC) units, airflow management, free cooling, liquid cooling, and thermosyphons, broadly categorized into airside and liquid-side systems [36]. Air-side cooling, particularly CRACs involving chillers, water pumps, fans, and cooling towers, is preferred for its reliability, lower initial costs, and maintenance ease. However, vapor compression systems, integral to CRACs, consume substantial energy due to year-round operation. Current practices also involve energy management, airflow optimization, single-phase, and phase change cooling to enhance efficiency, as detailed in **Figure 4**.

A review by Ma et al. [38] highlights the limited research on data center cooling systems incorporating cold energy storage for power outages. Simulations by Wang et al. demonstrated that PCM plates could maintain data center temperatures below

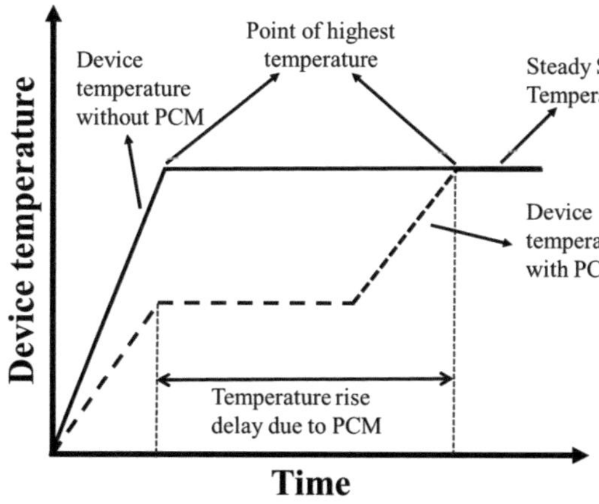

Figure 3.
Comparison of temperature response of a heat dissipating system with and without PCM.

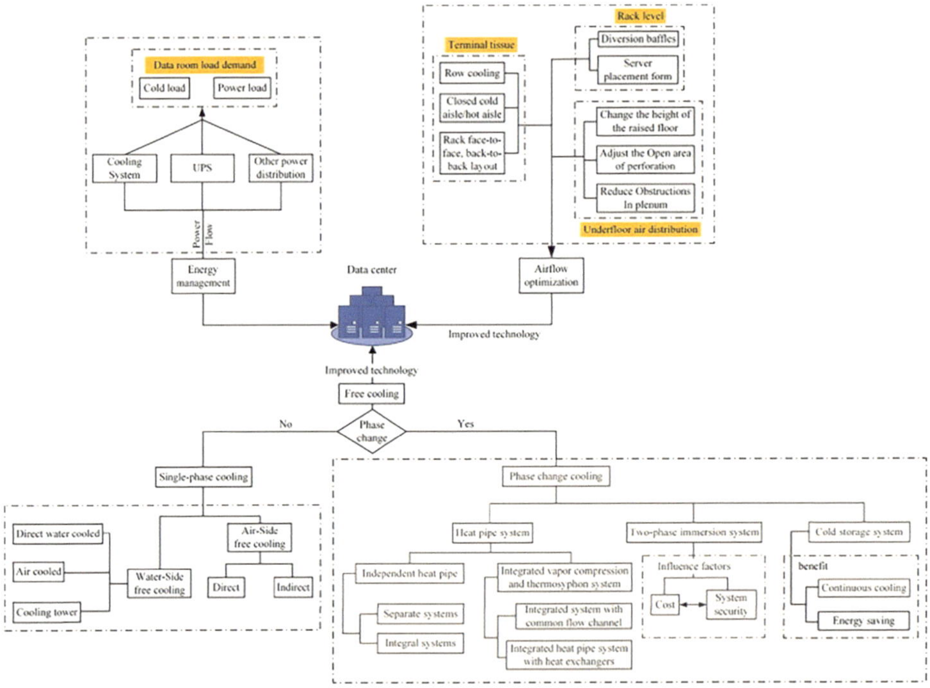

Figure 4.
Currently available high-performance and effective cooling technologies applied in data centers [37].

35°C for 9 hours during emergency power failures [39]. Another study optimized a latent heat storage system with a tube-in-tank design, finding that thermal conductivity enhancements significantly 0.2 to 1 W/(m·K) [40] increased its capacity effectiveness. Zheng et al. designed an air-based phase change material storage system for emergency cooling in data centers, showing through simulations that it could discharge over 2.5 kW for 30 minutes during power outages, offering a promising solution for maintaining critical temperature thresholds [41].

Phase change cooling (PCC) vastly surpasses traditional air-cooling methods, offering over a thousand times the cooling capacity [42]. It utilizes an under-evaporation technique, allowing for the use of various low-boiling electrolyte fluids and refrigerants [43], with the choice of refrigerant being key to its performance. Extensively researched and applied in data centers, PCC demonstrates exceptional thermal efficiency and energy conservation [44]. The incorporation of phase change materials (PCMs) further boosts PCC's efficiency by enhancing evaporation and condensation processes, enabling the temporary storage and release of excess heat [45, 46]. This prevents overheating in data centers, reducing dependence on conventional cooling methods and lowering energy consumption.

2.3 PCM technology for buildings

To create buildings with a comfortable indoor atmosphere, effective ventilation is key, achievable through the smart placement of diffusers and exhausts, alongside managing heating by circulating warm air and minimizing losses with PCM [47, 48]. The COVID-19 pandemic has further emphasized the importance of indoor space management to curb virus transmission. Building materials, like thick walls, are crucial for insulation, absorbing heat during the day and releasing it at night to maintain comfortable temperatures and mitigate temperature fluctuations in arid climates. Since the late 1980s, PCMs have been incorporated for efficient thermal management, absorbing excess heat to reduce temperature peaks.

This integration enhances energy density in building materials, reducing temperature spikes and HVAC energy needs, with PCM phase transition temperatures ideally between 20 and 30°C. Various PCM applications have been explored, including macro-encapsulated PCMs in gypsum boards, immersion in gypsum, mixing in concrete, and use in roofing, offering several strategies for improving thermal management in construction (**Figure 5**).

Through compression molding, Zhang et al. used microencapsulated PCM with gypsum and glass fiber. The results indicate that a mix of 50/50 composite gypsum board can absorb energy and moderate temperature rise for 48 mins and reduce peak temperature values to those of standard gypsum board. The major concerns with the use of PCM for buildings are availability, cost, additional weight, structural integrity, and suitable range of melting temperature afforded by PCM [49].

2.4 Supplemental cooling in air-cooled power plant condensers

Power plants, pivotal in electricity generation, necessitate efficient heat management, particularly in nuclear reactors, to ensure operational safety. Heat generated within a reactor core is dissipated into a water pool with submerged heat exchangers during maintenance. Without proper intervention, this can lead to thermal stratification, affecting the system's efficiency [50–52]. Utilizing adiabatic shrouds around heat exchangers can prevent such stratification [53, 54]. Moreover, incorporating PCMs in the pool's surrounding walls offers dual benefits: mitigating concrete damage and providing additional energy storage to postpone boiling. Given the extensive water and energy demands of conventional cooling systems, PCMs emerge as a vital innovation, exploiting their latent heat of fusion for significant thermal energy absorption during phase transition. This approach not only enhances cooling efficiency but also promotes sustainability by substantially reducing water consumption in power plant cooling operations.

2.4.1 Thermal power plant

Industrial cooling towers in thermal power generation are the largest consumers of freshwater. As energy demands continue to surge, the availability of freshwater for cooling purposes is expected to fall short, placing significant stress on freshwater resources. Consequently, there is a need for alternative technologies. One such alternative is dry cooling, utilizing air-cooled heat exchangers to eliminate the requirement for cooling towers. But the approach comes with trade-offs, including reduced operational efficiency, heightened capital and operational costs, and decreased reliability. In arid climates, air-cooled heat exchangers can become inoperative on scorching summer days when ambient air temperatures surpass the steam temperature at the turbine exhaust, resulting in power plant shutdowns and grid instability. To overcome this, supplemental cooling solutions like TES systems incorporate PCMs in LHTESS. It enhances performance and reliability, especially in areas with soaring daytime temperatures that exert backpressure on turbines and risk tripping failures.

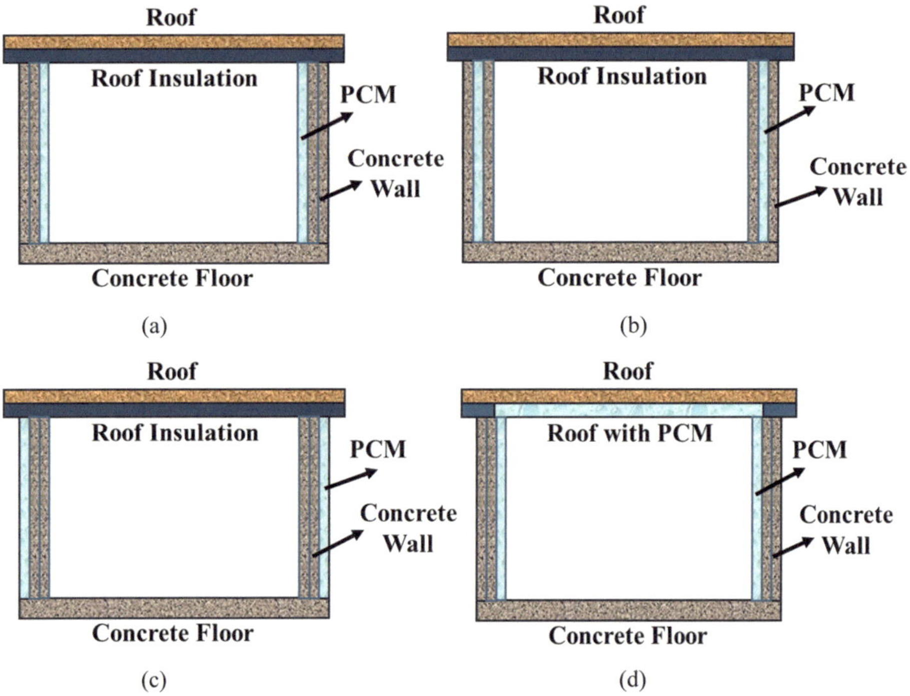

Figure 5.
Application of PCM in buildings: Encapsulated PCM (a) internally bonded, (b) sandwiched between concrete walls, (c) externally bonded, and (d) roofed and internally bonded room models.

2.4.2 Thermoelectric power generation

Thermoelectric power generation is a significant freshwater consumer in the United States, with a large part of the water used lost to evaporation. These power plants, operating at 35–55% efficiency, contribute to thermal pollution, endangering local ecosystems. While wet cooling systems, which utilize cooling towers, are more efficient and cost-effective due to water's superior heat transfer capabilities, their extensive use is increasingly seen as unsustainable due to the strain on freshwater resources. As a response, there is a growing need to improve dry cooling technologies. Latent Heat Thermal Energy Storage Systems (LHTESS), incorporating phase change material (PCM)-based heat exchangers, offer a promising solution. LHTESS enhances the efficiency and reliability of dry cooling systems, particularly beneficial in arid areas prone to significant temperature swings and power generation challenges [55].

2.5 Solar energy systems and power plants

The global population increase has escalated energy demands, largely fueled by economic and technological advancements. Traditional energy sources such as fossil fuels and nuclear energy, while crucial, are finite and contribute significantly to greenhouse gas emissions, posing a threat to the environment [56, 57]. Hence, there is an imperative shift toward renewable and sustainable energy sources like solar, wind, hydro, and bioenergy to minimize environmental impact. Solar energy, in particular, has gained global traction due to its universal availability and potential in power production, drying, and space heating [58–62]. Enhancements in Solar Air Heaters (SAH) efficiency through ribbed heating surfaces and PCM [63–65], alongside the development of computational and mathematical models for PCM and nanofluids [66–69], mark significant progress. Yet solar power systems face challenges such as irregular energy supply, efficiency fluctuations [70], and nighttime power generation gaps. Concentrated Solar Power (CSP) systems, which concentrate sunlight to heat a transfer fluid for power generation or storage in TES systems, offer solutions to extend operational hours and improve efficiency. Despite these advancements, residential solar thermal applications encounter distinct challenges, further explored in subsequent sections. The integration of TES in CSP, illustrated in the System Advisor Model (SAM) for a parabolic trough plant, is a step toward addressing these limitations, enhancing efficiency, and reducing costs (**Figure 6**).

Traditional TE systems use molten salts for sensible heat storage, preferred for their high operating temperature ranges and low vapor pressures. Innovations in molten salt formulations are being explored to extend these temperature ranges, aiming for cost reduction and enhanced thermal efficiency. Research by Robak et al. on using PCMs in CSP applications showed that a PCM-based system employing heat pipes could significantly reduce the volume and overall system cost by 65 and 15%, respectively, compared to traditional two-tank models [72]. Nithyanandam and Pitchumani supported this, finding an 11% cost reduction in electricity production with PCMs over conventional sensible heat storage. The adoption of PCMs in TES could lead to substantial savings, though challenges like limited high-temperature PCM availability, phase segregation, and low thermal conductivity need addressing [73].

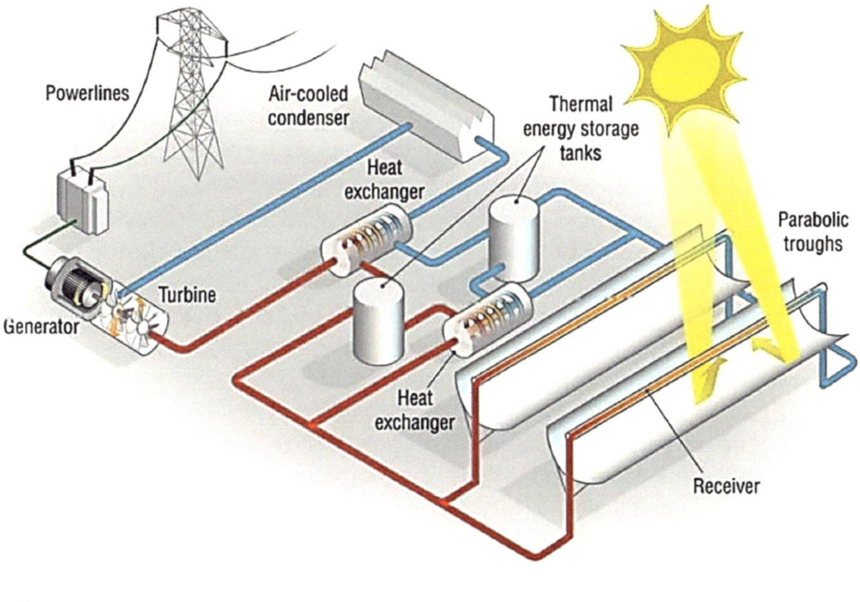

Figure 6.
System Advisor Model (SAM) sketch of a parabolic trough plant. Credit NREL/SAM [71].

3. Enhancing PCM reliability with additives and addressing challenges of supercooling and compatibility

To harness the full potential of the PCMs and their reliability, additives or specific material types can be used, which helps overcome challenges related to PCM stability, supercooling, and degradation, making PCMs more dependable for various applications. For the efficiency and effectiveness of PCM-based thermal energy storage systems, additives are added to enhance their thermal properties, stability, and overall performance. A list of additives includes graphite, carbon nanotubes, metal nanoparticles, expanded graphite, cellulose fibers, silicon dioxide, polymers, microencapsulated PCMs, salt additives, paraffin wax modifiers, phase change enhancers, nanoencapsulations, fibrous materials, inorganic materials, and organic materials. To improve the stability and reliability of the PCMs, stabilizers are added. The list of stabilizers includes encapsulation, nanoencapsulation, inorganic matrices, polymer matrices, thickeners, phase change enhancers, surfactants, crosslinking agents, anti-agglomerating agents, antioxidants, thermal stabilizers, chemical modifiers, hybrid materials, and composite materials.

Li et al. discuss the benefits of incorporating nanoparticles, such as metal oxides and carbon-based particles, into phase change materials (PCMs) to enhance thermal conductivity, reduce temperature fluctuations during phase transitions, and improve thermal cycling performance [74]. Organic polymer-based additives are used to ensure uniform phase distribution, preventing phase separation and leakage, common issues in certain PCMs. The employment of stabilizers enhances the long-term reliability of PCMs by preventing phase segregation and chemical degradation, crucial for thermal insulation and renewable energy applications. Additives significantly improve PCM performance, addressing challenges like supercooling and enhancing stability, durability, and thermal efficiency.

3.1 The PCM supercooling

Supercooling occurs when a phase change material (PCM) like water does not solidify at its equilibrium freezing point, staying liquid even below this temperature due to the absence of nucleation sites, which are necessary for phase change initiation. This state hampers the PCM's ability to store and release thermal energy efficiently, impacting the effectiveness, reliability, and performance of thermal energy storage (TES) systems. The degree of supercooling is influenced by factors such as impurities and surface conditions. To counteract supercooling, strategies like incorporating nucleators, using gelling agents, and applying the Cold Finger Technique are employed, enhancing the operational efficiency of TES platforms [75].

3.2 Nucleators as additives for reducing supercooling

Nucleators play a pivotal role in the nucleation process, acting as catalysts that trigger the initial solid crystal formation from a supercooled liquid, a crucial phase in various applications. These nucleators provide necessary sites for molecules or particles in a liquid or solution to begin clustering into solid crystals, significantly addressing the challenge of supercooling in PCMs. Research has explored the use of nucleation agents to mitigate supercooling, with typical nucleators including inorganic materials like talc or alumina nanoparticles and specific organic compounds. The choice of nucleator depends on the PCM used, its application, and the desired properties. For example, a study found that adding 20% by mass of NaCl to n-octadecane-based PCMs reduced supercooling by 6°C, demonstrating the impact of nucleators in enhancing PCM performance [76], provided in **Table 1**.

3.2.1 Mechanisms and benefits of nucleator

Nucleators are essential in phase change materials (PCMs) for providing sites where PCM molecules can organize and solidify. Without them, PCMs may not transition to a solid state efficiently, remaining disordered and liquid below the intended phase change temperature. By reducing the energy barrier, nucleators facilitate an organized crystalline formation, enabling a more immediate phase change near the desired temperature. This significantly enhances thermal energy storage efficiency by mitigating supercooling and ensuring predictable, consistent phase change behavior, crucial for reliable thermal energy storage applications [77].

3.2.2 Recent developments in the use of nucleators

Recent advancements in nucleator technology have significantly improved the effectiveness and adaptability of PCMs across various domains. The strategic incorporation of nanoparticles enhances nucleation, reducing supercooling and boosting the efficiency of thermal energy storage systems [77]. Custom-tailored nucleators now meet the specific needs of different PCMs, optimizing phase change performance. Microencapsulation techniques extend nucleation effects, enabling controlled thermal release in PCMs. This innovation integrates nucleators with renewable energy sources, enhancing energy management. Additionally, the use of nucleators in high-temperature PCMs expands their application in areas like solar power and electronics cooling. Advanced characterization methods offer insights into nucleator behavior, leading to the development of more efficient materials. The exploration of hybrid PCM-nucleator composites and eco-friendly nucleators highlights a commitment to sustainability and reduced environmental impact in thermal storage solutions [78].

PCM	Thickener	T_m (°C)	Nucleating agent (size, µm)	Supercooling	
				w/o nucleator	w/nucleator
$Na_2SO_4.10H_2O$	SAP	32	Borax (20 × 50–200 × 250)	15–18	3–4
$Na_2S_2O_3.5H_2O$	CMC	57	K_2SO_4	30	0–3
			$Na_2P_2O_7.10H_2O$		0–2
$CH_3COONa.3H_2O$	CMC	46	Na_2SO_4	20	4–6
			$SrSO_4$		0–2
			Carbon (1.5–6.7)		4–7
$Na_2HPO_4.12H_2O$	SAP	36	Borax (20 × 50–200 × 250)	20	6–9
			Carbon (1.5–6.7)		0–1
			TiO_2 (2–200)		0–1
			Copper (1.5–2.5)		0.5–1
			Aluminum (8.5–20)		3–10

Table 1.
Supercooling range of the thickened PCMs with the respective nucleating agents [45].

3.3 Gelling agents in PCMs

Gelling agents, or thickening agents, are added to PCMs to combat supercooling by altering their physical state to a semisolid or gel-like consistency, enhancing viscosity and cohesiveness. These agents help in stabilizing PCMs, preventing leakage, and improving handling. For instance, adding borax to Glauber's salt significantly reduces its supercooling from 15 to 3–4°C [24]. However, the effectiveness of these agents may decrease over time due to phase separation, agglomeration, or sedimentation. Common gelling agents include organic compounds, polymers, and inorganic materials like silica nanoparticles and clay minerals as given in **Table 2**.

3.3.1 Mechanisms and benefits of gelling agents

Gelling agents play a pivotal role in modifying the molecular arrangement and behavior of phase change materials (PCMs), rendering them more amenable to phase changes occurring at or near the desired temperature. These agents serve as nucleation sites within the PCM, initiating the crucial first step in the formation of solid crystals during the phase change process. By promoting nucleation, gelling agents encourage the PCM to initiate crystallization more readily, simultaneously increasing the viscosity of the PCM. This heightened viscosity hampers the movement of PCM molecules, compelling them to transition away from the supercooled state and toward the solid phase more expeditiously, effectively reducing the supercooling effect. Additionally, gelling agents lower the energy barrier necessary for nucleation and phase change processes to take place. Supercooling, a phenomenon where PCM molecules require a specific amount of energy to surmount this barrier and shift to the solid phase, is effectively mitigated by gelling agents. This reduction in the energy barrier ensures that the PCM solidifies more promptly, aligning with the intended phase change temperature. Moreover, certain gelling agents can augment the thermal conductivity of the PCM. This heightened thermal conductivity facilitates the

PCM component	Wt (%)	Function	T_m (°C)	ΔH (J/g)	C_{ps} (J/g·K)	C_{pl} (J/g·K)	K_s (W/m·K)
$Na_2SO_4.10H_2O$	95.0	PCM	32	227	3.4	4.6	0.64
SAP	3.0	Thickener					
Borax	2.0	Nucleator					
$CH_2COONa.3H_2O$	95.0	PCM	46	176	2.11	3.68	0.60
CMC	3.0	Thickener					
K_2SO_4	2.0	Nucleator					
$Na_2HPO_4.12H_2O$	92.8	PCM	36	104	3.7	4.1	0.40
SAP	3.5	Thickener					
TiO_2	3.7	Nucleator					
$Na_2S_2O_3.5H_2O$	92.0	PCM	57	206	2.2	—	—
CMC	3.0	Thickener					
$SrSO_4$	5.0	Nucleator					

Table 2.
Composition and thermophysical properties of PCM with thickener and nucleators [45].

efficient distribution and extraction of heat during the phase change process, further curtailing supercooling and enhancing the effectiveness of the PCM for thermal energy storage. Gelling agents also contribute to the creation of a more homogeneous PCM mixture within the material. This homogeneity significantly reduces the occurrence of regions within the PCM where supercooling can manifest, fostering uniform phase change behavior [67, 79, 80].

Gelling agents significantly alter the molecular arrangement of phase change materials (PCMs), facilitating phase changes at desired temperatures. Acting as nucleation sites, they initiate solid crystal formation, promoting crystallization and increasing PCM viscosity. This increased viscosity restricts molecular movement, accelerating the transition from a supercooled to a solid state, thereby reducing supercooling. Gelling agents also lower the energy barrier for nucleation, ensuring quicker solidification. Some agents enhance PCM thermal conductivity, improving heat distribution and extraction, further minimizing supercooling, and boosting thermal energy storage effectiveness. Additionally, they contribute to a more homogeneous PCM mixture, preventing uneven supercooling and ensuring uniform phase change behavior [69, 79].

Gelling agents enhance energy storage efficiency by diminishing supercooling, thereby reducing energy losses, and ensuring predictable phase change behavior through consistent nucleation. They improve thermal management by keeping the PCM within its intended temperature range and extend its lifespan by preventing phase separation across thermal cycles. Additionally, gelling agents boost heat transfer and thermal conductivity for quicker charging and discharging, minimize leakage risks by solidifying the PCM, and ensure compatibility with system components, reducing corrosion. Their use simplifies handling due to the semisolid state and offers versatility for various thermal energy storage needs.

3.3.2 Recent developments in gelling agents

Recent advancements in gelling agents have aimed to enhance performance, sustainability, and versatility across industries. The focus has shifted toward natural and renewable agents like agar-agar, carrageenan, and pectin, alongside bio-based alternatives to lessen environmental impacts of PCM-based systems. Microencapsulation techniques enable precise control over thermal energy release, improving efficiency and stability in PCM systems. Stable PCM emulsions created with gelling agents offer better stability and heat transfer. Hybrid gelling agents combine traditional and nanomaterials for superior thermal properties, while customization for specific PCM materials optimizes phase change temperature and viscosity. Sustainability is emphasized through biomass-derived or waste-material gelling agents, aligning with broader environmental goals [55].

3.4 Addressing supercooling challenges using the Cold Finger Technique (CFT)

The CFT is an innovative method aimed at mitigating supercooling in PCMs, ensuring their solidification at or near the intended phase change temperature. Praised for its simplicity and reliability over extended periods, CFT finds application in various sectors including energy storage, HVAC, and refrigeration, where precise temperature control and phase change efficiency are critical. Its versatility allows for customization to meet specific PCM temperature requirements and adapt to unique industry demands, highlighting its importance in improving thermal management and energy efficiency across a broad spectrum of applications [32].

3.4.1 Mechanisms and benefits of CFT

The cold finger is a device designed to prompt nucleation and crystallization in phase change materials (PCMs) at their phase change temperature. Functioning as a metal rod or plate cooled below the PCM's phase change point, it directly contacts the PCM, initiating cooling or solidification at its surface first. This method accelerates the PCM's shift to the solid phase by offering a surface for nucleation, thereby reducing the energy barrier for transition. Consequently, the Cold Finger Technique enhances the predictability and consistency of PCM phase changes, crucial for temperature stability and reliable performance in applications like thermal energy storage systems [79].

3.4.2 Recent developments in CFT

Researchers are innovating Cold Finger designs by integrating nanoparticles to enhance nucleation and combat supercooling, providing more effective nucleation sites. Focus has also been placed on refining temperature control within these setups to improve PCM efficacy. Customizing Cold Finger configurations to match the specific needs of different PCMs is key to optimizing phase change behavior. Hybrid approaches combining the Cold Finger Technique with other methods aim to offer comprehensive solutions. Efforts to adapt and optimize Cold Finger Techniques for thermal energy storage, incorporating renewable energy and ensuring grid stability, are underway. Advances in materials for Cold Finger components seek to improve durability and heat transfer, while developments in control and monitoring systems enhance understanding and optimization of PCM performance [81].

4. Issues of material compatibility with PCMs

Material compatibility with PCMs is crucial for applications in thermal energy storage and temperature control systems. As PCMs undergo phase transitions between solid and liquid states, absorbing or releasing thermal energy, potential interactions with surrounding materials may cause compatibility issues such as chemical reactions, corrosion, and leakage. Selecting materials that are chemically stable, corrosion-resistant, and can accommodate PCM volume changes is vital for the reliable and efficient operation of PCM-based systems. Proper compatibility ensures enhanced PCM reliability and addresses issues arising from interactions with additives and adjacent materials, preventing PCM degradation and ensuring system longevity [82].

4.1 PCMs with different metals and alloys

PCMs are strategically paired with different metals and alloys to augment storage capabilities. For instance, paraffin wax is blended with aluminum, preparing composite exhibiting improved thermal conductivity. Eutectic salt mixtures, such as LiCl-KCl, are integrated with copper to enhance thermal conductivity, for efficient heat transfer. For high-temperature TES systems, the combination of calcium sulfate hexahydrate ($CaSO_4{\cdot}6H_2O$) with copper proves invaluable. Copper enhances thermal conductivity, rendering it suitable for elevated temperature application. Additionally, sodium sulfate decahydrate ($Na_2SO_4{\cdot}10H_2O$) can be harmoniously merged with copper, ameliorating its thermal properties. Potassium nitrate (KNO_3), when paired with aluminum, becomes an option for low-temperature TES systems. Glauber's salt (Sodium sulfate decahydrate ($Na_2SO_4{\cdot}10H_2O$)) benefits from a copper infusion to boost thermal conductivity, critical for TES applications. In TES systems, the amalgamation of magnesium hydride (MgH_2) with nickel proves instrumental in augmenting hydrogen absorption and release properties. Moreover, sodium acetate trihydrate ($NaCH_3COO{\cdot}3H_2O$), combined with either copper or aluminum, finds utility in low-temperature thermal energy storage systems, fostering efficient heat exchange. Fatty acids, such as stearic acid, exhibit improved thermal conductivity and stability for thermal energy storage when intertwined with various metals like aluminum, copper, or graphite. Furthermore, organic PCMs like paraffins and polyethylene glycols stand to benefit from the inclusion of metal particles or nanoparticles, such as aluminum and copper, thereby enhancing their thermal properties [83, 84].

4.2 Importance of compatibility of PCMs in TES applications

Compatibility is crucial in PCM-based thermal energy storage (TES) systems, safeguarding against degradation and ensuring system reliability, efficiency, and safety. Incompatible materials can cause chemical reactions with PCMs, leading to deterioration and affecting system lifespan and performance. Compatibility reduces corrosion risks, which is vital since some PCMs can corrode container materials, risking structural integrity and causing leaks. It also prevents contamination, ensuring accurate temperature control and optimal energy storage. Safety is enhanced by preventing leaks and structural failures, especially important in high-temperature applications. Efficient heat transfer, crucial for system efficiency, relies on compatible materials. In critical applications like renewable energy integration, reliability is key, and compatibility supports this. Moreover, choosing environmentally responsible materials aligns with sustainability goals, reducing the ecological footprint and promoting responsible resource management. Compatibility not only ensures long-term cost-effectiveness by avoiding system failures but also contributes to sustainability by minimizing waste [85].

4.3 Recent findings on material interactions with PCM

Recent research on phase change material (PCM) interactions with various materials has significantly improved PCM-based system performance. Studies on compatibility between PCMs, container materials, additives, and surrounding components have been crucial in identifying and mitigating potential issues like chemical reactions, corrosion, and leakage. Advanced coating techniques and tailored additives have been developed to enhance compatibility, prevent phase separation, and improve phase change behavior. Analytical techniques such as spectroscopy and thermal analysis have provided deeper insights into PCM material interactions. Strategies like corrosion-resistant coatings and the exploration of innovative container materials, including composites and polymers, have improved resistance to chemical reactions and corrosion. Life cycle assessments and computational modeling have furthered the understanding of PCM behavior and system design, promoting more sustainable and efficient PCM solutions [82, 86].

5. Role of ML for TES and TMA systems and enhancing PCM reliability

ML has had a significant impact on enhancing the performance of TES platforms. It plays a pivotal role in predictive modeling for both TES and TMA. A study published by Li et al. in 2020 exemplifies the utilization of ML algorithms for predicting energy demand patterns with heightened precision. These ML models rely on diverse data sources such as weather conditions, user behavior, and historical data to anticipate energy requirements. ML empowers TES systems to store and release energy with pinpoint accuracy, effectively minimizing energy wastage and bolstering overall system efficiency. The outcome is a TES solution that is not only more sustainable but also cost-effective. Moreover, ML finds application in real-time temperature control within TES systems. It continuously monitors temperature fluctuations, fine-tuning the processes of energy storage and release accordingly. This dynamic adaptability ensures that TES systems can perform optimally even as conditions change, thereby enhancing their reliability and overall performance. ML modules can be strategically implemented to fortify TES systems, taking on essential roles in fault detection, resilience, and robustness.

In real time, ML algorithms are harnessed to identify potential faults or irregularities within TES systems. ML models are adept at detecting anomalies and can promptly trigger preventive maintenance or system adjustments. This, in turn, minimizes downtime and augments system resilience. Furthermore, ML models prove invaluable in the selection of the most suitable PCMs. Through an analysis of the thermal properties of various materials, ML-based optimization leads to an increase in energy storage capacity and an overall enhancement of TES system performance. The integration of TES with renewable energy sources, an integral aspect of modern energy systems, is seamlessly facilitated by ML. The ML models proficiently predict renewable energy generation patterns, enabling TES systems to efficiently adapt and store surplus energy. This integration ensures a reliable and uninterrupted energy supply, thereby promoting the sustainability of renewable energy sources. Lastly, ML models wield substantial influence in the management of TES systems within the broader energy grid. ML algorithms dissect grid data and make real-time decisions regarding energy charging and discharging. This grid management competency plays a pivotal role in balancing energy supply and demand while concurrently reducing stress on the energy grid [87, 88].

5.1 Reducing supercooling of PCM using CFT combined with ML

In addition to precise control of the TES system functionality, the ML models can also be used to solve the problem of supercooling. Integrating the CFT with suitable forecasting methods like ML has the potential to enhance the reliability and consistency of CFT. A previous study shows that the degree of supercooling was reduced to 1°C for $LiNO_3{\cdot}3H_2O$, which was demonstrated over 800 cycles of charging and discharging—with less than 6% degradation in energy storage capacity [89]. The successful implementation and execution of CFT techniques necessitates accurate forecasting models that need to provide accurate control of the melting process and robust predictions for different operating conditions. Hence, the history of PCM charging and discharging needs to be incorporated.

One potential approach for implementing CFT involves achieving a specific melting fraction of the PCM, such as 90% melt-fraction, with a high degree of accuracy and precision. To make this technique successful, it is crucial to predict in advance when a given mass of PCM will reach a target melt-fraction, like 90 or 95%, at which point the melting process should cease, and solidification should commence. The ability to predict the time required to reach the target melt-fraction should be robust and resistant to variations in environmental conditions, especially when dealing with multiple cycles of melting and solidification. Analytical and numerical models, which rely on energy and enthalpy balance approaches, often prove inadequate for providing reliable forecasts. These models can be overly sensitive to even minor fluctuations in environmental conditions, slight variations in process parameters, minor deviations in experimental procedures, and a narrow range of measurement uncertainties and experimental errors.

In such situations, ML algorithms prove to be well-suited for the precise prediction of melt fractions, a task that is often challenging for theoretical models due to their inherent complexity and sensitivity to transients. Theoretical models struggle to fully capture the system's behavior and predict the coupled and nonlinear thermal-hydraulic dynamics at the system level. ML techniques, on the other hand, rely on analyzing statistical variations within experimental data, especially during transient system states. The effectiveness of ML can be significantly enhanced when a substantial amount of "training data" is accessible for the algorithms to learn from [32, 90].

5.2 Artificial neural network principles as the basis of ML

The concept of artificial neural networks (ANNs) draws inspiration from the information processing in biological neural systems, such as the human brain [91]. Unlike structured programming methods, which rely on explicit instructions, ANNs operate by detecting patterns and relationships in data, learning from prior experiences or training data. This unique approach to forecasting allows ANNs to discern connections between parameters without requiring an in-depth understanding of the system [92]. One common implementation of ANN is the fully connected multilayer perceptron (MLP) model, consisting of nodes, often called "neurons," organized in sequential layers: input, hidden, and output. Neurons within a layer communicate through two-way connections, with the input layer having neurons equivalent to the input parameters and the output layer's neuron count determined by the number of output parameters. The hidden layer, positioned between the input and output layers, facilitates communication among neurons to solve problems. This mimics the functioning of the brain. The first layer is called the input layer (receiving input data vectors), while the last layer is called the output layer. A schematic of the MLP network is shown in **Figure 7**.

Depending on the output vector length, it may have more than one node. The middle layers are known as hidden layers. Each layer is connected to the previous layer using connectors. A single node is characterized by two entities—(1) bias, b, (2) and active function, f, while connectors have a weight, w. In each neuron, the activation function acts on the input a_n. received from the node in the preceding layer. The output value (a_{n+1}) from one layer serves as the input for the subsequent layer with a serial number (n + 2). This sequential relationship can be expressed mathematically as shown in the following equation.

$$a_{n+1} = f\left(w^T a_n + b\right) \tag{1}$$

The development of ANN models includes three major stages including data training, result validation, and efficacy testing [32]. Initially, during the data training phase, weights and biases are initialized randomly, and subsequently, outputs are generated based on the training data. Following this, in the result validation stage,

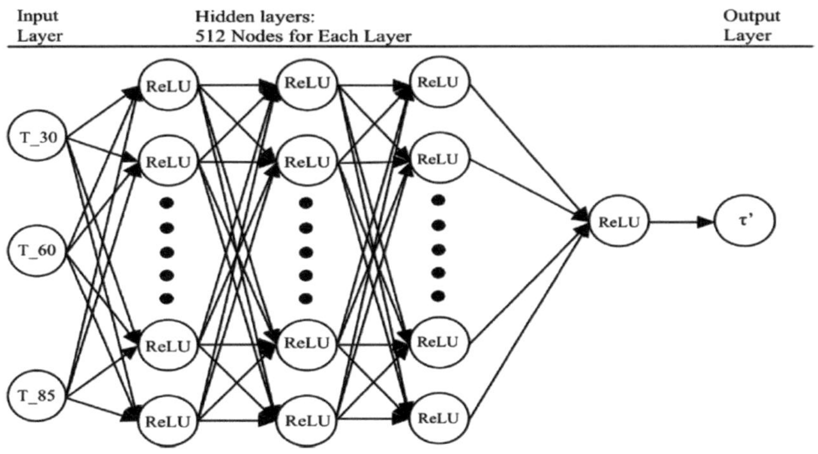

Figure 7.
Schematic of the MLP network [87].

the produced output is compared with the actual output, utilizing a cost function to compute the extent of disparity or error between them. The specific cost function employed during the result validation phase is the sum of squared error (SSE), as represented in the following equation.

$$SSE = \sum_{i=1}^{N} \left(p_i - a_i \right)^2 \tag{2}$$

Here, N is the number of result values, $\boldsymbol{p_i}$ is predicted value from ANN, and $\boldsymbol{a_i}$ is the actual value. The error denoted as SSE undergoes a backward propagation process from the result layer to the input layer, essentially implementing a gradient descent algorithm. This computational method is employed to adjust the biases and weights iteratively, to minimize the error between the predicted and actual values through multiple numerical iterations. This feedback mechanism continues until the desired level of error reduction is achieved. In the following stage, the efficacy of the ANN model is evaluated using a separate test dataset that was not utilized during the data training and result validation stages.

In previous applications, ANN techniques have been employed to predict the specific heat capacity of molten salt nanofluids used in TES applications, particularly in concentrated solar power (CSP) and solar thermal power generation [93]. Another category of ANNs, known as radial basis function neural networks (RBF-NNs), is particularly well-regarded for modeling material properties, including the thermo-physical properties of nanofluids. RBF-NNs, first introduced by Broomhead and Lowe, also consist of three layers: the input layer, hidden layer, and output layer [94]. The primary function of RBF-NN ANNs is function approximation. The key distinction between an MLP model and an RBF-NN model lies in the computation process within the neurons.

In the RBF-NN model, the activation function is a radial basis function ϕ, which operates based on the Euclidean norm between the neuron's center and the input vector. A connection weight is then applied to the result to produce outputs, as represented mathematically in the following equation.

$$y(x)=\sum_{i=1}^{n}w_i\text{ø}(x-x_i) \quad (3)$$

Prior research has demonstrated the viability of predicting melt fraction using RBF-NN models based on temperature measurements [95]. The effectiveness of a fully developed MLP model enables real-time predictions of the time required to achieve a predefined melt fraction (e.g., 90% melt fraction) based on temperature measurements at any given moment during a melting cycle. This predictive capability helps to optimize the utilization of thermal energy storage capacity while eliminating the need for supercooling to initiate nucleation during solidification. Additionally, ANN-based methods help predict the material properties of PCMs while also forecasting their transient performance, particularly their temperature responses for battery thermal management [96]. Therefore, the MLP/ANN-based predictive models can significantly enhance the reliability of PCM TES platforms while simultaneously improving their overall systemic and thermodynamic efficiencies.

5.3 ML approaches used for examining PCM reliability

ML techniques can be utilized to assess the reliability of PCMs to improve the efficiency of TES systems. **Table 3** presents various ML approaches and their respective purposes in this context [97, 98].

5.4 Recent developments in ML approaches for PCM-based TES systems

ML approaches offer versatile solutions adaptable to address future challenges in TES systems. Ongoing advancements involve the development of sophisticated ML models capable of predicting PCM behavior with increased precision, encompassing complex phase change processes and accounting for various influencing factors. The fusion of ML with artificial intelligence (AI) is being leveraged to tackle supercooling issues by creating intelligent control systems that actively manage phase transitions. Furthermore, the evolution of ML algorithms aims to provide comprehensive reliability assessments, including the prediction of long-term PCM system behavior and early detection of potential failure modes. Advanced ML systems enable autonomous and initiative-taking maintenance of PCM-based TES systems, detecting anomalies, scheduling maintenance tasks, and optimizing performance without human intervention. ML is also contributing to the rapid discovery and design of innovative PCM materials, enhancing thermal properties and system reliability. Additionally, ML plays a pivotal role in integrating PCM-based TES systems with smart grids and renewable energy sources, optimizing the interaction between energy storage and generation to bolster grid stability and energy efficiency. Evaluating the environmental impact of PCM materials and components and assessing the long-term sustainability of TES solutions are facilitated by advanced ML techniques. Finally, the development of real-time monitoring and control systems utilizing ML-based sensors empowers the effective management of PCM-based TES systems [99, 100].

5.5 Optimizing PCM selection and configuration with ML

The integration of ML with PCM optimization represents a significant advancement in the development of TES systems. By utilizing data science, this method

ML approaches	Intended function or application
Predictive modeling	• Anticipates PCM responses across varying conditions. • Utilizes models for projecting phase change behaviors, supercooling tendencies, and other attributes. • Contributes to the evaluation of reliability.
Anomaly detection	• Foresees unexpected PCM system behaviors. • Detects deviations from anticipated patterns, signaling potential reliability issues
Regression analysis	• Predicts material behavior in future scenarios. • Evaluate long-term reliability and degradation.
Failure Mode Analysis	• Envisions potential PCM failure scenarios. • Adopts a proactive stance in averting reliability issues.
Data-driven decision-making	• Offers guidance on maintenance, repair, or replacement to uphold reliability.
Optimization	• Projects optimal operating conditions for TES systems, reducing wear and tear and elevating reliability.
Probabilistic modeling	• Forecasts the probability of PCM failure or degradation over time, aiding in reliability assessments.
Pattern recognition	• Identifies patterns indicative of potential reliability concerns, allowing for timely interventions.
Life cycle assessment	• Conducts comprehensive life cycle assessments of TES systems, accounting for factors like environmental impact and long-term reliability.
Real-time monitoring	• Maintains ongoing assessments of PCM behavior through ML-based sensors and monitoring systems.

Table 3.
Machine learning approaches and their intended functions.

addresses the complex task of selecting the ideal PCM, factoring in thermal conductivity, melting point, latent heat capacity, and material compatibility [101]. Recent strides in ML, particularly through deep learning and reinforcement learning, have demonstrated substantial success in predicting PCM thermophysical behaviors in varying conditions, thus surpassing the limitations of traditional selection methods. The process involves training ML models on extensive datasets, including PCM properties and performance metrics across different scenarios, enriched with computational fluid dynamics (CFD) and finite element analysis (FEA) simulations [102]. These models, notably convolutional neural networks (CNNs) and recurrent neural networks (RNNs), excel at detecting intricate patterns and correlations, accurately forecasting PCM responses [103]. Additionally, the deployment of optimization algorithms, such as genetic algorithms (GAs) and particle swarm optimization (PSO), alongside ML models, meticulously identifies PCM combinations that optimize thermal efficiency while minimizing costs and environmental impact. This innovative approach not only boosts the thermal effectiveness of TES systems but also aligns with sustainable design principles, marking a forward leap toward a more sustainable, energy-efficient future in thermal management solutions.

6. Conclusion

PCMs are integral to TES and TMA, offering significant benefits across industries by storing and releasing energy to manage grid demands, improve cooling systems, and manage energy spikes. Widely applied in electronic cooling, data center regulation, energy-efficient architectural design, and renewable energy systems, the effectiveness of PCMs is dependent on their reliability and stability. Addressing supercooling and compatibility through additives is crucial for their broader application. ML models have become vital in enhancing PCM performance, aiding in selection, compatibility assessment, supercooling mitigation, and thermal behavior prediction. These models optimize TES system performance and facilitate the discovery of novel PCMs, driving sustainability in engineering systems. The collaboration between PCMs and ML offers potential for significant advancements in energy efficiency and environmental sustainability, marking a promising direction for research and development toward an eco-friendly future across various sectors.

Acknowledgements

This work received no funding.

Conflict of interest

The authors declare no conflict of interest.

Author details

Sunil Kumar* and Debjyoti Banerjee
Department of Mechanical Engineering, Texas A&M University,
College Station, Texas, USA

*Address all correspondence to: ksunil86.in@gmail.com

References

[1] Kumar S, Thyagarajan A, Banerjee D. An investigation on chevron plate heat exchanger filled with organic phase change material. In: Proceedings of the ASME 2020 Heat Transfer Summer Conference HT2020 July 13-15, 2020, Virtual, Online. American Society of Mechanical Engineers (ASME); 2020. pp. 1-6. DOI: 10.1115/HT2020-9122

[2] Kumar N, Banerjee D. Phase change materials. In: Handbook of Thermal Science and Engineering. Cham: Springer; 2018. pp. 2213-2275

[3] Solomon AD. Design criteria in PCM wall thermal storage. Energy. 1979;**4**:701-709. DOI: 10.1016/0360-5442(79)90092-6

[4] Theunissen P-H, Buchlin J-M. Numerical optimization of a solar air heating system based on encapsulated P.C.M. storage. Solar Energy. 1983;**31**:271-277

[5] Breen J, Rosner DE, Delgass WN, Nordine PC, Cibrian R, Krishnan NG. Catalysis Study for Space Shuttle Vehicle Thermal Protection Systems. NASA; 1973. Available from: https://ntrs.nasa.gov/citations/19740004385

[6] Humphries George WR, Griggs EI. A Design Handbook for Phase Change Thermal Control and Energy Storage Devices. NASA; 1977. Available from: https://ntrs.nasa.gov/citations/19780007491

[7] Hirano S. Thermal energy storage and transport. In: Handbook of Climate Change Mitigation and Adaptation, Chapter 2. 2nd ed. Vol. 2. Springer Publications; 2016. pp. 1433-1464. DOI: 10.1007/978-3-319-14409-2_20/COVER. Available from: https://link.springer.com/referencework/10.1007/978-3-319-14409-2

[8] Deb K, Pratap A, Agarwal S, Meyarivan T. A fast and elitist multiobjective genetic algorithm: NSGA-II. IEEE Transactions on Evolutionary Computation. 2002;**6**:182-197. DOI: 10.1109/4235.996017

[9] Adesina A. Use of phase change materials in concrete: Current challenges. Renewable Energy and Environmental Sustainability. 2019;**4**:9. DOI: 10.1051/rees/2019006

[10] Kumar S, Thyagarajan A, Banerjee D. Experimental investigation of thermal energy storage (TES) platform leveraging phase change materials in a Chevron plate heat exchanger. In: International Mechanical Engineering Congress and Exposition. Columbus, OH: Greater Columbus Convention Center; 2022. pp. 1-7. DOI: 10.1115/IMECE2022-96226

[11] Ren G, Chuttar A, Banerjee D. Exploring efficacy of machine learning (artificial neural networks) for enhancing reliability of thermal energy storage platforms utilizing phase change materials. International Journal of Heat and Mass Transfer. 2022;**189**:122628. DOI: 10.1016/J.IJHEATMASSTRANSFER.2022.122628

[12] Barzin R, Chen JJJ, Young BR, Farid MM. Application of PCM underfloor heating in combination with PCM wallboards for space heating using price based control system. Applied Energy. 2015;**148**:39-48. DOI: 10.1016/J.APENERGY.2015.03.027

[13] Jeon J, Lee JH, Seo J, Jeong SG, Kim S. Application of PCM thermal energy storage system to reduce building

energy consumption. Journal of Thermal Analysis and Calorimetry. 2013;**111**:279-288. DOI: 10.1007/S10973-012-2291-9/METRICS

[14] Chaturvedi R, Islam A, Sharma K. A review on the applications of PCM in thermal storage of solar energy. Materials Today Proceedings. 2021;**43**:293-297. DOI: 10.1016/J.MATPR.2020.11.665

[15] Qin Z, Ji C, Low ZH, Tong W, Wu C, Duan F. Geometry effect of phase change material container on waste heat recovery enhancement. Applied Energy. 2022;**327**:120108. DOI: 10.1016/J.APENERGY.2022.120108

[16] Douvi E, Pagkalos C, Dogkas G, Koukou MK, Stathopoulos VN, Caouris Y, et al. Phase change materials in solar domestic hot water systems: A review. International Journal of Thermofluids. 2021;**10**:100075. DOI: 10.1016/J.IJFT.2021.100075

[17] Ling Z, Zhang Z, Shi G, Fang X, Wang L, Gao X, et al. Review on thermal management systems using phase change materials for electronic components, Li-ion batteries and photovoltaic modules. Renewable and Sustainable Energy Reviews. 2014;**31**:427-438. DOI: 10.1016/J.RSER.2013.12.017

[18] Yahaya NA, Ahmad H. Numerical investigation of indoor air temperature with the application of PCM gypsum board as ceiling panels in buildings. Procedia Engineering. 2011;**20**:238-248. DOI: 10.1016/J.PROENG.2011.11.161

[19] Sánchez-Pérez JF, Mascaraque-Ramírez C, Moreno Nicolás JA, Castro E, Cánovas M. Study of the application of PCM to thermal insulation of UUV hulls using network simulation method. Alexandria Engineering Journal. 2021;**60**:4627-4637. DOI: 10.1016/J.AEJ.2021.03.058

[20] Akanda MAM, Shin D. A synthesis parameter of molten salt nanofluids for solar thermal energy storage applications. Journal of Energy Storage. 2023;**60**:106608. DOI: 10.1016/j.est.2023.106608

[21] Kuznik F, Johannes K, David D. Integrating phase change materials (PCMs) in thermal energy storage systems for buildings. In: Advances in Thermal Energy Storage Systems: Methods and Applications. Woodhead Publishing; 2015. pp. 325-353. DOI: 10.1533/9781782420965.2.325. Available from: https://www.sciencedirect.com/science/article/abs/pii/B9781782420880500134

[22] Hussein A, Abd-Elhady MS, El-Sheikh MN, El-Metwally HT. Improving heat transfer through paraffin wax, by using fins and metallic strips. Arabian Journal for Science and Engineering. 2018;**43**:4433-4441. DOI: 10.1007/S13369-017-2923-2/METRICS

[23] Tong X, Khan JA, Amin MR. Enhancement of heat transfer by inserting a metal matrix into a phase change material. Numerical Heat Transfer, Part A: Applications. 1996;**30**:125-141. DOI: 10.1080/10407789608913832

[24] Ryu HW, Woo SW, Shin BC, Kim SD. Prevention of supercooling and stabilization of inorganic salt hydrates as latent heat storage materials. Solar Energy Materials & Solar Cells. 1992;**27**:161-172

[25] Xu X, Cui H, Memon SA, Yang H, Tang W. Development of novel composite PCM for thermal energy storage using $CaCl_2{\cdot}6H_2O$ with graphene oxide and $SrCl_2{\cdot}6H_2O$. Energy and Buildings. 2017;**156**:163-172. DOI: 10.1016/J.ENBUILD.2017.09.081

[26] Liu Y, Yang Y. Form-stable phase change material based on Na_2CO_3 $10H_2O-Na_2HPO_4$12H_2O$ eutectic hydrated salt/expanded graphite oxide composite: The influence of chemical structures of expanded graphite oxide. Renewable Energy. 2017;**115(C)**:734-740. DOI: 10.1016/j.renene.2017.08.097

[27] Liu Y, Yang Y. Preparation and thermal properties of Na_2CO_3 Á10H_2O-Na_2HPO_4 Á12H_2O eutectic hydrate salt as a novel phase change material for energy storage. Applied Thermal Engineering. 2016;**112**:606-609. DOI: 10.1016/j.applthermaleng.2016.10.146

[28] Yinping Z, Yi J, Yi J. A simple method, the T-history method, of determining the heat of fusion, specific heat and thermal conductivity of phase-change materials. Measurement Science and Technology. 1999;**10**:201. DOI: 10.1088/0957-0233/10/3/015

[29] Mehling H, Cabeza LF. Chapter 9, Heat and Cold Storage with PCM. Springer Science & Business Media; 2008. DOI: 10.1007/978-3-540-68557-9. Available from: https://link.springer.com/chapter/10.1007/978-3-540-68557-9_9

[30] Meseguer J, Pérez-Grande I, Sanz-Andrés A. Spacecraft Thermal Control. Elsevier; 2012. DOI: 10.1533/9780857096081. Available from: https://www.sciencedirect.com/book/9781845699963/spacecraft-thermal-control

[31] Gunther Eva WM, Harald M. Melting and nucleation temperatures of three salt hydrate phase change materials under static pressures up to 800 MPa. Journal of Physics D: Applied Physics. 2007;**40**:4636. DOI: 10.1088/0022-3727/40/15/042

[32] Chuttar A, Thyagarajan A, Banerjee D. Leveraging machine learning (artificial neural networks) for enhancing performance and reliability of thermal energy storage platforms utilizing phase change materials. Journal of Energy Resources Technology, Transactions of the ASME. 2022;**144**:022001. DOI: 10.1115/1.4051048/1108736

[33] Fleischer AS. Chapter 2, Thermal Energy Storage Using Phase Change Materials: Fundamentals and Applications, SpringerBriefs in Applied Sciences and Technology. Springer; 2015. DOI: 10.1007/978-3-319-20922-7/COVER. Available from: https://link.springer.com/chapter/10.1007/978-3-319-20922-7_2

[34] Pai D, Joshi YK. Application of phase change materials to thermal control of electronic modules: A computational study. Journal of Electronic Packaging. 1997;**119**:40-50. DOI: 10.1115/1.2792199

[35] Vesligaj MJ. Transient thermal management of temperature fluctuations during time varying workloads on portable electronics. IEEE Transactions on Components, Packaging and Manufacturing Technology. 1999;**22**:541-550. DOI: 10.1109/6144.814970

[36] Cho J, Lim T, Kim BS. Measurements and predictions of the air distribution systems in high compute density (internet) data centers. Energy and Buildings. 2009;**41**:1107-1115. DOI: 10.1016/J.ENBUILD.2009.05.017

[37] Yuan X, Zhou X, Pan Y, Kosonen R, Cai H, Gao Y, et al. Phase change cooling in data centers: A review. Energy and Buildings. 2021;**236**:110764. DOI: 10.1016/J.ENBUILD.2021.110764

[38] Ma X, Zhang Q, Zou S. An experimental and numerical study on the thermal performance of a loop thermosyphon integrated with latent

thermal energy storage for emergency cooling in a data center. Energy. 2022;**253**:123946. DOI: 10.1016/J.ENERGY.2022.123946

[39] Wang J, Cao X, Yuan Y, Leng Z, Sun Y. Feasibility study on temperature control with phase change material in intensive heat-releasing space during emergency power failure: A case analysis of information system room. Energy and Buildings. 2021;**230**:110482. DOI: 10.1016/J.ENBUILD.2020.110482

[40] Fang Y, Niu J, Deng S. Optimizing LHS system using PCM in a tube-in-tank design for emergency cooling. Energy Procedia. 2017;**142**:3381-3387. DOI: 10.1016/J.EGYPRO.2017.12.474

[41] Zheng Z, Huang B, Lu G, Zhai X. Design and optimization of an air-based phase change cold storage unit through cascaded construction for emergency cooling in IDC. Energy. 2022;**241**:122897. DOI: 10.1016/J.ENERGY.2021.122897

[42] Berger RL, Jenkins LC. Electronic equipment thermal management. In: Proceedings of the Annual Reliability and Maintainability Symposium. San Francisco, CA: Annual Reliability and Maintainability Symposium; 1984. pp. 17-22. DOI: 10.1109/RAMS.1984.764256

[43] Liu M, Saman W, Bruno F. Review on storage materials and thermal performance enhancement techniques for high temperature phase change thermal storage systems. Renewable and Sustainable Energy Reviews. 2012;**16**:2118-2132. DOI: 10.1016/J.RSER.2012.01.020

[44] Dincer I. Refrigeration System and Application. Vol. 1377. John Wiley & Son; 2017. pp. 68-70. Available from: https://www.nec.com/en/global/techrep/journal/recommend_year/2014/02.html#:~:text=(1)%20Phase%20change%20cooling%20technology&text=The%20heat%20can%20be%20transported,the%20temperature%20of%20the%20fluids [Accessed: October 19, 2023]

[45] Ryu HW, Woo SW, Shin BC, Kim SD. Prevention of supercooling and stabilization of inorganic salt hydrates as latent heat storage materials. Solar Energy Materials and Solar Cells. 1992;**27**:161-172

[46] Safari A, Saidur R, Sulaiman FA, Xu Y, Dong J. A review on supercooling of phase change materials in thermal energy storage systems. Renewable and Sustainable Energy Reviews. 2017;**70**:905-919. DOI: 10.1016/J.RSER.2016.11.272

[47] Kumar S, Klassen M, Klassen D, Hardin R, King MD. Dispersion of sneeze droplets in a meat facility indoor environment-without partitions. Environmental Research. 2023;**236**:116603. DOI: 10.1016/j.envres.2023.116603

[48] Kumar S, King MD. Numerical investigation on indoor environment decontamination after sneezing. Environmental Research. 2022;**213**:113665. DOI: 10.1016/j.envres.2022.113665

[49] Zhang H, Xu Q, Zhao Z, Zhang J, Sun Y, Sun L, et al. Preparation and thermal performance of gypsum boards incorporated with microencapsulated phase change materials for thermal regulation. Solar Energy Materials & Solar Cells. 2012;**102**:93-102. DOI: 10.1016/J.SOLMAT.2012.03.020

[50] Kumar S, Vijayan PK, Kannan U, Sharma M, Pilkhwal DS. Experimental and computational simulation

of thermal stratification in large pools with immersed condenser by. Applied Thermal Engineering. 2017;**113**:345-361. DOI: 10.1016/j.applthermaleng.2016.10.175

[51] Kumar S, Grover RB, Vijayan PK, Kannan U. Numerical investigation on the effect of shrouds around an immersed isolation condenser on the thermal stratification in large pools. Annals of Nuclear Energy. 2017;**110**:109-125. DOI: 10.1016/j.anucene.2017.06.022

[52] Kumar S, Vijayan PK, Sharma M, Pilkhwal DS, Kannan U, Nayak AK. Experimental and numerical analysis of thermal stratification in large water pool. In: Proceedings of the International Thorium Energy Conference: Gateway to Thorium Energy. Mumbai, Maharashtra, India; 2015. Available from: http://inis.iaea.org/Search/search.aspx?orig_q=RN:47073982 [Accessed: January 12, 2021]

[53] Kumar S, Grover RB, Yadav H, Vijayan PK, Kannan U, Agrawal A. Experimental and numerical investigation on suppression of thermal strati fi cation in a water-pool: PIV measurements and CFD simulations. Applied Thermal Engineering. 2018;**138**:686-704. DOI: 10.1016/j.applthermaleng.2018.04.070

[54] Kumar S, Grover RB, Yadav H, Vijayan PK, Kannan U, Agrawal A, et al. Experimental investigations on thermal stratification in a large pool of water with immersed isolation condenser. International Conference on Nuclear Engineering Proceedings, ICONE. 2020;**3**:1-7. DOI: 10.1115/ICONE2020-16647

[55] Vasilyev G, Koifman N, Shuster M, Gishvoliner M, Cohen Y, Zussman E. Phase change material with gelation imparting shape stability. ACS Omega. 2022;**7**:11887-11902. DOI: 10.1021/ACSOMEGA.1C07376/ASSET/IMAGES/MEDIUM/AO1C07376_M001.GIF

[56] Rathore K, Kumar S. Net zero emissions for our future generations through renewables: A brief review. International Journal of Energy Resources Applications. 2023;**2**:39-56. DOI: 10.56896/IJERA.2023.2.1.004

[57] Kumar S, Rathore K. Renewable energy for sustainable development goal of clean and affordable energy. International Journal of Materials Manufacturing Sustainable Technology. 2023;**2**:01-15. DOI: 10.56896/IJMMST.2023.2.1.001

[58] Singh VP, Meena CS, Kumar A. Double pass solar air heater: A review. International Journal of Energy Resource Applications. 2022;**1**:22-43. DOI: 10.56896/ijera.2022.1.2.009

[59] Singh VP, Jain S, Karn A, Kumar A, Dwivedi G, Meena CS, et al. Recent developments and advancements in solar air heaters: A detailed review. Sustainability. 2022;**14**:12149. DOI: 10.3390/su141912149

[60] Kumar A, Singh VP, Meena CS, Dutt N. Chapter 2, Thermal Energy Systems: Design, Computational Techniques, and Applications. CRC Press; 2023. pp. 1-278. DOI: 10.1201/9781003395768. Available from: https://www.taylorfrancis.com/books/edit/10.1201/9781003395768/thermal-energy-systems-ashwani-kumar-varun-pratap-singh-chandan-swaroop-meena-nitesh-dutt

[61] Dutt N, Hedau AJ, Kumar A, Awasthi MK, Singh VP, Dwivedi G. Thermo-hydraulic performance of solar air heater having discrete D-shaped ribs as artificial roughness. Environmental

Science and Pollution Research. 2023;**1**:1-22. DOI: 10.1007/S11356-023-28247-9/FIGURES/26

[62] Dutt N, Kumar R, Murugesan K. Experimental performance analysis of multi-parabolic flat plate solar collector. In: Li X, Rashidi MM, Lather RS, Raman R, editors. Emerging Trends in Mechanical and Industrial Engineering. Lecture Notes in Mechanical Engineering. Singapore: Springer; 2023. pp. 159-172. DOI: 10.1007/978-981-19-6945-4_12/FIGURES/16

[63] Singh VP, Jain S, Kumar A. Establishment of correlations for the thermo-hydraulic parameters due to perforation in a multi-V rib roughened single pass solar air heater. Experimental Heat Transfer. 2022;**36**:597-616. DOI: 10.1080/08916152.2022.2064940

[64] Singh VP, Jain S, Karn A, Dwivedi G, Kumar A, Mishra S, et al. Heat transfer and friction factor correlations development for double pass solar air heater artificially roughened with perforated multi-V ribs. Case Studies in Thermal Engineering. 2022;**39**:102461. DOI: 10.1016/J.CSITE.2022.102461

[65] Singh VP, Jain S, Gupta JML. Performance assessment of double-pass parallel flow solar air heater with perforated multi-V ribs roughness—Part B. Experimental Heat Transfer. 2022;**35**:1059-1076. DOI: 10.1080/08916152.2021.2019147

[66] Singh VP, Jain S, Karn A, Kumar A, Dwivedi G, Meena CS, et al. Mathematical modeling of efficiency evaluation of double-pass parallel flow solar air heater. Sustainability. 2022;**14**:10535. DOI: 10.3390/SU141710535

[67] Awasthi MK, Dutt N, Kumar A, Kumar S. Electrohydrodynamic capillary instability of Rivlin–Ericksen viscoelastic fluid film with mass and heat transfer. Heat Transfer. 2023;**53**:115-133. DOI: 10.1002/HTJ.22944

[68] Shukla P, Awasti MK, Shukla AK, Agarwal S, Singh AK. Stability analysis of Rivlin-Ericksen fluid film with heat transfer through an annular porous medium. Special Topics & Reviews in Porous Media: An International Journal. 2023;**15**:25-39. DOI: 10.1615/SPECIALTOPICSREVPOROUSMEDIA.2023046399

[69] Awasthi MK, Dharamendra DY. Temporal instability of nanofluid layer in a circular cylindrical cavity. European Physical Journal Special Topics. 2022;**231**:2773-2779. DOI: 10.1140/EPJS/S11734-022-00599-2

[70] Dawahdeh A, Sharadga H, Kumar S. Novel MPPT controller augmented with neural network for use with photovoltaic systems experiencing rapid solar radiation changes. Sustainability. 2024;**16**(3):1021

[71] Boretti A, Nayfeh J, Al-Kouz W. Validation of SAM modeling of concentrated solar power plants. Energies. 2020;**13**:1949. DOI: 10.3390/EN13081949

[72] Robak CW, Bergman TL, Faghri A. Economic evaluation of latent heat thermal energy storage using embedded thermosyphons for concentrating solar power applications. Solar Energy. 2011;**85**:2461-2473. DOI: 10.1016/J.SOLENER.2011.07.006

[73] Nithyanandam K, Pitchumani R. Cost and performance analysis of concentrating solar power systems with integrated latent thermal energy storage. Energy. 2014;**64**:793-810. DOI: 10.1016/J.ENERGY.2013.10.095

[74] Li ZR, Hu N, Fan LW. Nanocomposite phase change materials for high-performance thermal energy storage: A critical review. Energy Storage Materials. 2023;**55**:727-753. DOI: 10.1016/J.ENSM.2022.12.037

[75] Garg VK, Rauthan DB. Choice of moon for G/T measurement. IETE Journal of Research. 1985;**31**:73-77. DOI: 10.1080/03772063.1985.11436496

[76] Yang Y, Luo J, Li S, Song G, Liu Y, Tang G. The experimental exploration of sodium chloride solution on thermal behavior of phase change materials. Solar Energy Materials & Solar Cells. 2015;**139**:88-94. DOI: 10.1016/J.SOLMAT.2015.03.009

[77] Beaupere N, Soupremanien U, Zalewski L. Nucleation triggering methods in supercooled phase change materials (PCM), a review. Thermochimica Acta. 2018;**670**:184-201. DOI: 10.1016/j.tca.2018.10.009ï

[78] Piquard L, Gagnière E, Largiller G, Mangin D, Bentivoglio F. Xylitol used as phase change material: Nucleation mechanisms of the supercooling rupture by stirring. Journal of Energy Storage. 2022;**48**:103922. DOI: 10.1016/J.EST.2021.103922

[79] Prabakaran R, Dhamodharan P, Sathishkumar A, Gullo P, Vikram MP, Pandiaraj S, et al. An overview of the state of the art and challenges in the use of gelling and thickening agents to create stable thermal energy storage materials. Energies. 2023;**16**:3306. DOI: 10.3390/EN16083306

[80] Cong L, Zou B, Palacios A, Navarro ME, Qiao G, Ding Y. Thickening and gelling agents for formulation of thermal energy storage materials – A critical review. Renewable and Sustainable Energy Reviews. 2022;**155**:111906. DOI: 10.1016/J.RSER.2021.111906

[81] Rashid FL, Al-Obaidi MA, Dulaimi A, Bernardo LFA, Redha ZAA, Hoshi HA, et al. Recent advances on the applications of phase change materials in cold thermal energy storage: A critical review. Journal Of Composites Science. 2023;**7**:338. DOI: 10.3390/JCS7080338

[82] Ostrý M, Bantová S, Struhala K. Tests on material compatibility of phase change materials and selected plastics. Molecules. 2019;**24**:1398. DOI: 10.3390/MOLECULES24071398

[83] Wu MQ, Wu S, Cai YF, Wang RZ, Li TX. Form-stable phase change composites: Preparation, performance, and applications for thermal energy conversion, storage and management. Energy Storage Materials. 2021;**42**:380-417. DOI: 10.1016/J.ENSM.2021.07.019

[84] Liu L, Niu J, Wu JY. Preparation of stable phase change material emulsions for thermal energy storage and thermal management applications: A review. Materials (Basel). 2022;**15**:121. DOI: 10.3390/MA15010121

[85] Devanuri JK, Gaddala UM, Kumar V. Investigation on compatibility and thermal reliability of phase change materials for low-temperature thermal energy storage. Materials for Renewable and Sustainable Energy. 2020;**9**:1-16. DOI: 10.1007/S40243-020-00184-4/TABLES/7

[86] Gu Y, Li Y, Ju G, Zheng T, Liang R, Sun G. Pcm microcapsules applicable foam to improve the properties of thermal insulation and energy storage for cement-based material. Construction and Building Materials. 2023;**409**:134144. DOI: 10.2139/SSRN.4435929

[87] Sudhir PS, Ren G, Chuttar A, Shettigar N, Banerjee D. Deploying machine learning (ML) for improving reliability and resiliency of thermal energy storage (TES) platforms by leveraging phase change materials (PCM) for sustainability applications and mitigating food-energy-water (FEW) nexus. In: Proceedings of the ASME 2022 International Mechanical Engineering Congress and Exposition. Vol. 8. Columbus, OH: Greater Columbus Convention Center; 2023. DOI: 10.1115/IMECE2022-97121

[88] Chuttar A, Banerjee D. Machine learning (ML) based thermal management for cooling of electronics chips by utilizing thermal energy storage (TES) in packaging that leverages phase change materials (PCM). Electronics. 2021;**10**:2785. DOI: 10.3390/ELECTRONICS10222785

[89] Kumar N, Von Ness R, Chavez R, Banerjee D, Muley A, Stoia M. Experimental analysis of salt hydrate latent heat thermal energy storage system with porous aluminum fabric and salt hydrate as phase change material with enhanced stability and supercooling. Journal of Energy Resources Technology, Transactions of the ASME. 2021;**143**:042001. DOI: 10.1115/1.4048122/1086267

[90] Chuttar AJ, Banerjee D. Machine learning (ML) based thermal management for cooling of electronics chips by utilizing thermal energy storage (TES) in packaging that leverage phase change materials (PCM). Electronics. 2021;**10**(22):2785. DOI: 10.20944/PREPRINTS202110.0253.V1

[91] Dongare A, Kharde RR, Kachare AD. Introduction to artificial neural network. International Journal of Engineering and Innovative Technology (IJEIT). 2012;**2**:189-194

[92] Agatonovic-Kustrin S, Beresford R. Basic concepts of artificial neural network (ANN) modeling and its application in pharmaceutical research. Journal of Pharmaceutical and Biomedical Analysis. 2000;**22**:717-727. DOI: 10.1016/S0731-7085(99)00272-1

[93] Hassan MA, Banerjee D. A soft computing approach for estimating the specific heat capacity of molten salt-based nanofluids. Journal of Molecular Liquids. 2019;**281**:365-375. DOI: 10.1016/J.MOLLIQ.2019.02.106

[94] Broomhead D, Lowe D. Radial basis functions, multi-variable functional interpolation and adaptive networks, R. In: Royal Signals and Radar Establishment Malvern (United Kingdom). UK: Royal Signals and Radar Establishment Memorandum 4148; 1988. Available from: https://apps.dtic.mil/sti/pdfs/ADA196234.pdf

[95] Shettigar N, Truong M, Thyagarajan A, Bamido A, Banerjee D. Application of machine learning (ML) for enhancing the transient performance of thermal energy storage (TES) platforms using radial basis function (RBF). Journal of Engineering Research and Reports. 2021:70-84. Available from: https://www.academia.edu/77823157/Application_of_Machine_Learning_ML_for_Enhancing_the_Transient_Performance_of_Thermal_Energy_Storage_TES_Platforms_Using_Radial_Basis_Function_RBF_ [Accessed: October 22, 2023]

[96] Kolodziejczyk F, Mortazavi B, Rabczuk T, Zhuang X. Machine learning assisted multiscale modeling of composite phase change materials for Li-ion batteries' thermal management. International Journal of Heat and Mass Transfer. 2021;**172**:121199. DOI: 10.1016/J.IJHEATMASSTRANSFER.2021.121199

[97] What Are The Top Machine Learning Methods? | Tableau. n.d. Available from: https://www.tableau.com/learn/articles/top-machine-learning-methods [Accessed: October 23, 2023].

[98] Geetha P, Ajitha S, Jyothirmayi M, Guha T, Chaturvedi A, Ganeshan P, et al. Smart operating range monitoring of solar PV cell with integrated phase change materials by using solar deep learning model. Electric Power Components and Systems. 2023:1-12. DOI: 10.1080/15325008.2023.2249882. Available from: https://www.tandfonline.com/doi/epdf/10.1080/15325008.2023.2249882?needAccess=true

[99] Anooj GVS, Marri GK, Balaji C. A machine learning methodology for the diagnosis of phase change material-based thermal management systems. Applied Thermal Engineering. 2023;**222**:119864. DOI: 10.1016/J.APPLTHERMALENG.2022.119864

[100] Rashid FL, Al-Obaidi MA, Dulaimi A, Bahlol HY, Hasan A. Recent advances, development, and impact of using phase change materials as thermal energy storage in different solar energy systems: A review. Designs. 2023;**7**:66. DOI: 10.3390/DESIGNS7030066

[101] Sharma A, Singh PK, Makki E, Giri J, Sathish T. A comprehensive review of critical analysis of biodegradable waste PCM for thermal energy storage systems using machine learning and deep learning to predict dynamic behavior. Heliyon. 2024;**10**(3):e25800. DOI: 10.1016/j.heliyon.2024.e25800

[102] Odoi-Yorke F, Opoku R, Davis F, Obeng GY. Optimisation of thermal energy storage systems incorporated with phase change materials for sustainable energy supply: A systematic review. Energy Reports. 2023;**10**:2496-2512. DOI: 10.1016/j.egyr.2023.09.044

[103] Ali M, Dewan A, Sahu AK, Taye MM. Understanding of machine learning with deep learning: Architectures, workflow, applications and future directions. Computers. 2023;**12**:91. DOI: 10.3390/COMPUTERS12050091

Chapter 5

Electrically Conductive Self-Healing Epoxy Composites for Flexible Applications: A Review

Van-Phu Vu and Tien Thuy Phan

Abstract

The rapid advancement in electronics, electric vehicles, and grid storage has created a growing need for cutting-edge electronic devices. Researchers are actively investigating innovative materials that can withstand mechanical stress, exhibit electrical conductivity, and self-repair to ensure top-tier performance in various scenarios. Traditional polymer materials, though versatile, often lack the necessary qualities to meet the rigorous demands of flexible electronic and energy applications. To address these limitations, a significant breakthrough has emerged with conductive self-healing epoxy composites. These composites incorporate conductive materials into the polymer matrix, providing them with electrical conductivity, making them ideal for use as active components in electronic systems. Moreover, the integration of self-healing mechanisms significantly enhances their durability, countering performance degradation from mechanical stress. This chapter explores the recent developments in electrically self-healing epoxy-based polymer composites, with a focus on intrinsic self-healing mechanisms, evaluation strategies, the design of the conductive network in these composites, and their diverse applications in electronic components. These applications encompass adhesives, anti-corrosion coatings, sensors, EMI shielding, soft actuators, and energy storage/harvesting devices such as supercapacitors, batteries, and nanogenerators. The chapter also addresses the challenges and prospects of advancing flexible devices, paving the way for more reliable and efficient electronic systems in the future.

Keywords: self-healing epoxy, conductive polymer composites, epoxy vitrimer, conductive materials, flexible electronics

1. Introduction

In the ever-advancing world of electronics, the demand for materials that combine exceptional mechanical properties with electrical conductivity is ever-increasing. Traditional epoxy materials are renowned for their excellent mechanical strength, chemical resistance, and adhesive properties, making them widely utilized in various applications [1–4]. However, their inherent insulating nature limits their usefulness in applications where electrical conductivity is required. The introduction of conductive materials, such as metallic particles, metallic nanowires, carbon nanotubes (CNTs), and graphene-based materials, has enabled the incorporation of electrical conductivity into epoxy composites (ECs). The conductive materials need to exhibit excellent conductivity while also demonstrating strong compatibility with the host polymer. Yet, these materials remain vulnerable to damage caused by repeated deformations, which can give rise to the formation of microcracks and fractures. Such structural impairments disrupt electrical pathways, yielding functional deficiencies that ultimately compromise performance and lifespan in electronic applications [5–8]. A captivating development aimed at addressing this demand is the fusion of self-healing properties with electrical conductivity in epoxy materials. Self-healing ECs endowed with electrical conductivity not only ensure the longevity and dependability of electronic components but also enable the development of advanced electronic systems that boast enhanced resilience and durability. This self-repairing capability is achieved through the incorporation of dynamic chemical bonds within the polymer matrix, including noncovalent bonds (hydrogen bonds [9, 10], metal-ligand coordination [11, 12], etc.) and reversible covalent bonds (hindered urea bonds [13, 14], disulfide bonds [15, 16], Schiff base [17, 18], etc.). **Figure 1** visually depicts the utilization of electrically functional epoxy composites in self-healing electronic components. The impetus for exploring the fusion of self-healing attributes and electrical conductivity in epoxy materials arises from the surging demand for robust materials possessing both mechanical vigor and electrical prowess. The amalgamation of these attributes opens avenues to augment the dependability and lifespan of electronic devices, while concurrently bolstering their resilience against mechanical and electrical adversities.

Within this chapter, we embark on a comprehensive exploration to present a panoramic overview of the substantial progress achieved in the application of epoxy composites (ECs) endowed with both self-healing capabilities and electrical functionality, specifically tailored for versatile and flexible applications. Our primary objective is to delve into their diverse implementation across various domains. These domains encompass adhesives, flexible coatings designed for anticorrosion purposes, sensors,

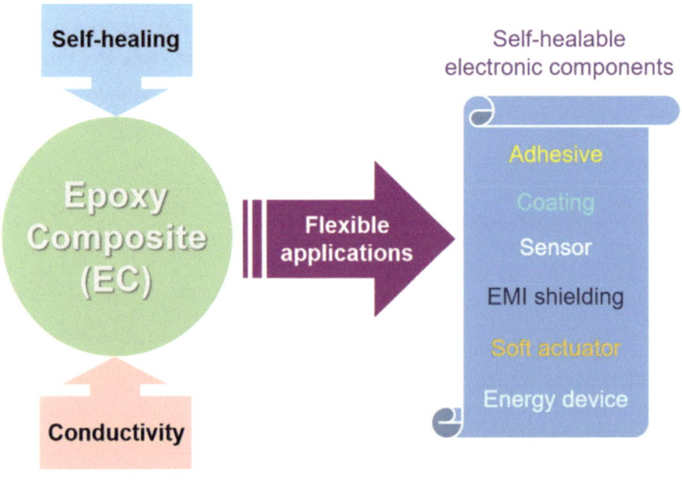

Figure 1.
Explore the electronic applications where epoxy composites with self-healing and electrical conductivity functionalities are utilized.

electronic components such as EMI shielding, soft actuators, and energy devices. We place distinct emphasis on meticulously unraveling the foundational differences between the mechanisms and strategies governing the healing of mechanical fractures and electrical breakdowns within ECs.

Furthermore, we endeavor to articulate the core principles, self-healing assessment, and design strategies of conductive networks that underpin self-healing EC-based devices. This highlighting showcases their inherent capacity not only to restore mechanical integrity but also to reinstate functionality and performance, even in the presence of simultaneous mechanical and electrical damage. While acknowledging the frontier that lies ahead, we also shed light on the challenges that await and the promising opportunities for future research. Through these efforts, this chapter aims to provide a comprehensive, up-to-date, and thorough understanding of the ever-evolving realm of conductive self-healing ECs within the landscape of flexible devices.

2. Fundamentals of conductive self-healing ECs

In the interest of providing comprehensive information, this section offers a concise overview of EC's self-healing strategies and materials design. However, the main emphasis of this chapter revolves around exploring the uses of conductive self-healing ECs in flexible electronic devices.

2.1 Self-healing chemistry strategies

The healing strategies for epoxy materials can be broadly categorized into two groups based on their structures and chemical compositions: intrinsic and extrinsic approaches. These two categories are distinguished by the mechanisms they employ to achieve healing functionality.

Intrinsic self-healing approaches do not rely on external healing agents. Instead, they possess an inherent ability to self-heal, which is facilitated by the thermal diffusion of molecules, reorganization of the polymer matrix, and the reformation of reversible/dynamic bonds or supramolecular interactions at the damaged interfaces. The efficiency of self-healing in epoxy polymers is dependent on critical factors such as the chain mobility, dissociation rate, and association rates. The majority of intrinsic recovery strategies are based on: (1) reversible covalent bonds (RCBs), which include Diels-Alder (DA) reactions [19], disulfide bonds (S—S) [20], boronated ester bonds [21], and imine bonds [22]. These RCBs offer a particularly favorable approach for designing self-healing epoxy polymers with enhanced mechanical properties, thanks to their robust bonding strength. According to Krishnakumar et al. [23], they demonstrated the potential of graphene oxide in facilitating healing of epoxy vitrimer nanocomposites, achieving approximately 90% restoration of the original strength through disulfide bonds. However, this healing process necessitates heating the cracked region to activate the bond exchange mechanism. At the earliest, Hao et al. [24] have showcased an innovative approach that involves a vanillin-based hyperbranched epoxy resin (VEHBP), integrating the synergistic effects of disulfide and imine dynamic covalent bonds (**Figure 2a**). This integration results in an epoxy resin that is both recyclable and malleable, while maintaining a notably high glass transition temperature (T_g), along with remarkable improvements in creep resistance and mechanical properties. Remarkably, the inclusion of 5% VEHBP in the dynamic

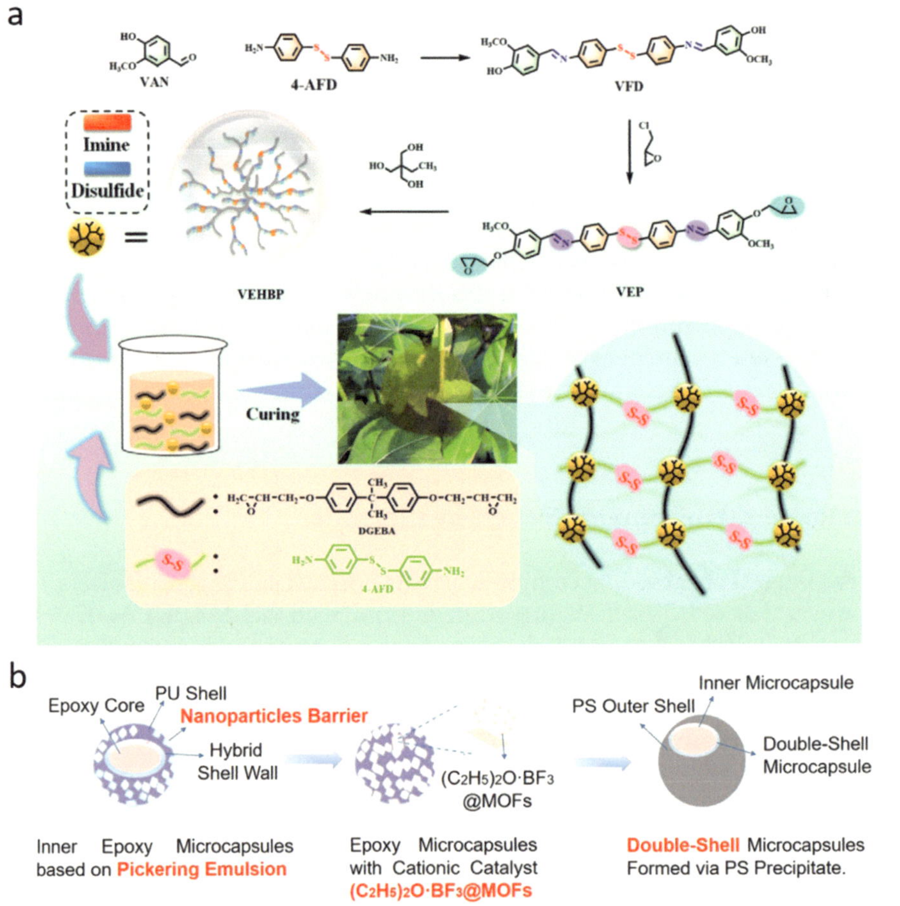

Figure 2.
(a) Synthetic routes of an epoxy resin incorporating a hyperbranched structure derived from vanillin, featuring disulfide and imine dynamic covalent bonds. Reprinted with permission from [24]. Copyright 2023 American Chemical Society. (b) Diagram illustrating the fabrication process of bilayered microcapsules encapsulating epoxy and $(C_2H_5)_2O{\cdot}BF_3$ dual agents through the utilization of MOF-stabilized Pickering emulsion and solvent precipitation techniques. Reprinted with permission from [25]. Copyright 2023 American Chemical Society.

covalent epoxy resin led to substantial enhancements in key characteristics. These enhancements encompassed a remarkable glass transition temperature of 175°C and a creep temperature of 130°C. Notably, the mechanical properties saw significant boosts, including a remarkable 34.1% increase in tensile strength, a 19.7% rise in storage modulus, and an impressive 173.3% surge in tensile toughness when compared to the pristine resin. Importantly, the distinctive hyperbranched architecture of VEHBP, coupled with the dual dynamic bonds, imbued these materials with exceptional self-healing capabilities, the potential for reprocessing, and a positive environmental footprint due to their degradability. These properties mark a significant stride forward in the realm of designing and crafting high-performance epoxy covalent adaptable networks. Liu and co-workers [26] conducted a study involving the synthesis of a range of ester-exchange catalysts, which were then analyzed for their potential in generating dynamic covalent bonds. In their investigation, they introduced metal-ion-free triethanolamine as an ester-exchange catalyst. This catalyst was utilized to develop epoxy resins possessing optimal characteristics suitable for the manufacturing of traditional electrical equipment. Furthermore, they combined 3,3′-dithiodipropionic acid (DTDPA), a compound containing carboxylic acid groups and S—S bonds, with MHHPA to create an epoxy vitrimer featuring dual dynamic bonds. By manipulating the ratio of the curing agents, they fine-tuned the structure of the dynamic covalent bonds within the crosslinking network. The results highlighted that the repair efficiency for surface scratches exceeded 90%. After fracture repair, the mechanical strength was rejuvenated to over 70%, and even instances of electrical damage exhibited some degree of restoration. Zou et al. [27] also investigated self-healing epoxy coatings utilizing reversible crosslinking network based on the DA reactions, which incorporated MXene flakes responsive to near-infrared (NIR) light. Their observations revealed that the presence of MXene enabled the epoxy to rapidly convert NIR light energy into heat, resulting in a self-healing capability of approximately 15 mins or/and (2) supramolecular chemistries primarily focus on reversible noncovalent interactions (RNBs), which encompass hydrogen bonds [28], metal-ligand coordination [29], ionic interaction [30], and host-guest interactions [31]. Liu et al. [32] developed a room-temperature self-healing epoxy coating with high elasticity in which the damaged coating can be repaired within 5 mins even under aqueous immersion thanks to reassociation of UPy hydrogen bonding units. In a study by Hu et al. [33], they introduced self-healable photocured epoxy acrylate resins that were formulated using photocured blends containing bisphenol-A epoxy diacrylate, a host-guest complex formed by 6-monomethacryl-substituted β-cyclodextrin and acrylamide-azobenzene, as well as butyl acrylate. The researchers demonstrated a unique approach to mending these light-responsive host-guest interactions. They achieved this by subjecting the material to UV light exposure at 365 nm, followed by a subsequent heat treatment at 120°C. This sequential process effectively enabled the damaged films to undergo self-healing. While not as strong as RCBs, these interactions have the advantage of creating mechanically dynamic and resilient systems, making them highly desirable for self-healing polymer designs. Due to their reversible nature, intrinsic self-healing epoxy materials can undergo multiple healing cycles through these interactions or reactions.

In contrast to intrinsic self-healing, extrinsic self-healing epoxy materials involve the incorporation of dispersed healing agents within an epoxy polymer matrix. These agents consist of reactive fresh precursors and catalysts. When damage occurs, these agents are released and initiate a process of spontaneous polymerization and reconstruction of the cross-linking network through chemical reactions, thereby repairing the affected areas [34]. Epoxy composites containing healing agent microcapsules have been also reported in several studies [35–38]. For instance, Zhu et al. [39] utilized a solvent evaporation technique to develop self-healing wave-absorption microcapsules, wherein epoxy resin cores were enclosed by hybrid walls composed of carbonyl iron powder and ethyl cellulose. As a result of this novel design, the material exhibited remarkable self-healing properties when subjected to electromagnetic irradiation. This improvement can be attributed to the increased mobility of the epoxy resin, achieved through the conversion of electromagnetic energy into thermal energy. This innovative approach shows significant potential for advancing self-healing epoxy materials, with promising applications across various fields. In a recent investigation, a novel approach was employed to achieve remarkable self-healing properties in epoxy resin. By incorporating 15 wt% microcapsules, the epoxy resin demonstrated an impressive 95% healing efficiency within a mere 10 mins at room temperature [25]. Simultaneously, the composite's interlaminar shear strength reached an outstanding 92% improvement within just 5 mins. This groundbreaking study marks the first instance of achieving rapid self-healing across diverse damage modes in both epoxy resin and carbon fiber/epoxy resin composites. This achievement was realized through the utilization of a unique self-healing system incorporating single-component double-shell microcapsules (**Figure 2b**). The innovation primarily revolved around the design of double-shell microcapsules, loaded with both epoxy resin and a catalyst ($(C_2H_5)_2O{\cdot}BF_3$). These capsules were ingeniously created by combining the principles of Pickering emulsion templating and solvent precipitation techniques. Noteworthy was the strategic use of metal-organic frameworks (MOFs) with precisely controlled sizes and hydrophilic properties. These MOFs acted as effective stabilizers for the Pickering emulsion, ensuring the stability and size control of the inner microcapsules. Furthermore, their distinctive properties provided an ideal platform for host-guest interactions, enabling the efficient capture, storage, and release of the cationic catalyst.

2.2 Self-healing performance assessment

Self-healing efficiency refers to the capacity of epoxy materials to restore their original functionalities and quality after experiencing damage. Typically, the evaluation of self-healing performance involves subjecting the materials to induced damage followed by conducting a mechanical test to initiate and quantify the self-healing process. This evaluation can be broadly categorized into three aspects: self-healing efficiency, healing rate, and damage volume.

2.2.1 Self-healing efficiency

The self-healing efficiency (η) is a crucial metric used to assess the restorative capabilities of a material following a damage and healing cycle. It quantifies the degree of functionality restoration, and there are several definitions of healing efficiency, with the most widely employed formula being $\eta = P_{healed}/P_{original}$. The parameter "*P*" can encompass a range of characteristics based on the research focus, including ultimate tensile strength, elongation, fracture toughness, Young's modulus values, and other properties that pertain to electrical, thermal, optical, or capacity-related indices. For instance, carbon/epoxy composites with self-healing capabilities, achieved through the incorporation of healing agents, exhibit a notable healing efficiency of 62% [40]. This efficiency is determined based on the changes observed in flexural strength. Benazzo et al. [41] engineered a high-performance epoxy vitrimer for deployment as the matrix phase within fiber-glass composites featuring the integration of Fiber Bragg Grating (FBG) sensors. The investigation showcased that laminates possessing a greater proportion of dynamic crosslinkers manifest enhanced reparative capabilities. Notably, during the initial healing cycle, the average and maximum recorded healing efficiencies, computed with respect to critical fracture toughness, were determined to be 89% and 95%, respectively. A higher healing efficiency enhances the material's self-healing ability, resulting in an extended service lifetime that aligns with specific application requirements.

2.2.2 Self-healing rate

The rate at which a material experiences damage is influenced by multiple factors, including strain rate, stress amplitude, and loading frequency. The effectiveness of self-healing processes relies on the healing rate exceeding the damage rate. This rate is influenced by various factors, including the concentration of broken bonds, the polymer chain mobility, and activation energies for bond exchange. These elements are intricately connected to the self-healing kinetics, encompassing temperature, mobility, and polymerization rate. To attain a high healing rate in self-healing materials, it is essential to customize the reaction kinetics by adjusting in temperature, repair time, or concentration. By carefully optimizing these parameters, the self-healing material can be designed to facilitate rapid and efficient healing processes. For example, the epoxy composite film, made using graphene-assisted host-guest chemistry, exhibits substantial improvement in healing rate at various time intervals. At 120°C, it achieves a 45.3% recovery after 5 mins and an 85.1% recovery after 30 mins [31]. In addition, raising temperatures typically enhances molecular diffusion, resulting in an increased healing rate and improved healing efficiency. For instance, Vertuccio et al. [42] investigated the impact of the curing temperature on a rubber-toughened bifunctional epoxy resin that was infused with self-healing molecules. This infusion aimed to confer an auto-repair capability to the resin. The study revealed that elevating the functionalization temperature to 160°C proved more effective in facilitating the interaction between the rubber phase and the epoxy precursor during the functionalization process. Notably, at 160°C, a healing efficiency of over 69% was achieved. The most remarkable outcome (88%) was observed in the sample incorporating the self-healing 1.3-dimethylbarbituric acid (DBA) filler. However, when the same sample underwent functionalization at 120°C, its healing efficiency dropped to 52%. Nevertheless, the practical application of thermal treatment is constrained, making room temperature self-healing highly desirable and a focal point of current research.

2.2.3 Damage volume

The capacity of a self-healing system to address damage volume plays a vital role in determining its ability to improve material reliability and lifespan. This capability relies on factors such as loading conditions, geometry, and polymer properties, and it can vary depending on the specific self-healing approach employed. Intrinsic self-healing systems are particularly effective for handling small damage volumes because they require the damaged surfaces to be in close proximity for re-bonding to occur. Wu et al. [43] employed epoxidized soybean oil (ESO) and natural glycyrrhizic acid to fabricate bio-based recyclable vitrimers that possess outstanding mechanical properties and thermal stability. Notably, crack widths measuring 100 microns gradually vanished within a span of 60 mins, highlighting the material's potential for recycling and chemical degradation. Microcapsule-based systems repair small to moderate damage volumes at limited microcapsule volume fractions. Microvessel-based systems offer the advantage of addressing a broader range of damage volumes since the healing agent can be replenished as required. In another study, it was observed that the microcrack located within the interface region possesses the ability for triple healing, showcasing its most remarkable self-healing performance at an impressive efficiency of 90.60% [44]. This noteworthy achievement was accomplished by developing an innovative interfacial self-healing system that utilizes carbon fiber and epoxy composites, anchored in the principles of the Diels-Alder reaction. Through the strategic integration of post-synthetic modified (PSM) metal-organic frameworks, these frameworks not only undergo in-situ grafting onto the carbon fiber (CF) surface but also serve as nanofillers, contributing to the dispersion of epoxy resin and enabling the fabrication of PSM/epoxy nanocomposites.

2.3 Design of electrically functional self-healing ECs

Combining self-healing capabilities with electrical properties offers a groundbreaking solution for overcoming material damage, enhancing durability, and opening new avenues for multifunctional applications. Within this subsection, we will discuss the design strategies employed for conductive networks in self-healing epoxy-based composites as well as explore the methods used for their manufacturing.

2.3.1 Percolation network by conductive filler reinforcement

Percolation theory states that introducing conductive fillers into a non-conductive matrix above the percolation threshold leads to the formation of a continuous conductive path in the composite. It is a common observation that the conductive filler concentrations (e.g., carbon-based, metal-based fillers, etc.) in the matrix exhibit a conductivity profile that can be represented by an S-shaped curve (**Figure 3**). This behavior indicates a substantial increase in conductivity within a small range of filler loading. For an insulating elastic matrix, the percolation threshold for conductive nanofillers, such as carbon blacks (CBs) with spherical morphology, typically falls within the range of 10–20% by volume fraction. This range is consistent with the predictions of classical percolation theory [46, 47]. An epoxy hybrid system was created by introducing multi-wall carbon nanotubes (MWCNTs) and graphene nanosheets (GNs) in two specific filler quantities: less than 0.1 wt% and greater than 0.5 wt% [48]. This was done using different ratios of MWCNTs to GNs. These hybrid epoxy systems demonstrated outstanding electrical capabilities, credited to the interactions between the π–π bonds of the multi-wall carbon nanotubes and the dispersed graphene layers within the epoxy resin matrix. When using high loading of stiff nanofillers, challenges arise due to increased viscosities, and inferior mechanical performance resulting from nanofiller aggregation at higher concentrations. Therefore, achieving an effective reduction in the percolation threshold while preserving electrical

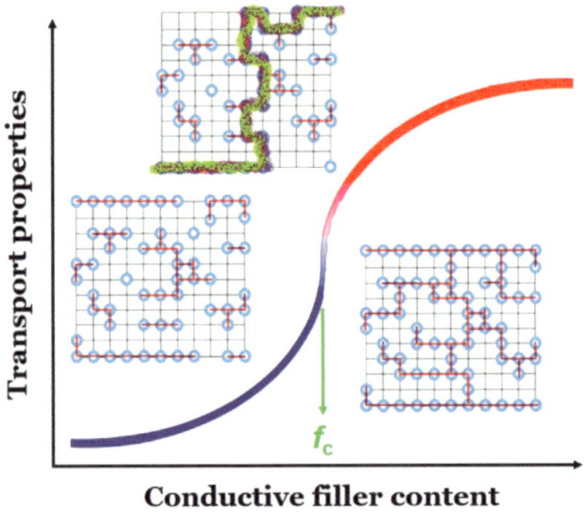

Figure 3.
The effective electrical conductivity exhibits an S-shaped curve near the percolation threshold f_c. This proximity to the threshold reveals notable phenomena of percolation theory and nonlinear changes in the transport properties of composites. The insets illustrate the geometric phase transition of fillers within the microstructures of the composites precisely at the percolation point. Reproduced from Ref. [45] with permission from the Royal Society of Chemistry.

properties becomes crucial for fabricating high-performance conductive composite materials. To achieve this objective, various parameters of conductive nanofillers can be tailored, including aspect ratio, dimensional size, size distribution, surface wettability, and contact resistance between individual nanofillers. This customization allows for fine-tuning the corresponding electrical properties to suit specific applications.

2.3.2 Segregated conductive filler network

By strategically forming segregated filler networks within an epoxy matrix, conductive epoxy composites can achieve remarkably improved conductivity even at lower nanofiller concentrations. Instead of being dispersed randomly throughout the entire matrix, conductive nanofillers are intentionally positioned at the interfaces of epoxy matrix domains, leading to a significantly lowered percolation threshold compared to conventional percolation networks. The establishment of these segregated conductive networks depends on the fabrication approach used during the construction of the composites [49, 50].

Through the implementation of an interface engineering strategy, branched CNTs were skillfully coated on the surface of catalyst-infused epoxy waste particles. Subsequently, vitrimerized epoxy/CNT composites were created via compression molding [51]. This innovative approach resulted in the development of a segregated filler structure at the interfaces of "vitrimerized" epoxy particles, achieving two important objectives simultaneously. Firstly, it enabled the recycling of epoxy waste, contributing to sustainability efforts. Secondly, it significantly enhanced the mechanical and electrical properties of the recycled epoxy, even with low carbon nanofiller content (below 1 wt%). The successful incorporation of branched CNTs facilitated the fabrication of highly electrically conductive composites, which hold great promise for applications such as electromagnetic interference (EMI) shielding and strain sensing. Zhang et al. [52] successfully accomplished the controlled dispersion of multi-walled carbon nanotubes (MWCNTs) within the distinct network of hexagonal boron nitride (h-BN) in the epoxy vitrimer matrix using compression molding. This method facilitated the creation of interfacial bonds through dynamic transesterification involving small molecules, particularly at elevated temperatures. As a result of this process, the epoxy vitrimer composites demonstrated improved thermal conductivity and electrical insulation even with a low concentration of fillers. This improvement is attributed to the collaborative effect of the well-organized nanofiller network. In a composite containing 1 wt% MWCNTs and 8 wt% h-BN, the recorded thermal conductivity and electrical resistivity values were 0.83 W $(m\ K)^{-1}$ and 1.92×10^{11} Ω cm, respectively. Furthermore, by increasing the portion of separated h-BN, it is possible to further enhance the electrical resistivity.

2.3.3 Production of electrically functional self-healing epoxy composites

Various strategies have been extensively explored to address the issue of inadequate electrical conductivity in epoxy composites. One commonly employed approach involves blending conductive single or hybrid fillers to utilize synergistic effects between epoxy polymers and various filler dimensions. For instance, Bian and his colleagues [53] introduced a remarkable self-healing epoxy composite capable of inhibiting electrical tree growth. They achieved this by incorporating SiO_2 particles into the epoxy matrix and creating hydrogen bonds between the SiO_2 particles and epoxy chains through solution blending. The results demonstrated excellent self-healing properties, with cracks healing upon the application of heat, and effective inhibition of electrical tree growth, a common cause of electrical breakdown in high-voltage applications.

Another approach, vacuum-assisted filtration, uses hierarchical filler architecture to prepare epoxy polymer composite films [54]. However, the time-consuming nature of this method has limited its practical application. On the other hand, the electrospinning coating method, known as the core-shell structure, provides a promising and simple means to create ordered composites [55–58]. Chen et al. [59] reported tough and rapidly self-healing carbon/epoxy composites using electrospun thermoplastic polyamide nanofiber (PAnf). The PAnf not only toughens the composite interlayer but also provides excellent self-healing capabilities. With just 1.2 wt% addition, the interlaminar shear strength and bending properties of composites increased by 17.6% and 14.7%, respectively. The entangled nanofiber structure enhances interlaminar adhesion, effectively suppressing microcrack and delamination propagation. The epoxy composites with PAnf demonstrated the best mechanical properties and healing efficiency after 3 h of electrospinning. Multiple healing cycles showed healing efficiencies of 70.1% for the third time. The inherent characteristic of thermoplastic PAnf allows it to repeatedly and steadily heal interlaminar fractures without depletion. The uniform distribution of porous PAnf promotes resin flow and impregnation without compromising the storage modulus and T_g of the composites, preserving their original properties.

Additionally, a novel approach involves generating a 3D highly electrically conductive network by dispersing the epoxy polymer matrix, ensuring high electrical conductivity [60]. However, complex template preparation and polymer filling have hindered its widespread use. Magnetic field forces have also been used to promote filler orientation or construct electrical conduction networks [61]. Additionally, hot pressing is sometimes employed after processes like filtration, electrospinning, or template methods to reinforce the structure [62]. This innovative technique holds promise for advancing electrical conductivity in epoxy composites.

3. Conductive self-healing ECs for flexible applications

Long-term operation of flexible electronic devices can lead to fatigue, corrosion, and damage, which ultimately culminate in significant device failure. To tackle this issue, researchers are actively investigating the potential of self-healing electronic devices. These devices possess the remarkable ability to automatically restore their electrical and mechanical performance in the face of mechanical damage, whether with or without external assistance. Integrating conductivity and self-healing functionality in flexible electronic devices is predicted to significantly enhance both their durability and performance. In this section, we emphasize the deployment of epoxy-based self-healing polymer composites in various components and devices of flexible electronics, encompassing adhesives, coatings, sensors, electromagnetic interference (EMI) shielding, soft actuators, and energy devices.

3.1 Adhesives

Epoxy resins find common applications as adhesives, with electrically conductive adhesives (ECAs) presenting a versatile solution in electronics due to their unique dual functionality: enabling electric current conductivity while mechanically bonding diverse substrates. Typically, conductive fillers like Ag, Au, Ni, carbon materials, or MXene are integrated into acrylic, silicone, or epoxy resins to create ECAs. Notably, epoxy resin-based ECAs have gained significant traction in electronics, serving purposes such as die attachment and solderless interconnection. This popularity stems from their exceptional attributes: a blend of chemical and thermal resistance, robust mechanical qualities, strong adhesion, and compatibility with various substrates and additives.

Of remarkable significance are epoxy-based ECAs with self-healing capabilities, which can autonomously repair minor damage in the conductive pathways of the adhesive matrix. These properties hold great promise for industries like flexible electronics, wearables, and automotive electronics, offering advantages through self-healing conductive adhesives. Their pliable nature makes them suitable for scenarios where traditional rigid soldering techniques are impractical. An example by Zhang et al. [63] involves a novel epoxy resin synthesized using the curing agent 1,4,5-oxadithiepane-2,7-dione (DSAA), containing reversible disulfide bonds. This epoxy resin demonstrated self-healing and malleability, showing potential as a reusable adhesive. However, effective self-healing required a relatively high level of DSAA (22.7 wt%).

Advancements in "green chemistry" aim to develop non-petroleum-based matrices for self-healing fiber-reinforced composites. Hu's group [64] obtained fully biobased vitrimers from camphoric acid (CPA) and epoxidized soybean oil (ESO), using them as matrices for a carbon fiber-reinforced composite (**Figure 4**). Network rearrangements via transesterification reactions (TERs) provided CF/CPA/ESO composites with reprocessing, self-adhesion, and repair capabilities at elevated

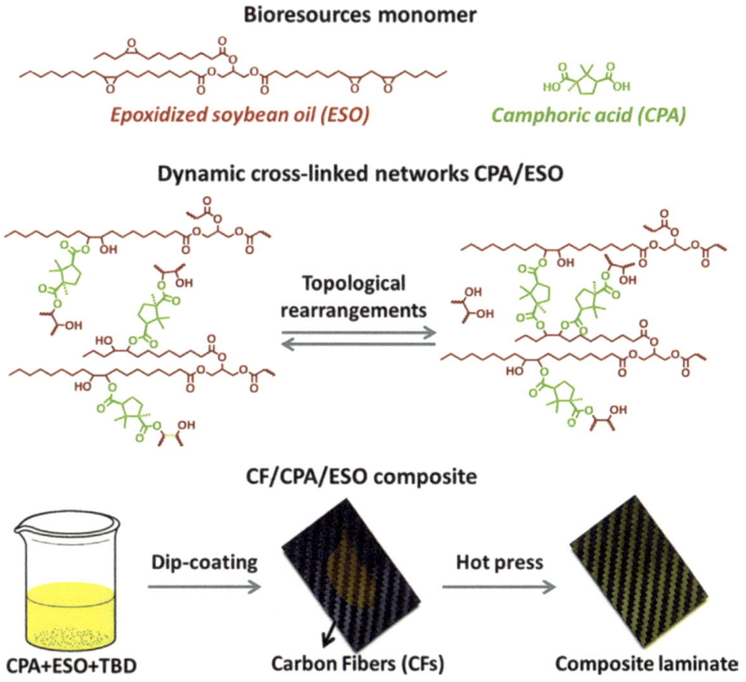

Figure 4.
Depictions showcasing the dynamic crosslinking rearrangements within CPA/ESO networks, and the creation of a composite laminate involving carbon fiber-reinforced composite. Reproduced from Ref. [64] with permission from the Royal Society of Chemistry.

temperatures (200°C). Carbon fibers could be fully recycled by degrading CF/CPA/ESO composites with ethylene glycol. In another study, Xu and colleagues [65] developed carbon fiber reinforced composites utilizing a degradable bio-based epoxidized menthane diamine-adipic acid (EMDA-AA) matrix. Notably, the EMDA-AA vitrimer exhibited topological rearrangement through dynamic transesterification reactions, facilitated by autocatalysis of tertiary amines, even in the absence of a catalyst. These developed vitrimers displayed self-healing behavior at 180°C for 30 mins (healing efficiency = 92.9%) and shape memory at 100°C, applicable to both EMDA-AA and EMDA-AA-CF systems. The reported epoxy vitrimer-based composite showcased commendable adhering properties (lap shear strength = 3.8 MPa) when two EMDA-AA-CF composites were bonded under a hot press at 180°C and 15 MPa.

Zhang et al. [66] introduced a fully bio-based epoxy vitrimer for recoverable adhesives, utilizing ozonated lignin and sebacic acid-derived epoxy cured with a zinc catalyst ($Zn(acac)_2$). This adhesive achieved a lap shear strength of 6.5 MPa when applied to coarse aluminum sheets. Moreover, separated halves of aluminum plates could be rebonded using catalyst-accelerated transesterification at 190°C for 1 h, a feat not achievable with traditional adhesives. This indicates potential for vitrimers as self-healing and recoverable adhesives from bio-based materials. Beyond practical advantages, these adhesives align with eco-friendly principles and cost-saving potential, positioning them as sustainable choices for future electronics. As research and development progress, self-healing conductive adhesives hold exciting promise for revolutionizing electronics manufacturing and maintenance paradigms.

3.2 Coatings

Coating represents one of the most prevalent applications for epoxy resins. Epoxy materials are extensively employed in coatings to safeguard surfaces and provide decorative attributes, owing to their exceptional mechanical properties and chemical resistance. However, when a coating becomes damaged, its intended functions are compromised. The removal of such coatings from substrates is challenging due to their robust interfacial bonding. To mitigate waste and extend the service life, the optimal approach likely involves repairing the coating and reinstating it for use. Yet, repairing epoxy coatings is intricate due to their permanent cross-linked structure, differing from conventional thermoplastics. A recent advancement by Han and colleagues involved the development of a catalyst-free epoxy vitrimer coating, exhibiting impressive self-healing characteristics [67]. This innovative coating was formulated by curing hyperbranched epoxy (HBE) prepolymers with succinic anhydride (SA). The abundance of hydroxyl groups in HBE facilitated rapid stress relaxation and efficient self-healing at elevated temperatures (>120°C). When applied as a vitrimer layer onto metal plates (with a thickness of approximately 80 μm), this coating not only demonstrated excellent adhesion and hardness but also displayed efficient self-healing at 150°C for 1 h. Furthermore, the healed coating effectively safeguarded the metal plates from electrolyte erosion, confirming its potent self-healing capability. Incorporating DGEBA (DER 331) into this system resulted in a repairable vitrimer coating that provided robust protection against salt and electrochemical corrosion. This type of epoxy coating holds potential for prolonging the lifespan of thermosets and reducing waste. Raspberry-like polydopamine@ polypyrrole (PDA@PPy) nanoparticles were synthesized to serve as a photothermal agent and subsequently incorporated into an epoxy resin, creating a self-healing coating [68]. Following 808 nm near-infrared (NIR) irradiation for 10 mins, the impact of varying amounts of PDA@PPy on scratch closure efficiency was explored. Elevated PDA@PPy content correlated with improved healing results, although excessive quantities hindered the repair process due to restrictions on the movement of epoxy resin molecular chains. Hence, precise incorporation of photothermal fillers is pivotal for optimizing the self-healing capacity of coatings. When evaluating corrosion resistance, the linear sweep voltammetry curve illustrated that the corrosion current of a healed coating diminished comparably to an intact coating, underscoring the effective safeguarding of X70 steel.

In a research endeavor undertaken by Hao et al. [69], they carried out a study where they modified Kraft lignin using esterification alongside 4-methylcylohexane-1,2-dicarboxylic anhydride. This chemical process resulted in a novel form of Kraft lignin referred to as L-COOH. The innovative aspect of this modification was its increased solubility in ethanol. By subsequently subjecting it to a reaction with poly(ethylene glycol) diglycidyl ether, utilizing a zinc catalyst, they successfully produced a vitrimer with a notable lignin content exceeding 47 wt%. The vitrimer's dynamic exchange mechanism, primarily involving ester bonds, was triggered at elevated temperatures beyond 140°C. This temperature-induced activation facilitated the efficient repair of coatings within a brief duration of 15 mins at 190°C, with the assistance of glycol. Notably, this coating also exhibited a distinctive swelling capability when exposed to alkaline aqueous solutions. This behavior was attributed to the conversion of phenol and carboxyl groups into sodium phenolate or sodium carboxylate, which caused a transformation in the chemical nature of the coating. As a result, the bonding between the coating and the substrate was weakened, leading to easier coating removal. Liu et al. [70] introduced an example of free radical polymerization based on a dynamic transesterification network. They synthesized a UV-curable oligomer (TMG) from tung oil using microwave technology, which was subsequently photopolymerized alongside a biobased-reactive diluent derived from malic acid (MA). This strategy yielded UV-curable coatings or materials featuring multiple hydroxyl and ester groups. These materials exhibited the capacity for repair and reshaping through dynamic transesterification reactions, activated at higher temperatures with the aid of a zinc catalyst. Importantly, the tensile strengths of welded (12.2 MPa) or reshaped materials (28.7 MPa) exhibited significant enhancement compared to the original material (7.1 MPa). This enhancement was attributed to an elevated crosslinking density resulting from C=C polymerization and esterification during high-temperature heating.

Feng and collaborators designed tung oil-loaded polyurethane (PU) microcapsules [71], with conductive polyaniline (PANI) impregnated into the microcapsule wall structure. Corrosion tests highlighted the exceptional anti-corrosion properties of the self-healing epoxy coating incorporating 10 wt% tung oil-loaded PU/PANI microcapsules. These microcapsules, with a tung oil core and PANI wall, acted as corrosion protectors, forming a self-healing film and a passivation layer. In a recent study by Alias et al. [72], linseed oil was encapsulated for corrosion protection of magnesium (Mg). Coatings containing microcapsules were applied to a bare Mg substrate and subjected to scratching to evaluate their self-healing capacity. Electrochemical measurements in a 3.5 wt% NaCl solution demonstrated that the epoxy coating combining linseed oil and urea-formaldehyde substantially reduced the corrosion current density (icorr) on the Mg sheet (1.552 A cm^{-2}) compared to the bare Mg sheet (109.8 μA cm^{-2}). Consequently, these self-healing coatings exhibited notable recovery in both self-healing and corrosion resistance on Mg alloys.

In another investigation, Haddadi et al. [73] reported the creation of amino-functionalized MXene (Ti_3C_2) nanosheets through an etching technique and a modification process involving 3-aminopropyltriethoxysilane. MXene's favorable interface interaction and stability with organic/polymeric materials position it as a promising anticorrosion nanofiller for enhancing the passive barrier properties of organic coatings [74]. It also serves as a suitable host for loading corrosion inhibitors [75]. Utilizing cerium (Ce^{3+}) cations as corrosion inhibitors, MXene nanosheets encapsulating these cations were employed to produce self-healing epoxy composite coatings, exhibiting robust corrosion protection performance (**Figure 5**). In a study by Niu et al. [76], a robust omniphobic slippery coating was employed to create self-healing coatings on ultrathin MXene multilayers. Upon exposure to just 10 mins of solar irradiation, the surface temperature of the coating rapidly rose to approximately 50°C. The successful fabrication of MXene multilayers on diverse metal substrates suggests the potential for developing a range of self-healing coatings. Consequently, the utilization of photothermal activation for self-healing emerges as a pivotal element within light-responsive coating technology. Wang and his colleagues [77] introduced a novel $Ti_3C_2T_x$/layered double hydroxide (LDH) pigmented epoxy composite hybrid coating that demonstrated remarkable corrosion resistance and self-healing capabilities on AZ31 magnesium alloy. The incorporation of environmentally friendly organic acid anions from cysteine acid (Cys) as corrosion inhibitors and the growth of $Ti_3C_2T_x$/MgAl-LDH (TML) heterostructure nanosheets via in-situ assembly led to the self-healing function. The TML heterostructure displayed excellent dispersion and compatibility with epoxy resin. The composite coating exhibited exceptional corrosion resistance, with a corrosion current density of 1.4×10^{-9} A cm^{-2}

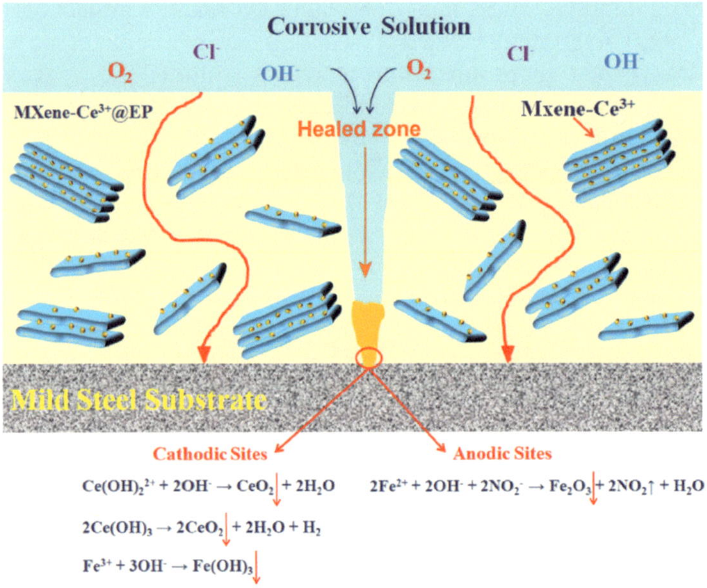

Figure 5.
Diagram illustrating the inhibition mechanism within the coating phase involving Ti_3C_2 MXene-Ce^{3+}. In the MXene-Ce^{3+}@EP coating containing a defect, Ce^{3+} ions have the potential to migrate out from Ti_3C_2 MXene-Ce^{3+} nanosheets. These ions subsequently get deposited onto the surface of the mild steel plate within the scribed zone, effectively obstructing cathodic sites. Reprinted with permission from [73]. Copyright 2021 American Chemical Society.

and an impedance value of $1.66 \times 10^7\ \Omega\ cm^{-2}$ at low frequency. Furthermore, the self-healing efficiency of the epoxy-TML composite coating reached 56.17% after 6 days of immersion in a corrosive environment.

Lorwanishpaisarn et al. [78] conducted a study with the goal of developing a self-healing epoxy vitrimer/CNT nanocomposite for use as a coating material. They employed two bio-based curing agents, namely cashew nut shell liquid (CNSL) and citric acid (CA), to create adaptable networks held together by covalent bonds. Within the epoxy/CNSL/CA matrix, they incorporated CNTs using probe sonication and agitation, spanning concentrations from 0 to 0.5 wt% (referred to as V-CNT0-0.5). The research outcomes indicated that the thermomechanical properties of the V-CNT nanocomposites improved as the CNT content increased. The study also found that a bond exchange reaction involving esterification could be triggered by near-infrared (NIR) light. Among the tested compositions, V-CNT0.5 exhibited the highest self-healing efficiency, as evidenced by a Shore D hardness of 97.34%. To evaluate corrosion resistance, coated steel samples containing V-CNT0 and V-CNT0.5 were immersed in a 3.5 wt% NaCl solution for a duration of 7 days. In this context, the corrosion rate of steel coated with V-CNT0.5 significantly decreased from 9.53×10^2 MPY to 3.12×10^{-5} MPY. Additionally, the protection efficiency surged by 99.99%. Capitalizing on these remarkable self-healing and anti-corrosion attributes, it is evident that V-CNT0.5 holds promise as an appealing material for organic anti-corrosion coatings. In a novel achievement, a thermally robust and high-performance pH-responsive anti-corrosive nanoreservoir system was developed [79]. This system is built upon a double-ligand zinc phosphate framework (ZPF) that decorates a graphene oxide skeletal template (G-ZPF) using the ligand exchange theory. The outcomes revealed that the G-ZPF nanoreservoirs loaded with structural inhibitors offer a combination of active (self-healing) and passive (ion-water barrier) attributes within the organic coating. As a result, the epoxy-polyamide coating embedded with G-ZPF nanoreservoirs demonstrated remarkable corrosion prevention capabilities and exhibited exceptional thermal and mechanical performance.

3.3 Sensors

A sensor is a sophisticated apparatus designed to identify and perceive external chemical or physical signals and subsequently translate the gathered data into the desired format. It plays a pivotal role in the advancement of wearable and adaptable intelligent electronic devices, which have vast applications in displays, artificial intelligence, and healthcare [80]. However, sensors equipped with multiple flexible functionalities are prone to encountering diverse damages during their operational lifespan, thereby impacting the overall longevity of these devices. Consequently, there has been a significant surge of interest in the development of self-healing sensors. The primary focus behind this research is to bolster the durability and sustainability of these electronic devices, ensuring they remain functional for an extended period.

In the earliest investigation conducted by Wang's research group [81], they developed a self-healing graphene/rubber-based supramolecular elastomer (GRSE) that capitalized on the synergistic interplay between dynamic boroxines and interfacial hydrogen bonds. The incorporation of graphene nanosheets not only elevated the conductivity and sensitivity of the sensors but also played a pivotal role in enhancing the overall performance. To further enhance the amino groups and augment adhesion among the graphene sheets and elastomers, the surface of the graphene nanosheet was modified through the absorption of 1-pyrenamine (PA) via $\pi-\pi$ conjugation. This strategic modification effectively improved the sensor's properties. This wearable sensor exhibited remarkable attributes, including high electrical conductivity (0.0029 S m^{-1}), rapid response time (250 ms), and a low detection threshold (1%). Additionally, the engineered sensor initiated its healing process under room temperature conditions, displaying impressive mechanical robustness (3.46 MPa) and a healing efficiency (η) of 91.1%. Leveraging its stress-sensing capabilities, this sensor holds promise for applications in human motion detection. Kai et al. [82] introduced photonic vitrimer-based electronics (PVBEs) as an innovative class of flexible and stretchable electronics that amalgamate the advantageous attributes of photonic crystals (PCs) and piezoresistive carbon textiles (CCT) with the intrinsic self-healing properties inherent in vitrimers. By addressing the limitations inherent in current flexible electronics, such as susceptibility to damage and constrained electrical output, the research employs synthesized poly(urethane-urea) vitrimer elastomers derived from specific constituents, encompassing polytetramethylene ether glycol (PTMG), poly(1,4-butanediol) bis(4-aminobenzoate) (PBDAB), isophorone diisocyanate (IPDI), and glycerin (GLY). These vitrimers exhibit robust mechanical characteristics, resilience, and a remarkable 93% self-healing efficiency attributed to dynamic covalent networks. The integration of photonic crystals and carbonized cotton textiles within the vitrimer matrix gives rise to PVBEs that manifest synchronized color alterations and electromechanical responses under conditions of mechanical stress or stretching. PVBEs showcase exceptional strain sensing capabilities, including swift synchronous electrical and optical responses (0.25 s), heightened sensitivity (with a gauge factor of 10.3), exceptional endurance (exceeding 10,000 cycles), mechanochromism, potential for self-healing in optical functionalities through dynamic covalent networks, stable electromechanical performance facilitated by a fully integrated structure, and the capacity for wireless transmission, enabling real-time monitoring of human movements with dual-signal feedback. This study pioneers a transformative platform in the realm of flexible electronics, promising applications in domains such as wearable devices, soft robotics, and the rapidly expanding landscape of the Internet of Things.

Yang et al. [83] also successfully developed a series of functional epoxy elastomer/carboxylated carbon nanotube composites, combining self-healing capabilities and degradability to create flexible and stretchable strain sensors (see **Figure 6a**). The epoxy elastomer was synthesized using carboxyl-terminated poly(ethylene glycol), 2,2′-dithiodibenzoic acid, and 1,4-butanediol diglycidyl ether as monomers, along with a bio-based epoxidized soybean oil as the cross-linker. These composites showcased exceptional mechanical properties, boasting a high tensile stress of 5.07 MPa and impressive stretchability, with a capacity to stretch up to 477%. The composite's remarkable self-healing performance of 92.5% at 80°C for 24 h (**Figure 6b**) was a result of the synergistic healing effect derived from the hydrogen bonds and disulfide bonds present in the epoxy matrix. Moreover, the strain sensor based on the elastomer composite with microstructure displayed a notably high gauge factor (GF) sensitivity of 176.7, along with rapid response and relaxation times of 60 ms and 100 ms, respectively. Additionally, it demonstrated exceptional repeatability withstanding 1000 cycles. Its versatile applications included successful detection of human motions and recognition of objects with various shapes, as depicted in **Figure 6c**.

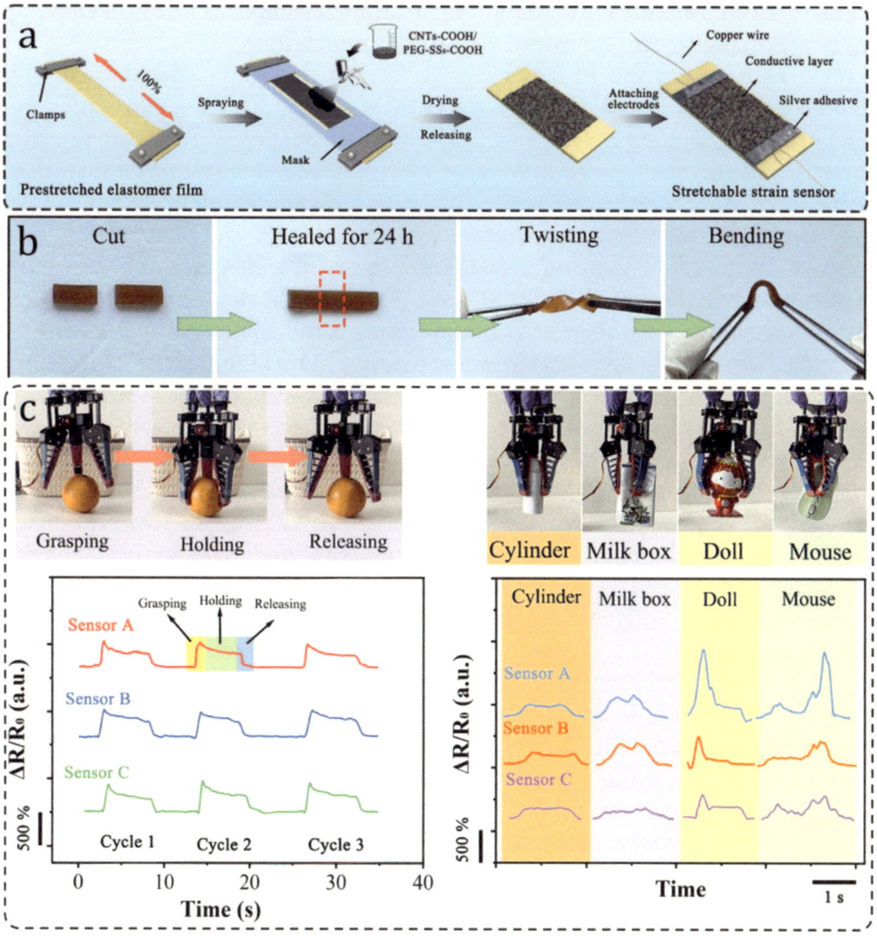

Figure 6.
Schematic illustrations of (a) the preparation process of the elastomer composites-based strain sensor. (b) Healing behavior after 24 h and mechanical deformations of epoxy elastomers. (c) The epoxy elastomer composite-based strain sensor finds practical use in detecting human motion and recognizing objects. Reprinted with permission from [83]. Copyright 2022 American Chemical Society.

Enhancing the stability of strain sensors is of utmost importance. However, the presence of scratches, mechanical damage, and harsh environments can significantly curtail their lifespan. To address this challenge, Lu and colleagues [84] have devised a remarkable solution—self-healing electronic sensors that boast exceptional stability and reproducibility. These sensors rely on the principles of metal-ligand coordination and possess a well-organized hierarchical structure. The preparation method for these self-healing sensors is quite straightforward: they employ polydopamine (PDA) to modify epoxidized natural rubber, which is then cross-linked using Fe^{3+}. Additionally, the incorporation of CNTs into the self-healing matrix further enhances their performance. As a result, the manufactured strain sensor exhibits outstanding mechanical properties and retains its high sensitivity even after undergoing extensive bending (over 50,000 times) or repetitive washing. A key highlight of these self-healing strain sensors is their remarkable ability to heal themselves at room temperature, making them stand out from conventional approaches. Given these exceptional attributes, these strain sensors are ideally suited for utilization in smart electronic devices, opening up new possibilities for advanced electronic applications.

A remarkable advancement in polymer technology lies in the development of bio-based epoxy vitrimers derived from soybean oil and tung oil. These novel polymer networks possess inherent advantages, including self-healing capabilities, repeatable processing, and recyclability. Additionally, when combined with carbon materials, these epoxy vitrimers can be utilized to create versatile multifunctional sensors, serving as energy converters, temperature warning sensors, and fire warning sensors. This breakthrough offers exciting new possibilities for a wide range of construction applications and sensor technologies [85–88]. Jia et al. [89] successfully developed bio-based polyschiff base vitrimer/graphene oxide composites with remarkable self-healing and reprocessability attributes, making them highly efficient for applications as temperature warning sensors and fire warning sensors. These polyschiff vitrimers were skillfully crafted using vanillin and tung oil as sustainable raw materials, resulting in a commendable tensile strength range of 1.20 MPa to 1.91 MPa. These vitrimer-based composites exhibited the ability to self-heal, with cracks effectively mending after being subjected to a temperature of 120°C in an oven for 120 mins. This self-healing phenomenon was attributed to a combination of sensitively dynamic imine covalent bonds and structural compatibility with the flexible aliphatic hydrocarbon chain of tung oil. Furthermore, the reprocessed composites displayed tensile strengths ranging from 1.21 MPa to 1.90 MPa, which closely resembled the tensile strengths of the original samples. This reprocessability feature adds to the practicality and durability of the bio-based vitrimer composites.

Cao et al. [90] utilized bio-derived carboxyl cellulose nanocrystals to create a novel nanostructured supramolecular sensor. This sensor was designed to interact with chitosan-decorated epoxy natural rubber latex, forming multiple H-bonding interactions. The results were impressive, as the sensor exhibited ultrafast self-healing (within 15 s) with remarkable repeatability. The healing efficiency of the sensor was exceptional, reaching 93% after the third healing cycle. Additionally, the team developed a highly sensitive strain sensor using a layer-by-layer technique, incorporating H-bonding between chitosan solutions and nanocomposite-assisted carbon nanotubes. This strain sensor demonstrated an impressively low strain detection limit of 0.2%. Even after undergoing cutting-healing cycles and being subjected to more than 20,000 bends, the sensor maintained its stability and provided reliable and repeatable response signals. Furthermore, the researchers took it a step further by integrating these sensitive and self-healable, flexible sensors into a human-machine interaction system. This system proved to be versatile, serving as a facial expression control system and an electronic larynx. The potential applications of this innovative technology are vast, and it opens up exciting possibilities in the field of flexible and interactive electronic devices.

3.4 Electromagnetic interference (EMI) shielding

As electronic technology and telecommunication-related industries continue to advance rapidly, the issue of electromagnetic radiation pollution has become increasingly prominent. Consequently, there is a growing need to explore new materials for electromagnetic interference (EMI) shielding. One class of high-performance EMI shielding materials that has gained significant attention is carbon-based materials, owing to their lightweight nature and ease of fabrication [91]. Among these materials, CF stands out as an appealing choice in electronic apparatus due to its exceptional mechanical and electrical properties [92]. To enhance the shielding effectiveness even further, CF is often coated with metal nickel, imparting magnetic properties to the material [93]. Yu and his coworkers [94] introduced a novel self-healing composite for EMI shielding, utilizing the Diels–Alder (DA) thermoreversible reaction system. This dynamic covalent bonding network was created by crosslinking furan-modified epoxy resin (FM-EP) with a bismaleimide at 60°C, resulting in an impressive self-healing efficiency of 92.5%. Within this self-healing thermoreversible epoxy resin matrix, nickel-coated carbon fiber (Ni/CF) was uniformly dispersed as a conductive filler to attenuate electromagnetic waves. The researchers achieved an EMI shielding effectiveness (EMI SE) of 40.5 dB by increasing the volume percentage of the conductive filler to 18%, while still maintaining excellent healing performance.

Through the ingenious utilization of associative dynamic bonding rearrangement within the epoxy vitrimer, Fang et al. [95] have achieved a remarkable breakthrough in developing segregated multiwalled-CNTs/epoxy composites. This innovative approach creates distinct pathways for conductive filler networks, allowing for outstanding performance across a broad range of compression temperatures and pressure conditions. These segregated composites exhibit an exceptionally low conductive percolation threshold of just 0.066 wt% and boast an impressive EMI SE of 22 dB, even with a mere 2 wt% loading of multiwalled-CNTs. Additionally, the incorporation of volatile ethylene glycol contributes to the mechanical robustness by enabling the decomposition and repolymerization of β-hydroxyl ester at the interface, even in the presence of numerous multiwalled-CNTs. Another significant advantage of this advancement is the preservation of the segregation structure and macroscopic properties after reprocessing. Moreover, the nanofillers can be effortlessly recovered using excessive EG at elevated temperatures, further demonstrating the versatility and practicality of this approach. With its exceptional combination of excellent conductive properties, high EMI shielding effectiveness, and mechanical resilience, this novel technique holds enormous promise for diverse applications in various fields.

3.5 Soft actuators

Soft actuators that can exhibit pre-set shapes and respond to various external stimuli such as heat, light, solvent, and electricity are attracting significant interest in diverse fields like soft robotics, energy generation, motors, and fluid propellers. The key attributes sought after in these soft actuators are multi-stimuli responsiveness and multi-stimuli-triggered self-healing capability, which are pivotal for their extensive applications. For practical utility in various scenarios, the integration of multi-stimuli responsiveness and multichannel self-healing ability is highly desirable. This feature would enable the soft actuators to be easily actuated or self-healed using a simple and convenient tool whenever required.

A notable contribution in this area is from Yang et al., who presented a groundbreaking advancement in the form of multifunctional, recyclable, thermosetting, and vitrimer-based soft actuators [96]. These unique actuators were developed through the hot-pressing of carbonized silk fabric (CSF) onto vitrimers, resulting in the formation of composite CSF-vitrimers. These actuators exhibit an impressive five-stimuli-triggered self-healing ability, including heat, light, electricity, electromagnetic waves, and solvent. Moreover, they demonstrate four-stimuli-triggered responsiveness to light, electricity, heat, and solvent. Additionally, the composites boast improved mechanical properties, prevent delaminations, and can be recycled and reprocessed due to their vitrimer feature. Furthermore, this strategy holds promise for expanding to other types of vitrimers (such as polyimine, polyurethane) and other fiber fabrics (such as cotton fabric, man-made-fiber fabrics), suggesting even broader industrial applications for this composite. Chen et al. [97] also developed a remarkable actuator utilizing a vitrimer liquid crystal elastomer equipped with exchangeable dynamic ester bonds. This actuator exhibited self-healing capabilities, enabling it to break old bonds and reform new covalent bonds even under challenging manipulative conditions. Additionally, the actuator displayed shape memory behavior, deforming appropriately in response to various stimulation conditions.

The integration of disulfide bonds in vitrimer actuating materials has proven highly effective and convenient for achieving self-healing properties under mild conditions, presenting a significant advantage over transesterification reactions that necessitate harsh conditions. Tang and colleagues [98] have devised a groundbreaking and accessible method to create multiple elastic-plastic shape-memory cycle soft actuators, capable of intricate movements. This remarkable achievement harnesses the potential of vitrimer liquid crystal elastomer (V-LCE) in combination with dynamic disulfide bonds and azobenzene chromophore functional groups. By incorporating these key components into the main chain through a curing reaction between epoxy and thiol compounds, the V-LCE actuators with dynamic disulfide bonds showcased an extraordinary ability to self-repair damaged areas under heating conditions. Additionally, the motion of the polymer chain was facilitated by changes in the configuration of the azobenzene chromophore moieties, leading to macroscopic shape morphing when exposed to UV light. The outstanding elastic-plastic shape-memory behavior and mechanical properties of V-LCE allowed for enhanced design flexibility, enabling the creation of diverse and complex 2D and 3D shape actuators. These soft "bionic" devices exhibited impressive functionalities such as grasping, transferring, and releasing objects, thanks to the exceptional mechanical robustness, photothermal response performance, and exceptional programability and reconfigurability of the covalent cross-linking system. The simple and efficient fabrication process of V-LCE, combined with its photothermal-responsive and extraordinary capabilities, opens up exciting possibilities for addressing complex tasks and provides a promising avenue for designing intelligent devices and bionic robotics with exceptional performance and versatility.

3.6 Energy devices

3.6.1 Energy storage devices

Supercapacitors (SCs) are devices used for energy storage, combining features found in traditional capacitors and rechargeable batteries. They possess various advantages, including high energy density, fast charge and discharge rates, reliable safety, long-lasting life cycles, low maintenance requirements, and stable performance [99]. In comparison to conventional capacitors, SCs can provide greater power delivery and energy storage capabilities. Nonetheless, their durability may be compromised by mechanical harm. As a result, SCs with self-healing properties have sparked significant interest in both mechanical and electrical research fields.

Typically, a SC device consists of an electrolyte placed between two electrodes. In the quest to develop a self-healing supercapacitor, it becomes essential to design electrodes or electrolytes that possess self-healing capabilities. One promising candidate for this purpose is the solid polymer electrolyte (SPE) known for its versatile properties, combining robust ionic conductivity and mechanical strength. To make it suitable for energy storage applications, the SPE is formed by blending a durable, cross-linked epoxy-based matrix with a fast-diffusing lithium salt/ionic liquid electrolyte. This blending process is carried out using a convenient one-pot curing method. In a successful endeavor, Kwon et al. [100] managed to create epoxy-based cross-linked SPEs by incorporating plasticized lithium salt, ionic liquid, and an inorganic slurry containing Al_2O_3 nanowires, all prepared in a single-pot process (illustrated in **Figure 7a**). According to their findings, when these SPEs were utilized alongside activated carbon electrodes, they exhibited excellent supercapacitor performance, demonstrating notably high-power density and energy density values. Specifically, the results showed a power density of 9.3 kW at 44 Wh kg^{-1} and an energy density of 75 Wh kg^{-1} at 382 W kg^{-1}.

Recently, Wang et al. [101] successfully fabricated liquid crystalline epoxy network (LCEN) composites with aligned CNT sheets, demonstrating electrically controlled actuation behavior (**Figure 7b**). The incorporation of CNTs into the vitrimer network directly promoted the photothermal conversion effect, resulting in higher thermal energy that activated transesterifications. This process facilitated Joule heating and played a crucial role in introducing structural anisotropy to the composite film. As a result, the composite film exhibited spontaneous bending behavior when powered by electricity, eliminating the need to induce macroscopic liquid crystalline orientation. Furthermore, the film displayed self-healing properties, which allowed it to recover not only through exposure to light but also by using electricity after being broken under a voltage of 1.18 V mm^{-1}. Additionally, the authors showcased the potential application of the conductive LCEN composite as self-healing supercapacitors (**Figure 7c**).

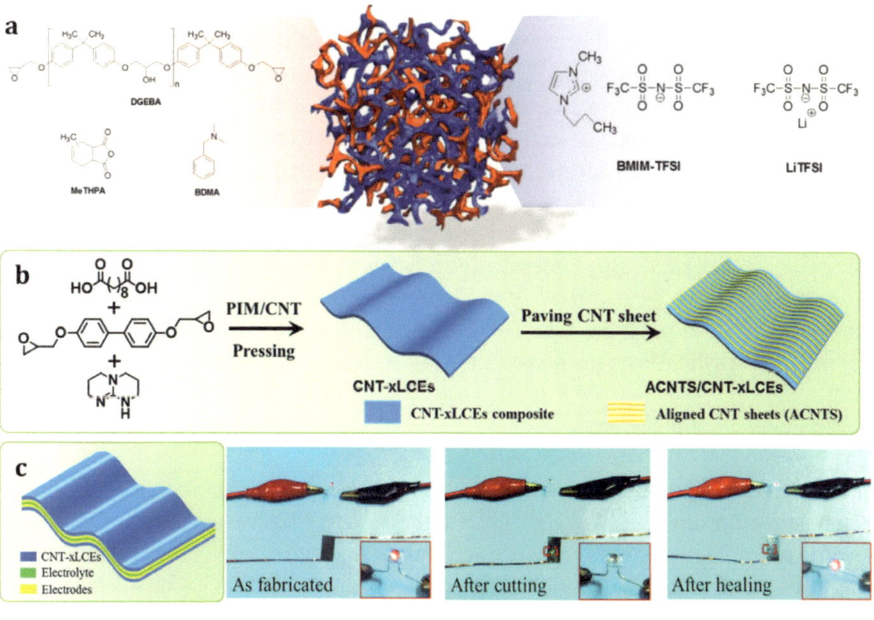

Figure 7.
(a) Chemical structures of epoxy-based solid polymer electrolytes containing an ionic liquid, a lithium salt, and inorganic Al_2O_3 nanowires. Reprinted with permission from [100]. Copyright 2018 American Chemical Society. (b) Fabrication of aligned CNT sheet-reinforced LCEN composite. (c) Flexible LCEN composite supercapacitor with self-healing behavior. Reprinted with permission from [101]. Copyright 2020 American Chemical Society.

Additionally, there is a growing interest in epoxy composite-based rechargeable batteries with self-healing capabilities. Sun et al. have achieved significant breakthroughs in the field of highly conductive self-healing SPEs [102]. Their innovative approach involved incorporating disulfide dynamic covalent bonds into an epoxy matrix to create a unique and recyclable electrolyte named RFSPE-3. To fabricate this self-healing and recyclable electrolyte, the researchers strategically combined diglycidyl ether of bisphenol A (DGBE) and poly(ethylene glycol) diglycidyl ether (PEGDGE) as the matrix components. The crosslinker used was 2-aminophenyl disulfide (2-AFD), which provided the disulfide bonds crucial for the exceptional self-healing ability and recyclability of the electrolyte. The resulting RFSPE-3 exhibited impressive characteristics. The inclusion of epoxy resin endowed the electrolyte with superior mechanical strength, boasting a tensile strength exceeding 20 MPa. Meanwhile, the dynamic exchange of disulfide bonds in 2-AFD contributed to enhanced self-healing capabilities, achieving a remarkable healing efficiency of over 95% within just 2 h at room temperature. Moreover, RFSPE-3 demonstrated exceptional ion conductivity, reaching 10^{-3} S cm^{-1}, and even after multiple healing processes, there were no changes observed in its ionic conductivity. Furthermore, this electrolyte exhibited an inhibitory effect on the growth of lithium dendrites, ensuring excellent cycling stability for up to 1800 h at 1 mAh cm^{-2}. These promising results pave the way for the development of advanced and long-term rechargeable batteries, with potential applications in various fields requiring durable and efficient energy storage solutions. The combination of high mechanical strength, excellent self-healing properties, and outstanding ion conductivity positions RFSPE-3 as a promising candidate for next-generation energy storage devices.

Sun and Wu [103] adopted a similar strategy by incorporating disulfide dynamic networks into an epoxy matrix. Their approach involved introducing the disulfide bond through a disulfide-containing aliphatic polyamine as the epoxy-curing agent. The resulting SPEs exhibited remarkable properties, being optically transparent and capable of self-healing above the T_g. An intriguing observation was the substantial decrease in curing time, from 102 to just 1.4 mins, achieved by raising the temperature from 40 to 100°C. Among various formulations tested, the SPE with 15% LiTFSI content demonstrated the highest ion conductivity. Specifically, it exhibited values of 3.35×10^{-6} S cm^{-1} at 80°C and 8.31×10^{-6} S cm^{-1} at 100°C. Although this conductivity was slightly lower compared to other self-healing SPEs, it still marked a significant advancement in the field. By successfully incorporating disulfide dynamic networks into the epoxy matrix, Sun and Wu's work represents a notable step forward in the development of solid polymer electrolytes with enhanced properties, showing potential for various applications in the realm of materials science and beyond.

3.6.2 Energy-harvesting devices

Conductive self-healing ECs find another crucial application in the realm of energy harvesting devices. The advancement of energy harvesting technologies represents a significant stride towards achieving an environmentally friendly society. Among these technologies, triboelectric nanogenerators (TENGs) have emerged as a highly regarded and eco-conscious power source. They offer numerous benefits, including cost-effectiveness, simple production processes, abundant material choices, impressive conversion efficiency, and the ability to design them in various versatile configurations [104–108]. These advantages have drawn considerable attention to TENGs as a sustainable and green solution. These TENGs capitalize on the phenomena of triboelectrification and electrostatic induction, making them a highly promising option for efficiently converting ambient mechanical energies into electricity, particularly in situations where low frequencies are involved [109]. To enhance the output power of conventional TENGs, researchers have employed patterned triboelectric layers with micro-nanostructures or self-assembled monolayers (SAMs) with different functional groups, effectively increasing the contact surface area [110, 111]. Nevertheless, enduring challenges persist, such as the decline in overall performance and the restricted lifespan of TENG devices, primarily attributable to interface friction between dissimilar materials and frequent mechanical deformation, including twisting, stretching, bending, and compressing. A viable approach to address these challenges involves incorporating a "self-healing" capability into the TENGs. This self-healing feature enables the repair of broken micro-nanostructures in the triboelectric layer, thereby restoring the overall performance of the nanogenerators [105, 112].

Ye et al. [113] developed a self-healable vitrimer-based triboelectric nanogenerator (VITENG) by employing a rapid thiol-Michael reaction with diacrylate poly(dimethylsiloxane) (AA-PDMS) elastomer. They introduced 2,3-dihydroxypropyl methacrylate (DHPMA) to generate hydroxy end dangling side chains within the matrix, which enhanced the flexibility and transesterification efficiency of the PDMS network. This innovation resulted in an impressive healing efficiency of 84% within a mere 15 mins for the developed vitrimer elastomer. The research findings revealed that the dangling hydroxy end chains not only enhanced the material's flexibility and stretchability but also significantly improved its self-healing ability. The low T_g system facilitated efficient collisions during the dynamic bond exchange process, contributing to this enhancement. An important outcome of their work was that VITENG exhibited a remarkable healing effectiveness of 100% and demonstrated a stretchability of 125%. Moreover, the VITENG showcased excellent output performance, producing 135 V of energy at low frequencies and applied forces. This remarkable performance makes the developed VITENG a highly efficient energy harvester with great potential for powering a wide array of soft devices.

4. Conclusions and outlooks

In this chapter, we delve into the fundamental mechanisms and healing strategies of epoxy composites, highlighting their crucial distinctions. Next, we explore the key principles underpinning self-healing epoxy composites-based devices, which enable them to restore mechanical integrity and regain functionality even when faced with mechanical and electrical damage. This exploration is complemented by a comprehensive overview of recent research progress in epoxy composites, providing a broader understanding of their applications. Subsequently, we present a thorough examination of the latest advancements in the use of conductive self-healing epoxy composites within the realm of flexible electronic components. These applications span from sensors, EMI shielding, and soft actuators to energy storage and harvesting electronic devices. While remarkable achievements have been made, we also recognize the existence of challenges and research gaps that necessitate ongoing and dedicated efforts for resolution.

Firstly, the demand for self-healing devices calls for the design of epoxy composite materials capable of repetitive rapid healing while maintaining stable filler conductivity post-damage. Secondly, current self-healing conditions often necessitate high temperatures or prolonged light induction, making it imperative to explore milder experimental conditions that involve simple manipulation. Thirdly, improving the self-healing efficiency becomes crucial to ensure the restoration of the original performance of the materials. Fourthly, the healing efficiency of electrical conductivity and mechanical behavior in conductive self-healing epoxy composites presents a contradiction, necessitating future research to achieve synergistic characteristics of high conductivity and excellent mechanical properties.

Despite significant progress in developing epoxy composites with favorable self-healing and mechanical properties, their applicability in energy electronics remains limited. One of the challenges lies in the mutual influence of different components in some composite functional materials. For instance, in self-healing electrodes, the introduction of self-healing polymers may replace some of the active material, leading to a reduction in electrode capacity. Additionally, conductive fillers can impact the original interface compatibility, mechanical properties, transparency, and other critical factors. To address these limitations, it is crucial to focus on creating self-healing conductive or semiconducting epoxy-based polymers that possess inherent conductivity. Moreover, there is a compelling need to explore the potential research direction of developing epoxy-based materials capable of automatically adapting various device properties based on specific demands. By pursuing these avenues, we can pave the way for advanced and versatile materials that overcome current constraints and unlock new opportunities in energy electronics.

Beyond these challenges, multifunctional electronic and energy systems with diverse interfaces, such as those between devices and substrates or between layers, can encounter delamination due to differences in surface chemistry and mechanical properties. Addressing this issue, self-healing polymer networks can be utilized to enhance mechanical robustness, facilitating seamless integration and self-bonding at layer interfaces, thereby enabling the efficient fabrication of three-dimensional electronic devices. Observing the wide range of applications for electrically conductive self-healing epoxy composites in adhesives, coatings, sensors, electronic components, and energy storage/harvesting, it is clear that these materials will play an increasingly vital role in shaping the future of various industries. Of utmost importance, these self-healable materials possess the capability to significantly diminish electronic waste by employing their autonomic repair or semi-autonomic functions. As a result, they play a pivotal role in propelling our interconnected world towards enhanced sustainability and, potentially, facilitating the growth of space exploration endeavors.

Conflict of interest

The authors declare no conflict of interest.

Author details

Van-Phu Vu[1]* and Tien Thuy Phan[2,3]*

1 School of Semiconductor and Chemical Engineering, Jeonbuk National University, Jeonju, Korea

2 Center for Cognition and Sociality, Institute for Basic Science, Daejeon, Korea

3 University of Science and Technology, Daejeon, Korea

*Address all correspondence to: vanphu.itims@gmail.com and thuytienphan0604@ust.ac.kr

References

[1] Li L, Zhang Y, Shi Y, Guo F, Yang X, Shi W. A hydrophobic high-crystalline g-C3N4/epoxy resin composite coating with excellent durability and stability for long-term corrosion resistance. Materials Today Communications. 2023;**35**:105692

[2] Moreira VB, Alemán C, Rintjema J, Bravo F, Kleij AW, Armelin E. A biosourced epoxy resin for adhesive thermoset applications. ChemSusChem. 2022;**15**(7):e202102624

[3] Xiang Q, Xiao F. Applications of epoxy materials in pavement engineering. Construction and Building Materials. 2020;**235**:117529

[4] Zhuo H, He Z, Liu J, Ma G, Ren Z, Zeng Y, et al. Bulk polymerization of thermoplastic shape memory epoxy polymer for recycling applications. Polymers. 2023;**15**(4):809

[5] Zhang F, Zhang L, Yaseen M, Huang K. A review on the self-healing ability of epoxy polymers. Journal of Applied Polymer Science. 2021;**138**(16):50260

[6] Pandey A, Sharma AK, Shukla DK, Pandey KN. Effect of self-healing by dicyclopentadiene microcapsules on tensile and fatigue properties of epoxy composites. Materials. 2023;**16**(14):5191

[7] Zhang H, Zhou L, Zhang F, Yang Q, Chen M, Chen Z, et al. Aromatic disulfide epoxy vitrimer packaged electronic devices: Nondestructive healing and recycling. Polymer. 2022;**255**:125163

[8] Kim G, Caglayan C, Yun GJ. Epoxy-based catalyst-free self-healing elastomers at room temperature employing aromatic disulfide and hydrogen bonds. ACS Omega. 2022;**7**(49):44750-44761

[9] Guadagno L, Vertuccio L, Naddeo C, Calabrese E, Barra G, Raimondo M, et al. Self-healing epoxy nanocomposites via reversible hydrogen bonding. Composites, Part B: Engineering. 2019;**157**:1-13

[10] Guadagno L, Raimondo M, Naddeo C, Vertuccio L, Russo S, Iannuzzo G, et al. Rheological, thermal and mechanical characterization of toughened self-healing supramolecular resins, based on hydrogen bonding. Nanomaterials. 2022;**12**(23):4322

[11] Cao L, Gong Z, Liu C, Fan J, Chen Y. Design and fabrication of mechanically strong and self-healing rubbers via metal-ligand coordination bonds as dynamic crosslinks. Composites Science and Technology. 2021;**207**:108750

[12] Li CH, Zuo JL. Self-healing polymers based on coordination bonds. Advanced Materials. 2020;**32**(27):1903762

[13] Chen B, Liu X, Liu J, Feng Z, Zheng X, Wu X, et al. Intrinsically self-healing, reprocessable and recyclable epoxy thermosets based on dynamic reversible urea bonds. Reactive and Functional Polymers. 2022;**172**:105184

[14] Zhang J, Zhang C, Song F, Shang Q, Hu Y, Jia P, et al. Castor-oil-based, robust, self-healing, shape memory, and reprocessable polymers enabled by dynamic hindered urea bonds and hydrogen bonds. Chemical Engineering Journal. 2022;**429**:131848

[15] Li B, Zhu G, Hao Y, Ren T. An investigation on the performance of epoxy vitrimers based on disulfide bond. Journal of Applied Polymer Science. 2022;**139**(5):51589

[16] Shan S, Mai D, Lin Y, Zhang A. Self-healing, reprocessable, and

degradable bio-based epoxy elastomer bearing aromatic disulfide bonds and its application in strain sensors. ACS Applied Polymer Materials. 2021;**3**(10):5115-5124

[17] Atta AM, Ahmed MA, Al-Lohedan HA, El-Faham A. Multi-functional cardanol triazine schiff base polyimine additives for self-healing and super-hydrophobic epoxy of steel coating. Coatings. 2020;**10**(4):327

[18] Jiang L, Tian Y, Wang X, Zhang J, Cheng J, Gao F. A fully bio-based Schiff base vitrimer with self-healing ability at room temperature. Polymer Chemistry. 2023;**14**(7):862-871

[19] Zhang Y, Ye J, Qu D, Wang H, Chai C, Feng L. Thermo-adjusted self-healing epoxy resins based on Diels–Alder dynamic chemical reaction. Polymer Engineering and Science. 2021;**61**(9):2257-2266

[20] Eidi M, Zamani PM. Thermal induced intrinsic self-healing in epoxy based elastomer coatings provided by disulfide metathesis reactions. Journal of Applied Polymer Science. 2022;**139**(22):52239

[21] Wang X, Cao L, Xu C, Fan B, Lin Z, Li W, et al. Novel dual dynamic boronate ester bond regulated bio-based polymer with rapid self-healing and multiple recyclability. Industrial Crops and Products. 2022;**189**:115855

[22] Rashid MA, Zhu S, Jiang Q, Wei Y, Liu W. Developing easy processable, recyclable, and self-healable biobased epoxy resin through dynamic covalent imine bonds. ACS Applied Polymer Materials. 2022;**5**(1):279-289

[23] Krishnakumar B, Sanka RSP, Binder WH, Park C, Jung J, Parthasarthy V, et al. Catalyst free self-healable vitrimer/graphene oxide nanocomposites. Composites, Part B: Engineering. 2020;**184**:107647

[24] Hao Y, Zhong L, Li T, Zhang J, Zhang D. High-performance recyclable and malleable epoxy resin with vanillin-based hyperbranched epoxy resin containing dual dynamic bonds. ACS Sustainable Chemistry & Engineering. 2023;**11**(30):11077-11087

[25] Yuan W, Shen Y, Wang B, Li J, Liu L, Huang Y, et al. Ultrafast self-healing fiber/matrix composite with single-component microcapsules loaded with cationic catalyst. ACS Applied Polymer Materials. 2023;**5**(3):2016-2025

[26] Liu H, Sun Z, Wei L, Liu Y, Zhou S, Ge Q, et al. Double-dynamic crosslinked epoxy vitrimer resin prepared using transesterification and dynamic disulfide bonds: High-performance, degradable, self-healing, environment-friendly. Polymer Testing. 2023;**126**:108145

[27] Zou Y, Fang L, Chen T, Sun M, Lu C, Xu Z. Near-infrared light and solar light activated self-healing epoxy coating having enhanced properties using MXene flakes as multifunctional fillers. Polymers. 2018;**10**(5):474

[28] Wang N, Feng X, Pei J, Cui Q, Li Y, Liu H, et al. Biobased reversible cross-linking enables self-healing and reprocessing of epoxy resins. ACS Sustainable Chemistry & Engineering. 2022;**10**(11):3604-3613

[29] Huang QS, Zhao PC, Lai JC, Zhang XP, Li CH. A healable, recyclable and thermochromic epoxy resin for thermally responsive smart windows. Polymer Chemistry. 2022;**13**(15):2178-2186

[30] Boumezgane O, Suriano R, Fedel M, Tonelli C, Deflorian F, Turri S. Self-healing epoxy coatings with microencapsulated ionic PDMS

oligomers for corrosion protection based on supramolecular acid-base interactions. Progress in Organic Coating. 2022;**162**:106558

[31] Hu Z, Zhang D, Lu F, Yuan W, Xu X, Zhang Q, et al. Multistimuli-responsive intrinsic self-healing epoxy resin constructed by host–guest interactions. Macromolecules. 2018;**51**(14):5294-5303

[32] Liu T, Zhao H, Zhang D, Lou Y, Huang L, Ma L, et al. Ultrafast and high-efficient self-healing epoxy coatings with active multiple hydrogen bonds for corrosion protection. Corrosion Science. 2021;**187**:109485

[33] Hu Z, Liu Y, Xu X, Yuan W, Yang L, Shao Q, et al. Efficient intrinsic self-healing epoxy acrylate formed from host-guest chemistry. Polymer. 2019;**164**:79-85

[34] Kosarli M, Bekas DG, Tsirka K, Baltzis D, Vaimakis-Tsogkas DT, Orfanidis S, et al. Microcapsule-based self-healing materials: Healing efficiency and toughness reduction vs. capsule size. Composites, Part B: Engineering. 2019;**171**:78-86

[35] Pittala RK. Effect of epoxy resin healing agent viscosity on the self-healing performance of capsules reinforced polymer composite. Journal of Polymer Research. 2021;**28**:1-19

[36] Peñas-Caballero M, Chemello E, Grande AM, Hernández Santana M, Verdejo R, Lopez-Manchado MA. Poly (methyl methacrylate) as healing agent for carbon fibre reinforced epoxy composites. Polymers. 2023;**15**(5):1114

[37] Ebrahimnezhad-Khaljiri H, Eslami-Farsani R, Mirzamohammadi S. The effect of capsulated nanosilica-epoxy healing agents on the self-healing ability of glass fibers-epoxy composites under mechanical loading. Journal of Industrial Textiles. 2022;**52**:15280837221119833

[38] Ahangaran F, Hayaty M, Navarchian AH, Pei Y, Picchioni F. Development of self-healing epoxy composites via incorporation of microencapsulated epoxy and mercaptan in poly (methyl methacrylate) shell. Polymer Testing. 2019;**73**:395-403

[39] Zhu J, Wang X, Wang X, Lei Y, Li Y. Carbonyl iron powder/ethyl cellulose hybrid wall microcapsules encapsulating epoxy resin for wave absorption and self-healing. Composites Science and Technology. 2021;**214**:108960

[40] Prabhakar M, Hassim MT, Song JI. Development of self-healing carbon/epoxy composites with optimized PAN/PVDF Core–Shell Nanofibers as Healing Carriers. ACS Omega. 2022;**7**(46):42396

[41] Benazzo F, Rigamonti D, Bettini P, Sala G, Grande AM. Interlaminar fracture of structural fibre/epoxy composites integrating damage sensing and healing. Composites, Part B: Engineering. 2022;**244**:110137

[42] Vertuccio L, Calabrese E, Raimondo M, Catauro M, Sorrentino A, Naddeo C, et al. Effect of temperature on the functionalization process of structural self-healing epoxy resin. Aerospace. 2023;**10**(5):476

[43] Wu J, Yu X, Zhang H, Guo J, Hu J, Li MH. Fully biobased vitrimers from glycyrrhizic acid and soybean oil for self-healing, shape memory, weldable, and recyclable materials. ACS Sustainable Chemistry & Engineering. 2020;**8**(16):6479-6487

[44] Li Y, Jiang B, Huang Y. Interfacial self-healing performance of carbon fiber/epoxy based on postsynthetic modification of metal-organic

frameworks. Composites Science and Technology. 2022;**227**:109564

[45] Jiang Q, Yang J, Hing P, Ye H. Recent advances, design guidelines, and prospects of flexible organic/inorganic thermoelectric composites. Materials Advances. 2020;**1**(5):1038-1054

[46] Mamunya E, Davidenko V, Lebedev E. Percolation conductivity of polymer composites filled with dispersed conductive filler. Polymer Composites. 1995;**16**(4):319-324

[47] Pang H, Xu L, Yan DX, Li ZM. Conductive polymer composites with segregated structures. Progress in Polymer Science. 2014;**39**(11):1908-1933

[48] Guadagno L, Naddeo C, Sorrentino A, Raimondo M. Thermo-mechanical performance of epoxy hybrid system based on carbon nanotubes and graphene nanoparticles. Nanomaterials. 2023;**13**(17):2427

[49] Doan VC, Vu MC, Thieu NAT, Islam MA, Park PJ, Kim SR. Copper flake-coated cellulose scaffold to construct segregated network for enhancing thermal conductivity of epoxy composites. Composites, Part B: Engineering. 2019;**165**:772-778

[50] Bae YH, Yu MJ, Vu MC, Choi WK, Kim SR. Synergistic effects of segregated network by polymethylmethacrylate beads and sintering of copper nanoparticles on thermal and electrical properties of epoxy composites. Composites Science and Technology. 2018;**155**:144-150

[51] Yuan D, Guo H, Ke K, Manas-Zloczower I. Recyclable conductive epoxy composites with segregated filler network structure for EMI shielding and strain sensing. Composites Part A: Applied Science and Manufacturing. 2020;**132**:105837

[52] Zhang F, Ye W, Zhou W, Gao X, Fang H, Ding Y. Endowing thermally conductive and electrically insulating epoxy composites with a well-structured nanofiller network via dynamic transesterification-participated interfacial welding. Industrial & Engineering Chemistry Research. 2022;**61**(9):3320-3328

[53] Bian W, Wang W, Yang Y. A self-healing and electrical-tree-inhibiting epoxy composite with hydrogen-bonds and SiO2 particles. Polymers. 2017;**9**(9):431

[54] Zeng X, Sun J, Yao Y, Sun R, Xu JB, Wong CP. A combination of boron nitride nanotubes and cellulose nanofibers for the preparation of a nanocomposite with high thermal conductivity. ACS Nano. 2017;**11**(5):5167-5178

[55] Cescato R, Rigotti D, Mahmood H, Dorigato A, Pegoretti A. Thermal mending of electroactive carbon/epoxy laminates using a porous poly (ε-caprolactone) electrospun mesh. Polymers. 2021;**13**(16):2723

[56] Kotrotsos A. An innovative synergy between solution electrospinning process technique and self-healing of materials. A critical review. Polymer Engineering & Science. 2021;**61**(1):5-21

[57] Liu Y, Liu X, Liu P, Chen X, Yu DG. Electrospun multiple-chamber nanostructure and its potential self-healing applications. Polymers. 2020;**12**(10):2413

[58] Neisiany RE, Lee JKY, Khorasani SN, Ramakrishna S. Self-healing and interfacially toughened carbon fibre-epoxy composites based on electrospun core–shell nanofibres. Journal of Applied Polymer Science. 2017;**134**(31):44956

[59] Chen B, Cai H, Mao C, Gan Y, Wei Y. Toughening and rapid self-healing for

carbon fiber/epoxy composites based on electrospinning thermoplastic polyamide nanofiber. Polymer Composites. 2022;**43**(5):3124-3135

[60] Yuan D, Delpierre SB, Ke K, Raquez JM, Dubois P, Manas-Zloczower I. Biomimetic water-responsive self-healing epoxy with tunable properties. ACS Applied Materials & Interfaces. 2019;**11**(19):17853-17862

[61] Pan D, Yang G, Abo-Dief H, Dong J, Su F, Liu C, et al. Vertically aligned silicon carbide nanowires/boron nitride cellulose aerogel networks enhanced thermal conductivity and electromagnetic absorbing of epoxy composites. Nano-Micro Letters. 2022;**14**(1):118

[62] Xie Z, Cai Y, Wei Z, Zhan Y, Meng Y, Li Y, et al. Robust and self-healing polydimethylsiloxane/carbon nanotube foams for electromagnetic interference shielding and thermal insulation. Composites Communications. 2022;**35**:101323

[63] Zhang Y, Yuan L, Liang G, Gu A. Developing reversible self-healing and malleable epoxy resins with high performance and fast recycling through building cross-linked network with new disulfide-containing hardener. Industrial and Engineering Chemistry Research. 2018;**57**(37):12397-12406

[64] Zhang W, Wu J, Gao L, Zhang B, Jiang J, Hu J. Recyclable, reprocessable, self-adhered and repairable carbon fiber reinforced polymers using full biobased matrices from camphoric acid and epoxidized soybean oil. Green Chemistry. 2021;**23**(7):2763-2772

[65] Xu Y, Dai S, Bi L, Jiang J, Zhang H, Chen Y. Catalyst-free self-healing bio-based vitrimer for a recyclable, reprocessable, and self-adhered carbon fiber reinforced composite. Chemical Engineering Journal. 2022;**429**:132518

[66] Zhang S, Liu T, Hao C, Wang L, Han J, Liu H, et al. Preparation of a lignin-based vitrimer material and its potential use for recoverable adhesives. Green Chemistry. 2018;**20**(13):2995-3000

[67] Han J, Liu T, Hao C, Zhang S, Guo B, Zhang J. A catalyst-free epoxy vitrimer system based on multifunctional hyperbranched polymer. Macromolecules. 2018;**51**(17):6789-6799

[68] Dong Y, Gong M, Huang D, Gao J, Zhou Q. Shape memory, self-healing property, and NIR photothermal effect of epoxy resin coating with polydopamine@ polypyrrole nanoparticles. Progress in Organic Coating. 2019;**136**:105232

[69] Hao C, Liu T, Zhang S, Brown L, Li R, Xin J, et al. A high-lignin-content, removable, and glycol-assisted repairable coating based on dynamic covalent bonds. ChemSusChem. 2019;**12**(5):1049-1058

[70] Zhang J, Huang J, Zhu G, Yu X, Cheng J, Liu Z, et al. Self-healing, recyclable, and removable UV-curable coatings derived from tung oil and malic acid. Green Chemistry. 2021;**23**(16):5875-5886

[71] Feng Y, Cui Y, Zhang M, Li M, Li H. Preparation of Tung oil-loaded PU/PANI microcapsules and synergetic anti-corrosion properties of self-healing epoxy coatings. Macromolecular Materials and Engineering. 2021;**306**(2):2000581

[72] Alias J, Johari NA, Zanurin A, Alang NA, Zain MZM. Self-healing epoxy coating with microencapsulation of linseed oil for the corrosion protection of magnesium (Mg). Journal of Physics Conference Series. 2021;**2129**(1):012008

[73] Haddadi SA, Hu S, Ghaderi S, Ghanbari A, Ahmadipour M, Pung SY, et al. Amino-functionalized MXene nanosheets doped with Ce (III) as potent nanocontainers toward self-healing epoxy nanocomposite coating for corrosion protection of mild steel. ACS Applied Materials & Interfaces. 2021;**13**(35):42074-42093

[74] Amin I, Brekel HVD, Nemani K, Batyrev E, De Vooys A, Van Der Weijde H, et al. Ti3C2T x MXene polymer composites for anticorrosion: An overview and perspective. ACS Applied Materials & Interfaces. 2022;**14**(38):43749-43758

[75] Yan H, Fan X, Cai M, Song S, Zhu M. Amino-functionalized Ti3C2Tx loading ZIF-8 nanocontainer@ benzotriazole as multifunctional composite filler towards self-healing epoxy coating. Journal of Colloid and Interface Science. 2021;**602**:131-145

[76] Niu W, Chen GY, Xu H, Liu X, Sun J. Highly transparent and self-healable solar thermal anti-/Deicing surfaces: When ultrathin MXene multilayers marry a solid slippery self-cleaning coating. Advanced Materials. 2022;**34**(10):2108232

[77] Wang Z, Fang L, Wu F, Ruan H, Tang Y, Hu J, et al. Anti-corrosion, self-healing and environmental-friendly Ti3C2Tx/MgAl-LDH@ epoxy composite organic coating for Mg alloy protection. Journal of Materials Science. 2023;**58**(7):3283-3306

[78] Lorwanishpaisarn N, Srikhao N, Jetsrisuparb K, Knijnenburg JT, Theerakulpisut S, Okhawilai M, et al. Self-healing ability of epoxy vitrimer nanocomposites containing bio-based curing agents and carbon nanotubes for corrosion protection. Journal of Polymers and the Environment. 2022;**30**(2):472-482

[79] Ramezanzadeh M, Ramezanzadeh B, Mahdavian M. Graphene skeletal nanotemplate coordinated with pH-responsive porous double-ligand metal-organic frameworks (DL-MOFs) through ligand exchange theory for high-performance smart coatings. Chemical Engineering Journal. 2023;**461**:141869

[80] Vu VP, Sinh LH, Choa SH. Recent progress in self-healing materials for sensor arrays. ChemNanoMat. 2020;**6**(11):1522-1538

[81] Ting Y, Dajiang K, Weiyi H, Yunjie Y, Chaoxia W. Room-temperature self-healing graphene/rubber-based supramolecular elastomers utilized by dynamic boroxines and hydrogen bonds for human motion detection. Colloids and Surfaces, A: Physicochemical and Engineering Aspects. 2023;**657**:130411

[82] Zhao K, Wang Y, Guo J, Zhang S, Niu W. Photonic vitrimer-based electronics with self-healing and ultrastable visual-digital outputs for wireless strain sensing. Chemical Engineering Journal. 2022;**450**:138285

[83] Yang Z, Li H, Zhong Y, Lai X, Ding J, Chen Z, et al. Functional epoxy elastomer integrating self-healing capability and degradability for a flexible stretchable strain sensor. ACS Applied Materials & Interfaces. 2022;**14**(39):44878-44889

[84] Han Y, Wu X, Zhang X, Lu C. Self-healing, highly sensitive electronic sensors enabled by metal–ligand coordination and hierarchical structure design. ACS Applied Materials & Interfaces. 2017;**9**(23):20106-20114

[85] Hu Y, Tong S, Sha Y, Yu J, Hu L, Huang Q, et al. Cardanol-based epoxy vitrimer/carbon fiber composites with integrated mechanical, self-healing, reprocessable, and welding properties

and degradability. Chemical Engineering Journal. 2023;**471**:144633

[86] Wu Y, Wei Y, Ji Y. Carbon material/vitrimer composites: Towards sustainable, functional, and high-performance crosslinked polymeric materials. Giant. 2022;**13**:100136

[87] Liu Y, Tang Z, Chen Y, Wu S, Guo B. Programming dynamic imine bond into elastomer/graphene composite toward mechanically strong, malleable, and multi-stimuli responsive vitrimer. Composites Science and Technology. 2018;**168**:214-223

[88] Hu Y, Jia P, Lamm ME, Sha Y, Kurnaz LB, Ma Y, et al. Plant oil-derived vitrimers-graphene composites with self-healing ability triggered by multiple stimuli. Composites, Part B: Engineering. 2023;**259**:110704

[89] Jia P, Shi Y, Song F, Bei Y, Huang C, Zhang M, et al. Bio-based and degradable vitrimer-graphene/graphene oxide composites with self-healing ability stimulated by heat, electricity and microwave as temperature and fire warning sensors. Composites Science and Technology. 2022;**227**:109573

[90] Cao J, Lu C, Zhuang J, Liu M, Zhang X, Yu Y, et al. Multiple hydrogen bonding enables the self-healing of sensors for human–machine interactions. Angewandte Chemie. 2017;**129**(30):8921-8926

[91] Kumar R, Sahoo S, Joanni E, Singh RK, Tan WK, Kar KK, et al. Recent progress on carbon-based composite materials for microwave electromagnetic interference shielding. Carbon. 2021;**177**:304-331

[92] Zhang HY, Li JY, Pan Y, Liu YF, Mahmood N, Jian X. Flexible carbon fiber-based composites for electromagnetic interference shielding. Rare Metals. 2022;**41**(11):3612-3629

[93] Qiao Y, Wang Y, Yang J, Li Q, Gu J. High performance porous Ni@ Cf paper with excellent electromagnetic shielding properties. Composites Part A: Applied Science and Manufacturing. 2023;**172**:107618

[94] Yu B, Feng Y, Zhu W, Wang R, Qi S. Self-healing electromagnetic interference shielding composite based on Diels–Alder chemistry. Journal of Materials Science: Materials in Electronics. 2019;**30**:19994-20001

[95] Fang H, Ye W, Yang K, Song K, Wei H, Ding Y. Vitrimer chemistry enables epoxy nanocomposites with mechanical robustness and integrated conductive segregated structure for high performance electromagnetic interference shielding. Composites, Part B: Engineering. 2021;**215**:108782

[96] Yang Y, Wang H, Zhang S, Wei Y, He X, Wang J, et al. Vitrimer-based soft actuators with multiple responsiveness and self-healing ability triggered by multiple stimuli. Matter. 2021;**4**(10):3354-3365

[97] Chen Q, Yu X, Pei Z, Yang Y, Wei Y, Ji Y. Multi-stimuli responsive and multi-functional oligoaniline-modified vitrimers. Chemical Science. 2017;**8**(1):724-733

[98] Tang D, Zhang L, Zhang X, Xu L, Li K, Zhang A. Bio-mimetic actuators of a photothermal-responsive vitrimer liquid crystal elastomer with robust, self-healing, shape memory, and reconfigurable properties. ACS Applied Materials & Interfaces. 2021;**14**(1):1929-1939

[99] Vu VP, Mai VD, Lee SH. Hybrid carbon nanofiller/polymer composites as self-healable current collector electrodes

for use in high-performance flexible metal-free supercapacitors. Journal of Alloys and Compounds. 2023;**933**:167823

[100] Kwon SJ, Kim T, Jung BM, Lee SB, Choi UH. Multifunctional epoxy-based solid polymer electrolytes for solid-state supercapacitors. ACS Applied Materials & Interfaces. 2018;**10**(41):35108-35117

[101] Wang H, Yang Y, Zhang M, Wang Q, Xia K, Yin Z, et al. Electricity-triggered self-healing of conductive and thermostable vitrimer enabled by paving aligned carbon nanotubes. ACS Applied Materials & Interfaces. 2020;**12**(12):14315-14322

[102] Sun Z, Wu J, Yuan H, Lan J, Yu Y, Zhu Y, et al. Self-healing polymer electrolyte for long-life and recyclable lithium-metal batteries. Materials Today Energy. 2022;**24**:100939

[103] Sun S, Wu T. Preparation and properties of self-healable solid-state polymer electrolytes based on covalent adaptive networks enabled by disulfide bond. Journal of Polymer Science. 2022;**60**(17):2582-2590

[104] Zhang W, Wang J, Zhang W, Cui D, Cheng T, Qiu D, et al. Scalable micropatterned epoxy vitrimer films by thermo-triggered bond exchange for repairable and recyclable triboelectric nanogenerators. Applied Materials Today. 2023;**32**:101789

[105] Li C, Guo H, Wu Z, Wang P, Zhang D, Sun Y. Self-healable triboelectric nanogenerators: Marriage between self-healing polymer chemistry and triboelectric devices. Advanced Functional Materials. 2023;**33**(2):2208372

[106] He W, Li S, Bai P, Zhang D, Feng L, Wang L, et al. Multifunctional triboelectric nanogenerator based on flexible and self-healing sandwich structural film. Nano Energy. 2022;**96**:107109

[107] Xiao X, Xiao X, Nashalian A, Libanori A, Fang Y, Li X, et al. Triboelectric nanogenerators for self-powered wound healing. Advanced Healthcare Materials. 2021;**10**(20):2100975

[108] Deka BK, Hazarika A, Kwak MJ, Kim DC, Jaiswal AP, Lee HG, et al. Triboelectric nanogenerator-integrated structural supercapacitor with in situ MXene-dispersed N-doped Zn–Cu selenide nanostructured woven carbon fiber for energy harvesting and storage. Energy Storage Materials. 2021;**43**:402-410

[109] Dzhardimalieva GI, Yadav BC, Tat'yana VL, Uflyand IE. Polymer chemistry underpinning materials for triboelectric nanogenerators (TENGs): Recent trends. European Polymer Journal. 2021;**142**:110163

[110] Zargari S, Koozehkanani ZD, Veladi H, Sobhi J, Rezania A. Cost-effective fabrication approaches for improving output performance of triboelectric energy harvesters. Journal of Electrostatics. 2022;**115**:103640

[111] Zhou J, Gao M, Choi J. Enhancement of output power and durability of DLC-based sliding TENGs modified with self-assembled monolayers. ACS Applied Electronic Materials. 2023;**5**(5):2853-2861

[112] Zhu J, Zhu M, Shi Q, Wen F, Liu L, Dong B, et al. Progress in TENG technology—A journey from energy harvesting to nanoenergy and nanosystem. EcoMat. 2020;**2**(4):e12058

[113] Ye C, Li W, Voet VS, Folkersma R, Pei Y, Loos K. Flexible Vitrimers for self-healable triboelectric nanogenerators. Advanced Materials Technologies. 2023;**8**(8):2201670

Chapter 6

Recent Progress on Metal Hydride and High Entropy Materials as Emerging Electrocatalysts for Energy Storage and Conversion

Andile Mkhohlakali, Nonhlahla Ramashala, Sivuyisiwe Mapukata, Sanele Nyembe and Lebohang Hlatshwayo

Abstract

The global energy demand and energy crisis such as the use of fossil fuel for energy conversion and storage have created a need for the development of clean and sustainable renewable energy sources such as fuel cells, batteries, supercapacitors, solar. However, commercialization of renewable energy devices relies heavily on exploring and devising highly functional and stable materials. High entropy materials are emerging, high-performing electrocatalysts due to their intrinsic tenability; hence, these materials may result in earth-abundant catalysts for efficient electrochemical energy storage and conversion. In this chapter, advancements in the energy storage and conversion efficiencies of emerging materials, i.e. high entropy and metal hydrides, as well as their counterparts, i.e. PGMs and MOFs, respectively are discussed. Their applications in fuel cells, hydrogen and oxygen evolution reactions, hydrogen storage, and batteries are deliberated. Furthermore, computer modeling (density functional theory) and machine learning are factored in to supplement the catalytic processes in energy generation and storage reactions.

Keywords: high entropy materials, PGMs electrocatalysts, metal-hydrides, metal-organic framework, electrocatalysis, energy storage, computational modeling

1. Introduction

The new and advanced materials such as high entropy materials, metal hydrides and MOFs enables the realization of renewable energy technology (i.e. Fuel cell, zinc-air batteries, green hydrogen technology, oxygen reduction) [1]. The high entropy alloy's (HEA's) unique structural and morphological characteristics, tuneable chemical composition and functional properties have attracted considerable interest in the field of renewable energy technology [1]. In order to develop environmentally sustainable energy sources, innovative methods for designing catalytic nanoparticles are required [2].

IntechOpen

Fuel cells and hydrogen generation have gained substantial attention as an alternative to conventional energy conversion technologies to overcome the consequences that arise with the utilization of fossil fuels. These issues are in connection with the depletion of fossil fuel reserves, demand for carbon-neutral energy sources, climatic changes concerns and economic consideration. Furthermore, energy storage has become the research of interest for renewable energy sources such as metal-organic frameworks, and metal hydride to store hydrogen and batteries materials.

This chapter presents the emerging materials (i.e. high entropy, metal hydride), their counterparts (i.e. binary and ternary PGMs and PMGs-MOFs) and the application of these materials for energy storage and conversion. The synthetic/fabrication approach of high entropy alloys (HEAs) and the computational (i.e. machine learning and density functional theory) approach for energy storage and conversion catalyst screening process are discussed. In addition, the intrinsic properties, geometric properties, and challenges associated with electrocatalysts in energy conversion, current status, conclusions, and future perspective are also fully elucidated.

2. Platinum group metal catalysts

Platinum group metals consist of six (6) principal elements and they are also known as precious metal, and these metal includes Ruthenium (Ru), Rhodium (Rh), Palladium (Pd), Osmium (Os), Iridium (Ir) and Platinum (Pt) [3]. They are often classified into two categories namely, light (Ru, Rh, Pd) and heavy PGMs (Os, Ir and Pt) based on whether they are based on 4d electron shells or 5d electron shells based on their atomic number [4]. The PGMs are used in various applications across different applications; however, in this section, we will focus on their fabrication methods and use in electrocatalysis [5]. PGMs are applied in numerous energy conversion and storage as electrocatalysts and as energy carriers. PGMs are key electrocatalyst materials that are employed in numerous electrochemical energy conversions, including fuel cell and hydrogen generation to name a few. In electrocatalysis, the PGMs are applied in the proton exchange membrane, and they are mainly responsible for oxygen reduction and hydrogen oxidation reactions at the anode and cathode of the fuel cell [6]. However, recently there has been a drive to substitute PGM-based catalysts with non-PGM catalysts, and this was due to the relatively high costs associated with the former [7]. The use of these relatively non-expensive, high abundance and good activity has further progressed the improvement of energy materials [8].

2.1 Fabrication methods of PGMs electrocatalysts

Fabrication of PGMs nano-electrocatalysts has been the search of interest for many years to improve the properties for a wide range of applications. In nanotechnology, nanostructure PGMs electrocatalysts are fabricated using two (2) synthetic approaches, i.e. "Top-down" and "Bottom-up" techniques [9]. The top-down mainly focuses on forming the nano-scale materials from bulk *via* physical crushing milling, plasma etching and lithographic methods (i.e., photo-lithology, soft lithology, colloidal lithology) to name a few. Whereas the "Bottom-up" counterpart implies the fabrication of PGMs electrocatalysts, starting from molecular/atomic scale *via* self-assembly of atoms and molecules. Bottom-up techniques include chemical vapour deposition, molecular beam epitaxy, sol-gel method, green synthesis, hydrothermal and chemical reduction to name a few. Bottom-up has shown many advantages compared to its counterpart that as the control over shape size, cost-effective, short methods with low consumption. The later approach usually involves the use of a metal salt which is reduced in the liquid medium in the presence of a stabilizing agent to form relatively stable nanoparticles [10]. Examples include the use of platinum chloride salt as a Pt source, sodium borohydride as the reducing agent and citrate moiety as the stabilizing agent [11]. An example would include the use of electron beam lithography to produce metal nanoparticles with the desired dimensions [12]. In addition, excellent electrochemical activity demands a robust and unified synthetic approach to design phases of catalyst materials.

2.2 Intrinsic, geometric properties of PGMs-M electrocatalyst

Generally, the use of metallic nanocatalysts that are not anchored onto a support structure is relatively unstable. Furthermore, we will highlight the effects each method has on the geometric properties and the intrinsic performance of the catalysts in application. They tend to undergo Ostwald ripening and coalesce producing larger particles with reduced surface area and eventually loss of performance as electrocatalysts hence the use of various mesoporous structures as support materials [13]. The synthetic approach is found to control the selectivity, mass activity, agglomeration, surface morphology, atomic scale and size of nanomaterials-based catalysts. Zhang et al. have reported the electrodeposition (self-supporting electrode) of Pt/C for the oxidation of liquid fuels, such as ethanol and methanol. The method showed uniformly dispersed nanoparticles due to weak Vander Waal force between Pt and substrate and compared with the traditional method [14]. Notably, the influence of geometric properties such as shape, size and morphology influences the electrochemically active surface area due to increasing active sites which further improves the catalytic activity. Mkhohlakali et al. argued that the best descriptor for catalytic activity of PdTe is best described by geometric features. Furthermore, AFM and DFT showed rugged surface morphology and High O* binding energy respectively, which enhanced the oxidation of ethanol reaction intermediates [15]. Mesoporous structures have been used successfully to fabricate electrocatalysts with unique structures and speciality morphology. However, great care is required in controlling the pore sizes and pore distribution and their specific surface area as these factors affect the catalyst properties and their performance during application [16]. In the past decade, various studies have been conducted in template development for electrocatalysts [17]. The high adsorption capacity often directly relates to better metal loading and enhanced electrocatalysis performance.

3. Recent progress of PGMs electrocatalysts

Over the past decades, PGMs-based electrocatalysts have been mostly studied using various approaches for wide-range applications such as in fuel cell technology (HOR, AOR and ORR), hydrogen evolution reactions, and oxygen evolution reactions. These applications' behaviour relies on the best choice of catalysts, thus far PGMs-based electrocatalysts are taking the lead. The most commonly used PGM electrocatalysts over the years for fuel cells, HER and OER are platinum-based in acidic medium, Pd and Ru-based electrocatalysts in alkaline medium, while for OER they are iridium, Pt-based as well as PGMs free [18]. Traditionally, these electrocatalysts have significantly improved the catalytical performance with lower overpotential and faster reaction kinetics. In addition, one of the most significant roles played by PGM-based and PGM-free electrocatalysts was to enhance the mass activity by turning the physicochemical properties, such as structural or electronic effects and change the adsorption/desorption of the intermediates on the surface of the catalysts. In this book chapter, we summarize the recent progress of PGMs and PGMs free electrocatalysts that are classic and believed to have potential applications in HER, OER, ORR and HOR for sustainable energy conversion.

3.1 Pt and Pd alloys-based electrocatalysts

Since Pt-based electrocatalysts were first coined for the development of fuel cells, HER and OER are seen as cost reductions and possible commercialization over pure Pt catalysts. Therefore, adding two or more non-noble metals onto to Pt surface can be of great importance to enhance its desorption as well as complete oxidation of CO intermediates and accelerate the reaction kinetics [19]. Since electrocatalysts are surface effect dependent, they have the propensity to agglomerate on the surface leading to a decrease in surface area and affecting the catalytical performance [20]. Therefore, the use of advanced supporting materials such as carbon black, carbon nanotubes, graphene, carbon nanofibers, carbon nano-horns and heteroatom-doped carbon material will allow the increase in surface area, high porosity, high electrical conductivity and high dispersion of Pt-based electrocatalysts [19].

Although Pt-based electrocatalysts have been chosen as the best candidate thus far, their ability to operate only in an acidic medium hinders the development of efficient and practical AOR, HOR, HER and ORR [21] in the future. In acidic environments, most electrocatalysts are unstable, easily decomposed and corrosive. Therefore, an urgent needs to develop new electrocatalysts that can easily operate in alkaline medium to mitigate these obstacles. To date, Pd-based electrocatalysts have previously been indicated as an alternative to replace Pt due to their excellent properties [22, 23]. These different approaches have resulted in high-efficiency AOR, HOR, OER and ORR. Various types of Pt-based electrocatalysts were investigated for the development of HER, HOR, HOR and ORR in an alkaline environment. The investigation has revealed the higher exchange current density or mass activity at a low overpotential of 0.05 or 0.1 V. Developing electrocatalysts with high activity and durability is of the most significance. Recently, the development of PGMs-free electrocatalysts for the application of fuel cells, OER and HER has become the current topic of interest. In the recent 5 years, extensive work has been published regarding the utilization of PGMs free in alkaline conditions. Designing highly efficient and long-term stable non-noble metals with advanced support material to increase the surface area will result in high electrochemical activity and excellent performance of HER, HOR, ORR and OER.

3.2 PGM-based electrocatalysts for application in ORR and OER

Oxygen reduction reaction (ORR) as a cathodic reaction for fuel cells has gained a lot of attraction for the development of next-generation energy conversion [24]. The ORR plays a huge role in fuel cells as it controls the whole device's performance. ORR reaction mechanism has two reaction pathways that occur during the reduction of oxygen with the aid of electrocatalysts. The proposed mechanism involves the direct four electrons ($4e^-$) step where the reduction of O_2 to water/OH^- (acidic or basic electrolyte) occurs without H_2O_2 formation and the two electrons ($2e^-$) transfer is where O_2 is firstly reduced directly to H_2O_2 as intermediates and later to H_2O [24]. The former reaction is ideally applicable in fuel cells and the latter reaction produces H_2O_2 which is good for the green synthesis process, for the treatment of wastewater and disinfection [25]. The good ORR performance should obey the Sabier principle where the adsorption strength of these intermediates (OOH*, OH* and O*) should neither be too weak nor too strong to bind the active site. In theory, according to the volcano plot the different electrocatalysts are demonstrated based on their catalytical activity performance. The use of the model assists in knowing why Pt and Pd with their alloys are better cathode electrocatalysts over other metals before performing the experimental analysis [26]. Then from now on, Pt and Pd-based electrocatalysts have greatly been investigated as ORR catalysts due to their electronic structure for O^- (from dissociation of O_2) and OH^- binding energies [27].

3.2.1 PGM-based electrocatalysts for application in OER in acidic and alkaline environments

Oxygen evolution reaction is based on the 4-electrons process and their reaction mechanism is found to be difficult whether conducted in the alkaline or acidic medium leading to high overpotential. The high overpotential causes the efficiency and the performance of OER to be sluggish for the formation of hydrogen from the water-splitting process. On that note, suitable electrocatalysts are needed to break down the O-H bond to form the O – O bond, accelerate the kinetics of OER and enhance the overall efficiency. Ideally, a catalyst should be able to follow the Sabatier principle whereby it is stated that a catalyst should not be too weak nor too strong to bind oxygen. The computational studies were performed on the potential catalysts and the volcano curve gives the relationship between the catalytical activity of the catalysts and bond strength. Therefore, Trassati et al. have reported that RuO_2 has shown a high OER activity on the volcano shape between the oxidizing enthalpy and OER activity [28]. An extensive search has been conducted for RuO_2 base electrocatalysts, ranging from the monometallic (Ru, Ru, and Ir, Ru, and the mixed IrO_2 and RuO_2) in both acidic and basic environments. The development of these materials from fundamental to commercialization is hindered by the slow kinetic and high cost of Ru and Ir [29–35]. In the past decade, researchers have reported the use of second or third metal oxides (IrO_2-SnO_2, RuO_2-TiO_2, IrO_2-MnO_2, RuO_2-ZrO_2, IrO_2-Ta_2O_5 and IrO_2-IrO_2-Ta_2O_5) to design binary and ternary electrocatalysts not only to minimize the noble metal content but also to reduce the overpotential towards OER. In earlier studies, it was reported that the binary catalysts/multi-metal oxides outperform monometallic catalysts because of the ability to promote bifunctional (bonding interaction (M-OH)) activity and lower the overpotential kinetics [36, 37]. Lee et al. have reported the use of highly active metals and stable three-dimensional mesoporous Ir-Ru binary based on mixed metal oxides and metal/metal support as OER catalysts. Their structural and electrochemical behaviour was investigated for the OER activity and mesoporous IrO_2/RuO_2 with a molar ratio of (1:10 Ir/Ru) gave a lower overpotential of 300 mV at 10 mA cm^{-2} as compared to IrO_2/RuO_2 with a high overpotential of 340 mV. After the stability test was performed on both catalysts ran for over 2 h, the overpotential was as low as 22 mV for mesoporous IrO_2/RuO_2 and 44 mV for IrO_2/RuO_2 [33]. In addition, Huang et al. have also affirmed that coupling RuO_2 with oxides such as TiO_2 with a high valence state can enhance the intrinsic stability under an acidic medium without compromising the performance. The electrochemical behaviour revealed that the RuO_2/HTI/TI composite has low overpotential of 220 and 265 mV with the current densities of 10 and 50 $mAcm^{-2}$ and gave the mass activity of 1760 ± 60 mA $g_{Ru}{}^{-1}$ at 290 mV overpotential, which is 7.5 times higher than pure RuO_2 nanoparticles. The study also revealed the high stability of composite for 11 hr. with a current density of 500 $mAcm^{-2}$ and mass loading of 0.1 mg cm^{-2} [38]. These findings highlight that the incorporation of metal oxides into noble metals is promising electrocatalyst for OER [39]. These results were better as compared to amorphous Ir atomic clusters, and IrO_2 nanoneedles [40]. More work has been conducted on the OER electrocatalysts and interestingly zirconium oxides were also investigated because they are less likely to be corroded and possess a homogenous dispersion property among other reported oxides when dispersed on noble metals. Liu et al. studied the ternary metal oxides Ti/IrO_2-RuO_2-ZrO_2 for OER activity and the electrochemical investigation shows the high electrocatalytic activity and good stability when the Ru content is 21 wt%, which emphasizes that Ru content plays a critical role in designing the efficient OER catalyst [41]. Furthermore, it has been reported earlier that the catalysts for OER applications in acidic environments should have high Ru content for better performance [42]. However, a pioneering study by Niu et al. demonstrates that the addition of transition metals onto Ru can minimize the noble metal content in an acidic medium and also contributes to regulating the electronic structural interaction. They aimed to investigate the low loading RuO_2 (2.51 wt%) unto $(Mn, Co)_3O_4$ as highly efficient catalysts succeeded for OER application. The $RuO_2/(Co, Mn)_3O_4$ catalysts show a high performance with an overpotential reported to be 270 mV at a current density of 10 mA cm^{-2}, superior to commercial RuO_2, and RuO_2/Co_2O_3/CC catalysts [42]. Though PGM electrocatalysts have exhibited remarkable catalytical performance for OER, high cost and poor chemical stability in alkaline environments impede the development of tenable applications [43]. Therefore, the electro-design of non-noble electrocatalysts such as Co-based, Ni-based metals, and multi-metal oxides (spinel, layered hydroxides, perovskites, etc.) as active OER showed the best alternative is an alkaline environment [37, 44]. In general, the spinel metal oxides are expressed as follows AB_2O_4, whereby cation A has a charge of 2+, occurred at the tetrahedral sites (A M^{+2}) and cation B with M^{+3} occurs at the octahedral sites of the close-packed structure [37]. Among them all, spinel ferrites, such as $NiFe_2O_4$ and $CoFe_2O_4$ have been widely studied as efficient and promising catalysts for OER activity and other applications [36]. In addition, layered double hydroxides (LDHs) were considered a class of synthetic layered terrestrial 3d transition metals with high electrocatalytical activity for OER application. **Table 1** summarizes the electrocatalytic activity of mono-metallic, bimetallic and trimetallic towards different electrocatalytic reactions.

3.2.2 PGM-based electrocatalyst for hydrogen evolution reaction

Hydrogen evolution is the production of hydrogen gas through a water-splitting process and the reverse is utilizing that Hydrogen gas as fuel in fuel cell applications [50]. Both the HER and HOR processes occur with the aid of electrocatalysts at the cathodic and anodic sides during electrochemical reactions, respectively. Thus far, PGM electrocatalysts such as Pt, Ru, Pd, Ir and Rh are the most efficient, and more active and have advanced tremendously in the past years to lower the overpotential while enhancing the reaction rate [51, 52]. The ultimate goal in the development of electrocatalysts under acidic conditions for both HER and HOR reactions is to understand the concept of adsorption-free energy of hydrogen (ΔG_H) because the theory can describe intensively the binding strength of intermediate H*on the catalyst

Catalysts	Electrolytes	Mass loading	Mass activity	Stability	Over potential	Refs
IrO_2-SnO_2	0.1 M H_2SO_4	1.2 mg cm^{-2}				[45]
Ru-RuO_2-NC	0.5 M H_2SO_4		1.35 A g^{-1}	40 h at 10 $mAcm^{-2}$	176 mV	[46]
Ru-NiO/ Co_3O_4					269 mV at 100 $mAcm^{-2}$	
Ir-Co_3O_4	0.5 M H_2SO_4	2.88 wt% Ir		130 h at 10 $mAcm^{-2}$	225 mV at 10 $mAcm^{-2}$	[47]
Ti/IrO_2-RuO_2-ZrO_2						
RuO_2(Mn, Co)3O4	0.5 M H_2SO_4	2.51 wt% Ru	366.5 A/g Ru	1000 cycles between 0.867 to 1.667 V	270 mV at 10 mA cm^{-2}	[42]
Ru_3MoCeO_x	0.5 M H_2SO_4		1469.5 A/g		164 mV at 10 mA cm^{-2}	[48]
NiFe LDH/ CB	1 M KOH	0.16 mg cm^{-2}		12 h with	300 mV at 10 mA cm^{-2}	[49]

Table 1.
Summary of electrocatalysts and their electrocatalytic activity.

surface [53]. The surface of the electrocatalysts should neither be too strong nor too feeble to bind hydrogen, so it enables the surface to form hydrogen gas easily by improving the kinetics of HER as expressed in equation [50, 54]. This is according to the empirical rule obtained from the Sabatier principle where it is expected that ΔG_H of the electrocatalysts should be close to zero [53]. Nørskov and his coworkers were inspired by these experimental findings and later decided to conduct the computational work utilizing density functional theory (DFT) on various metallic surfaces and the outcomes were the same as the one attained from the volcano shape. The volcano curve investigates a linear relationship between the hydrogen evolution reactions's (HER) exchange current density and the hydrogen binding energy (HBE) on the metallic surface [54]. This is basically to show that the variation of the measured exchange current densities is well comprehended by using a facile kinetic model. The volcano shape in Figure displayed the role PGMs electrocatalysts play in HER reactions and those electrocatalysts gave the best performance in enhancing the kinetics in an acidic medium. Thus far, Pt and Pd (which are closer to the highest peak at the volcano) are leading efficient HER catalysts, with low overpotential and showing rapid kinetics [55]. It is desirable to introduce the support materials as well as the non-noble metals to eliminate the problem while maintaining the high catalytic activity [55]. Alloying PGMs with non-noble metals (Fe, Au, Co, Ni, Cu, etc.) and with their corresponding metal oxides/hydroxides by turning intrinsic and extrinsic properties towards HER [51]. The table summarizes the electrocatalysts used for HER. The development of non-noble metals as an alternative to Pt and Pd has gained attractive attention to date as bifunctional electrocatalysts owing to their low cost and possessing good corrosive resistance in alkaline medium [56]. So far electrocatalysts, such as cobalt, Ni, Mn, Fe based, were widely explored and the performance in HER gives high electrocatalytical behaviour. Even though sluggish kinetics and lower ECSA in most of the electrocatalysts were observed, however, alloying and the use of support materials combat the problems. The support materials also play a crucial role in PGMs-free alloys as they can increase the surface area.

3.2.3 Reaction mechanism for HER in acidic and alkaline media

To comprehend the application of electrocatalysts towards HER reaction, it is advisable to first understand the catalytic reaction mechanism evolved throughout. The reaction can take place in two conditions: acidic and alkaline conditions and are described as follows. The hydrogen evolution reaction can be expressed in the following equation, where the hydrogen intermediates are formed throughout the process to produce hydrogen gas: the process includes three major steps in an acidic environment. This reaction step involves the adsorption and desorption of H* intermediates on the surface. The initial step involves the adsorption of a proton into the electrocatalyst surface – M-H_{ads} where M represents the active material with one electron migrating (R1), followed by the second step which can be derived from either Heyrovsky or Tafel reaction because this particular step-dependent on the coverage of hydrogen intermediates on the surface of the active site, subsequently, the final step involves the coverage of two hydrogens adsorbed on the active site to form hydrogen atom [57, 58]. In an alkaline environment, the HER reactions occur in two steps mainly, the Volmer and Heyrovsky reactions, whereas the Tafel reaction remains the same with the acidic step [59]. Briefly, the initial step involves the formation of M-H_{ads} and OH^- on the active site followed by the interaction of water with the adsorbed hydrogen and electron to form a hydrogen atom (**Table 2**) [60].

3.3 Binary and ternary electrocatalysts towards electro catalytic reaction

Researchers have approached the drawback associated with sluggish kinetic and 2 electron pathways during ORR by introducing the second and third elements. The enhancement of the catalyst activity is ascribed to the synergistic effect and functional activity of two elements and the O^- O-species formed by exophilic species. Lankiang et al. reported the binary and ternary electrocatalysts such as $Pt_{70}Pd_{15}Au_{15}$ using the micro-emulsion method, and the electrocatalysts were tested for ORR in in O2-saturated 0.1MHClO4 by rotating disk electrode. Recent research on PtPd-based catalysts has shown an enhancement in the electroactivity towards ORR in acidic medium, which is explained by a synergetic effect between Pd and Pt.

3.4 PGM-based electrocatalysts for application in HOR

The PGMs electrocatalysts for application in hydrogen-oxygen reaction (HOR) have gained much attention in the past decades. Thus far, the widely studied PGM

Reaction mechanism	Acidic	Alkaline
Volmer reaction	$H^{++e^{-\to M-H_{ads}}}$ (R1)	$H_2O{+}e^{-\to M-H_{ads}} + OH^-$
Heyrovsky reaction	$H^{++e^{-+M-H_{ads}\to H_2}}$ (R2)	$H_2O{+}e^{-+M-H_{ads}\to =OH^{-+H_2}}$
Tafel reaction	$M - H_{ads} + M - H_{ads} \to H_2$ (R3)	

Table 2.
Summary of reaction mechanism in both acidic and alkalinity.

electrocatalysts are Pt, Pd, Ru, Ir, Re and Rh on HOR in fuel cells for electrochemical conversion purposes. Inspired by Sabatier principles it states that the best electrocatalysts for HOR should possess a moderate adsorption strength. The HOR reaction pathway in the alkaline medium can be obtained following the elementary steps: The R1 and R2 which are Tafel and Heyrovsky are reactions where the H_{ads} intermediates are chemically absorbed on the active site and where M represents the active site for hydrogen adsorption. Later, the absorbed hydrogen together with hydroxyl ions forms water and therefore electrons are released from the surface. In addition, the Tafel reaction step as differs from Heyrovsky ought two vicinal active sites to allow the adsorption of two hydrogen intermediates (bond length between H-H is approximately 0.74 Å. Therefore, it is generally accepted that in a basic electrolyte, the kinetics of HOR is derived from hydrogen binding energy (HBE). HBE is a very important factor as it affects the efficiency of HOR, sluggish kinetics results in weak or strong adsorption interaction between binding PGMs adsorbate and intermediates. As it was observed from the volcano curve when the pH (from 0 to 14) is increased the use of PGMs aids in binding the hydrogen stronger hence, the PGMs are found on top of the curve [56].

4. High entropy materials

High entropy materials (HEMs) are emerging routes for the production of high-performing electrocatalysts due to their intrinsic tenability and the coexistence of several possibilities and these materials may result in earth-abundant catalysts for efficient electrochemical energy storage and conversion. High-entropy materials were first proposed in 2004 and have been applied in a range of systems and applications. They are a promising class of disordered multicomponent materials with tailorable properties/functionalities (and maybe unparalleled performances). HEMs are a new class of materials that have just been invented and exhibit extraordinary properties that outperform those of conventional alloy (i.e., binary, ternary alloys) material. This concept was first introduced recently by Cantor et al. and Yeh et al. in 2004 [61, 62] based on the composition, and HEAs/HEMs materials are defined as a single phase (alloy) multi-component system with 5 or more major metal element of which each having a near equiatomic (equimolar ratio) and each element has at least concentration between 5 and 35% content [1, 63]. These materials are also known as multi-principal component alloys, in contrast to traditional alloys which have 1 principal metal element dominant. Another definition is based on the thermodynamic configurational entropy of mixing (ΔS_{mix}, as shown in Eq. (2)), the following formula can be used to express the aforementioned definition

$$N_{major} \geq 5, 5 \in \%, c_i \leq 35, 35 \in \% \wedge n_{minor \geq} 0 c_j \leq 5\% \quad (1)$$

Where ci_{major} *and* in_{minor} are the number of primary and secondary elements, respectively, and $c_i \wedge c_j$ are the atomic percentages of primary and secondary elements, respectively.

$$\Delta S_{mix} = -R \sum_{i=1}^{n} c_i ln c_i = -R \sum_{n} \frac{1}{n} ln \frac{1}{n} = R ln n \quad (2)$$

Where R is the gas constant, n is several major elements, and C_i is the concentration of the components. Compared to conventional compounds, HEAs with more than

5 elements trigger the formation of a single-phase solution (structure) with lattice distortion (which is beneficial for gas sorption and hydrogen storage). There are various factors affecting the formation of HEAs (1) Enthalpy of mixing (ΔS_{mix}), (2) Atomic size difference, (3) Electronegativity difference, (4) Valence of the electron concentration, (5) Effect of kinetics on phase formation, (6) Geometric parameter and (7) excess of entropy. Based on the high entropy effect phenomenon, the high entropy of mix ΔS_{mix} must be higher than 1.5R.

$$\Omega = T_m \frac{\Delta S_{mix}}{\Delta H_{mix}} \tag{3}$$

The enthalpy entropy relationship during solidification is symbolized as (magma). This parameter is calculated based on Eq. (3). Where Tm and H denote the melting temperature and the enthalpy of mixing, respectively. When $\Omega > 1$, the solid single solution phase is expected in the alloy structure, while in the case of $\Omega < 1$ is attributed to the formation of an intermetallic compound or phase separation.

According to Yeh et al., another attractive property of multiple elements (large HEAs) with different characteristics that enhance the performance in diverse applications are (i) sluggish diffusion, (ii) high entropy effect and (iii) cocktail effect (which describes the synergistic response, beneficial for energy conversion) [64]. As illustrated in **Figure 1(a)** and **(b)** HEAs exist in various structures namely (i) face-cantered cubic, (ii) body-cantered cubic (bcc) and (iii) hexagonal closely packed (hcp). The characteristic features, generation and classification of HEMs will be discussed in the following section.

4.1 Generations and category of HEAs

Refractory element-based HEAs are frequently categorized into the first, second and third generations based only on the time of their formation [65]. In contrast to traditional alloys (i.e. binary and ternary, with 1–2 principal elements), the first generation of high entropy alloys which comprised of a single phase with at least 5 principal elements at equal concentration with BCC single-solid phase. These typical HEAs were established in 2004 as Canter alloys such are AlCoCrNiMn, FeCrMnNiCo, and $Mo_{25}Nb_{25}Ta_{25}W_{25}$. Due to the thermal stability (i.e. resilience in high temperatures), these HEAs were being applied in aerospace that is wind turbine blades. The second generation consists of at least four principal elements with dual phases at

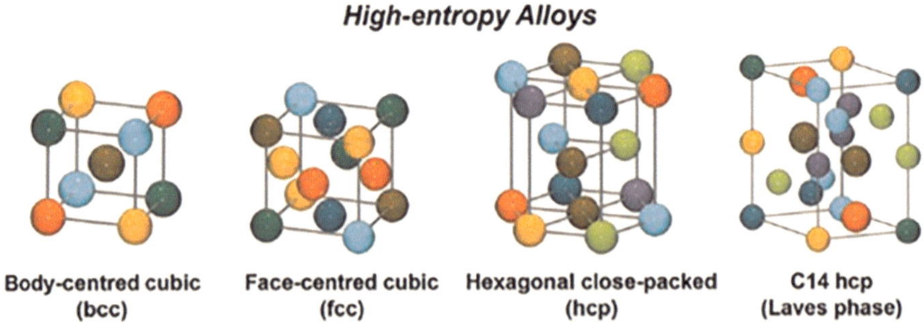

Figure 1.
Schematic representation of high entropy alloy. Reproduced with permission from reference [1].

non-equiatomic started to emerge because of the high density and low corrosion resistance of the first-generation HEAs. The metals with a high density such as Ta, Nb and Mo in the first generation were substituted with light and corrosion-resistant elements such as Al, Fe and Cr to name a few and transitioned from single to dual phases such as $Fe_{50}Mn_{30}Co_{10}Cr_{10}$ [66]. The HEAs formation strategy is reviewed and discussed in the following section.

4.2 Preparation methods

Ever since the discovery of HEAs in 2004, there has been an extensive search for suitable methods to form a single-solid solution of such materials [67]. The common methods to prepare 3D and low-dimensional structures of HEAs are abundant and mainly include (i) Vacuum induction melting, (ii) Vacuum arc furnace melting, (iii) Top-down chemical dealloying, (iv) bottom carbothermic shock, (v) Solvothermal co-reduction, (vi) Ball milling and (vii) Sputtering. In the following section, the various preparation methods will be discussed.

4.2.1 Vacuum arc melting method

Guler et al. have reported another second generation of HEMs using vacuum arc melting followed by a heat treatment process. The materials were composed of AlCoCrFeNiTix with (x = 0.0, 0.2, 0.4, 0.6, 0.8 and 1.0) stoichiometry [68]. Ti was incorporated in the most studied HEAs (AlCoCrNi). The materials were heated at various temperatures that is 500, 700 and 1000°C. Zhang et al. reported the equimolar NbZrTiCrAl HEAs prepared by the vacuum arc melted. The equimolar HEAs were prepared with oxidation response (treated) vacuum arc melted NbZrTiCrAl HEA at 800–1200°C (**Figure 2**).

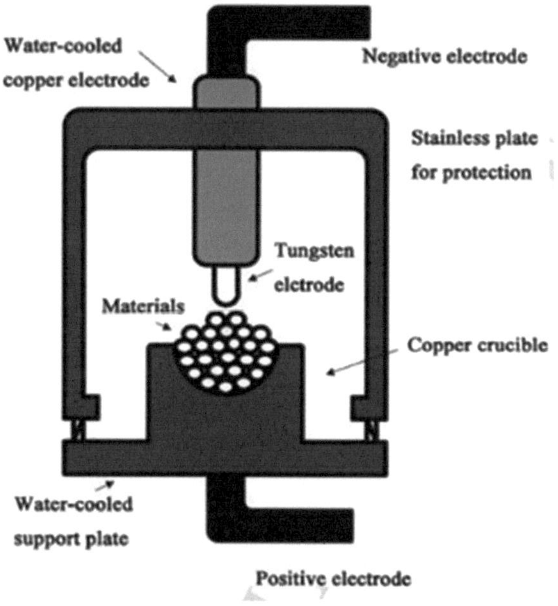

Figure 2.
Schematic illustration of vacuum arc method. Reprinted with permission from literature [69].

The HEAs prepared using this type of method show fine and conformal (uniform) grains, uniform chemical composition and high density [70]. However, it is difficult and tedious to evenly combine five or more metal species into a single nanostructured phase and precisely shape the HEAs into low dimensional ($_1$D, $_0$D) structures to obtain diverse applications (owing to their intrinsic and thermodynamic instability) using the tradition wet chemical method. To overcome this shortcoming, there have been many approaches developed for the fabrication of HEAs to low dimensional structure, including boot-up and top-down to fabricate HEAs. Some of these methods include the sputtering technique, solvothermal deposition, carbothermal shock method and chemical dealloying method [71]. Recently, HEAs have been prepared successfully by employing the carbothermal shock and chemical dealloying method. The section below will discuss the methods for the fabrication of HEAs.

4.2.2 The bottom-up carbothermal shock method

The carbothermal shock (CTS) method relies on thermally shocking the metal salt covered with carbon nanofibers at around 1730°C, followed by rabid quenching [72–74]. As illustrated in **Figure 3**, this method is a facile throughput to fabricate HEAs on carbon support by rapid heating and ultrafast cooling at about 105–2000 $K.S^{-1}$. The rapid increase of temperature from ambient temperature to high temperature (≥1700°C) and cooling fast to 25°C is the promising solution to dictate the alloying of multicomponent at ultrafast kinetics [76]. This method was first introduced in 2018 by Yao et al. by developing the HEAs nanoparticles with PtPdRhRuCe composition for application in ammonium oxidation [72]. Through optimizing the method parameters namely, the concentration of the metal precursor, the shock duration, substrate and cooling rate a uniform dispersion and narrow particle size distribution can be achieved (**Figures 3** and **4**) [67].

4.2.3 Solvothermal co-reduction technique

So far, several methods for the preparation of HEA have been investigated to address the drawbacks associated with high-temperature preparation methods for HEAs. Solvothermal co-reduction which is analogous to hydrothermal has attracted considerable interest in the preparing HEAs and HEAs oxides. The word "solvothermal" was first introduced in the early 1980s to distinguish this specific

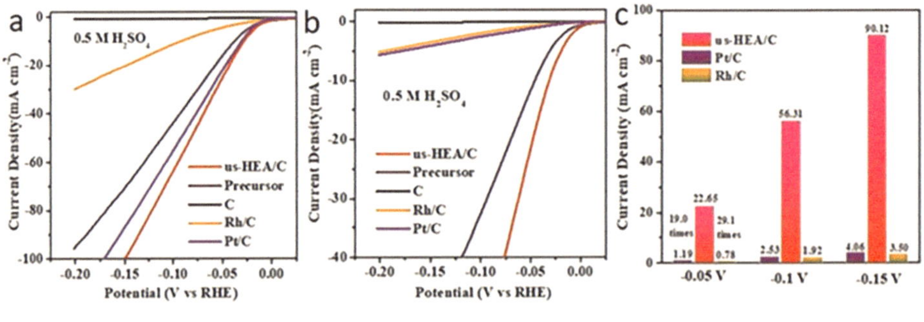

Figure 3.
Comparative electrocatalytic performance showing HER polarization curve of commercial electrocatalysts Pt/C and Rh/C and HEAs towards HER in 0.5 M H_2SO_4 solution. Reprinted with permission from literature [75].

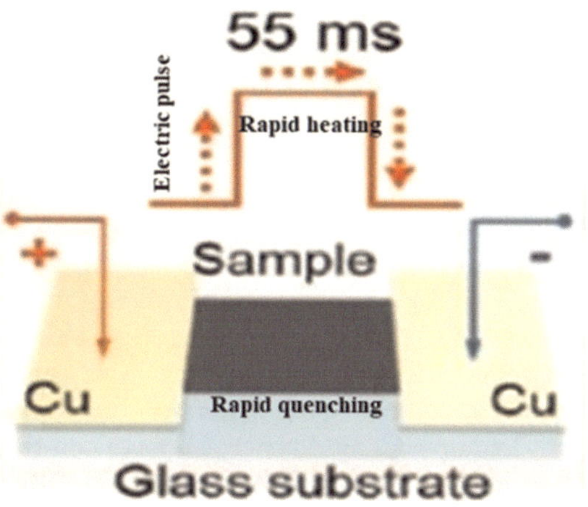

Figure 4.
Schematic illustration of carbothermal strategy using carbon substrate.

High entropy alloy composition	Fabrication method	Size (nm)	Catalytic application	Ref
PtPdRhRuCe	Carbothermal		Ammonium oxidation	[73]
Al-Cu-Ni-Pt-Mn	Top-down de-alloying	15	ORR	[74]
np- p-PtRuCuOsIr	Chemical dealloying	60	ORR	[79]
FeNiMoCrAl	Sputter-deposition	10	OER	[80]
$(Co,Cu,Fe,Mn,Ni)_3O_4$	Solvothermal	5	OER	[78]
[78]	Vacuum arc	25	-	-
NiCoFePtRh	Facile chemical coreduction	1.62	HER	[75]

Table 3.
Summary of various HEAs prepared by different synthetic methods, size and their catalytic application.

reaction from others in other solvents [77]. The solvothermal method is known to be the most effective synthetic route for HEAs with controlled sizes and morphology. Wang et al. prepared small-size (5 nm) HEAs for application in water oxidation reactions, particularly OER and HER [78]. In this method, an auto-clave is a closed system where the mixture is reacted with solvent (i.e. organic or non-aqueous) under a controlled temperature (i.e. between 29 and 98°C) and pressure. The process uses straightforward autoclaves that have been moderately heated. There are various HEAs reported that are prepared by the solvothermal co-reduction method. A summary of different HEAs, with different properties and fabricated using methods, is listed in **Table 3**.

4.3 Application of high entropy materials: Recent progress

The cock-tail effect of high entropy alloys promotes the synergistic effect which is beneficial in energy conversion and electrocatalysts and increases the investigation of

HEAs for ORR, OER, HEA and ethanol oxidation reaction [61]. The discovery of emerging efficient electrocatalyst materials such as high-entropy alloys for use in renewable energy sources such as fuel cells, batteries and water electrolysis is imperative to the development of energy conversion because it is the carbon-neutral approach. Having oxygen reduction reaction (ORR) and oxygen evolution reaction as a key reaction in fuel cell and zin-air batteries respectively are a good example of an electrochemical energy conversion reaction, which requires better electrocatalysts. In addition, hydrogen production through hydrogen evolution on the cathode and oxygen evolution on the anode during electrochemical splitting into chemical energy is an intriguing pathway to convert chemical energy stored in water into electricity [81]. Electrocatalysis is a key asset in the transition towards sustainability because it enables the net-zero-carbon synthesis of value-added chemicals and chemical fuels, including power-to-X approaches. Examples include energy conversion through fuel cells, hydrogen evolution, hydrogen oxidation of fuel cells, ORR, OER reaction in zin-air batteries and water electrolysis. The energy conversion system that occurs in fuel cells, water electrolysis (H_2O splitting) and metal-air batteries is driven/formed by ORR and OER has a cell reaction. The efficiencies of this electrochemical energy conversion system are hampered by sluggish/slow reaction kinetics, which requires the development of active electrocatalysts to overcome these bottlenecks [1]. The complexity associated with these reaction intermediates requires promising bifunctional (i.e. with multiple active sites) electrocatalyst materials such as HEAs [64, 82]. Many attempts have been made to develop HEAs, specifically, FCC structured type of HEAs has become a current subject of intense investigation as emerging electrocatalysts for OER and ORR. The active catalyst sites of HEAs and the unique physico-chemical properties of HEAs have attracted great interest in OER and ORR. In the section below, the continuous work on the development of HEA as electrocatalysts towards electrochemical conversion, particularly OER and ORR will be reviewed and discussed.

4.3.1 Emerging HEMs towards ORR

A major issue that prevents the practical use of fuel cells and metal-air batteries is the oxygen reduction reaction's (ORR) sluggish kinetics. Pt-based catalysts have exhibited high activity for ORR; however, the high cost of Pt limits their application []. Therefore, it is still critically necessary to improve the ORR activity of Pt-based nanostructures and reduce the Pt loading quantity. Thus, developing noble metal with Pt activity by alloying with other metals provides *OH to enhance ORR performance. Alloying with 5 or more metals in one solid phase solution makes HEAs rather than binary, ternary. HEAs are increasingly being investigated concerning ORR; this is because of their many unique mechanical and physical properties, which make them particularly suitable as structural materials [74]. The higher stability and activity of PdCuMoNiCo HEA [83] compared to 20% Pt/C was due to the synergistic effect of its hollow structure. Recently, Chen et al. reported the remarkable 15.8 folds of higher mass activity (1.738 A mg^{-1}), 0.90 V and higher stability of PtFeCoNiCu than the commercial Pt/C towards ORR in polymer electrolyte membrane fuel cell (PEMFCs)- [84], as indicated in **Figure 5**.

Al-Cu-Ni-Pt-Mn np-HEA exhibited (with low Pt loading 20%Pt) exhibited a higher electrochemical activity towards ORR as compared to the commercial Pt/C [74].

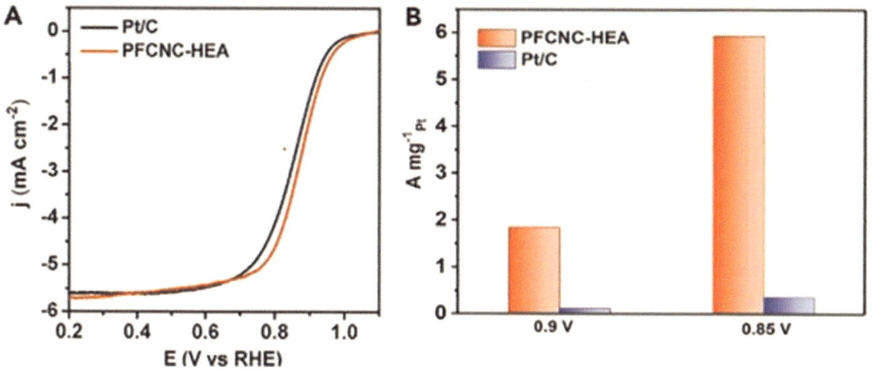

Figure 5.
Schematic representation of ORR polarization curve of PtFeNiCu and commercial Pt/C electrocatalysts scanned in oxygen saturated 0.1 M $HClO_4$ at 25 C at a scan rate of 10 mVs^{-1} and rotation rate of 1600 rpm. Reproduced with permission from the literature [84].

4.3.2 Progress of HEAs on OER

The oxygen evolution reaction (OER), one of the most crucial steps in water splitting and other renewable energy storage and conversion techniques like metal-air batteries and fuel cells, has a four-electron transfer process, which exhibits slow reaction kinetics and a high reaction energy barrier, resulting in a high overpotential. Noble metal oxides such as RuO_2/IrO_2 have been ideal/active candidates for OER. In the search for highly active electrocatalysts, transition metal-based electrocatalysts emerged as potential candidates for OER electrocatalysts. With the synergistic effect possessed by HEAs and their oxide, there is an extensive search for HEAs for OER application. Qiu used the dealloying method to prepare the quaternary Ir-based HEAs such as AlNiCoIrX (i.e. X = M0, Cu, Cr, V, Nb) for OER and Li et al. reported the FeNiMoCrAl thin film electrocatalysts deposited using sputter deposition, and HEAs electrocatalyst showed superior electrochemical performance with low potential and higher current density as compared to binary and ternary counterparts [**10.1002/smll.202106127**].

4.3.3 Advances of HEAs towards green (HER)

The challenge associated with hydrogen production is a sufficiently effective catalyst towards hydrogen production [4]. In contrast to traditional binary and ternary alloy, the HEAs possess the advantage of lattice distortion, which often provide the diffusion pathways and interstitial sites for hydrogen atoms, leading to intriguing hydrogen properties. Feng et al. reported the HEA NiCoFePtRh denoted at us-HEA is the smallest size HEA ever reported [82]. The high activity was characterized by the mass activity is about 28.3 Amg-1 at −0.05 V for HER in acidic solution (0.5 M H2SO4) which is 74.5-folds higher than the commercial Pt/C. These later findings are attributed to the lattice distortion of single-solid solution HEAs which promote the synergistic effect.

4.3.4 PGMs-HEAs towards EOR

In 2022, Fan et al. reported on the HEAs composed of PdPtCuAgAu nanowires for methanol oxidation reaction (MOR) for fuel cell application. Combining 5–6 PGMs to

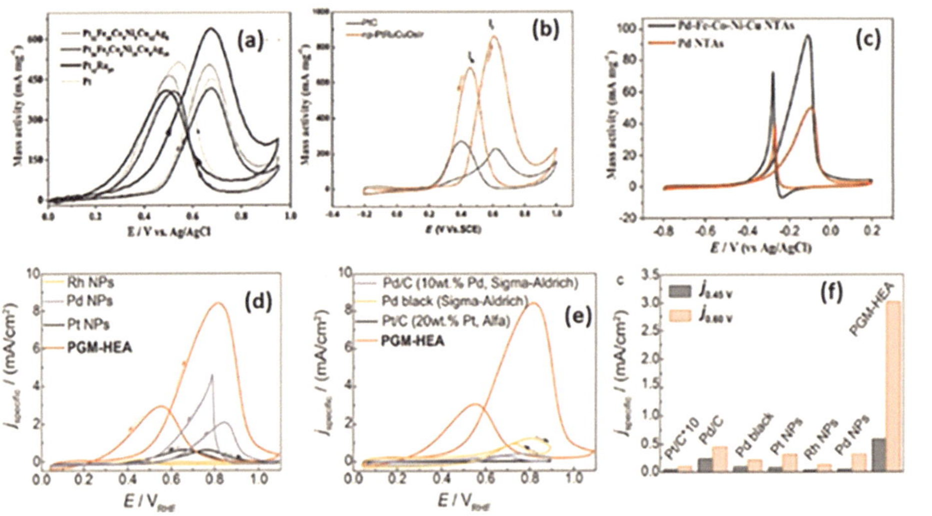

Figure 6.
Illustrate the comparison of current density and mass activity using cyclic voltammetry (CV) plot of the HEAs and commercial Pt/C in 1.0 M MeOH in 0.5 M H2SO4 solution at 5mVs-1 (a-c) and of PGM-HEAs with commercial catalyst (Pd/Pt) (d-f) towards EOR. Reproduced with permission from literature [85, 86].

make large PGMs-HEAs, creates a variety of active sites on their surface to catalyze the multistep reactions. Recently, Wu et al. have reported the 6 principal PGM elements to catalyze ethanol oxidation reaction, and PGMs-HEAs exhibited higher activity by 12.8-folds than the commercial Pt/C in terms of current density, mass (intrinsic) activity as shown in **Figure 6(c-d)** [85]. This increase is attributed to the scissoring of the ethanol C-C bond which is the key step for the 12e- pathway for EOR by adding the second and third, fourth elements.

4.4 Challenges and advantages of PGMs and high entropy for energy conversion

As much PGMs-based electrocatalysts have been a leading material for energy conversion, PGMs occurs in trace amount in the earth crust and this scarcity inhibits the commercialization of PGMs. In addition, the high possibility of poisoning by reaction by-products on the surface of electrocatatalyst has become the biggest challenge. For instance, the (i) H-H (heron sky) forms Pt-H and decreases the electrochemical activity on the catalyst surface during the hydrogen evolution reaction. (ii) The acetate by-which is the EOR reaction product (COOH*) on the catalysis surface poison the catalysts surface and subsequently decreases the stability, selectivity and catalysts overall performances. (iii) Furthermore, the O-O (metal atom (M-OO) to form superoxo-metal complex which results to yielding of two electrons triggering the Pauling's model in oxygen reduction reaction instead of 4 electron pathway why which is an undesirable electrochemical reaction. During the review of the literature, the researcher's approach's trend been found that the are numerous efforts to find a simple and facile methodthat can selectively form the low-dimension catalyst materials (1D to 2D) that possess highly active sites for electrocatalytic activity. The bottom-up fabrication methods has been promising to fine-tune the development of active electrocatalysts with high selectivity. Nonetheless, this is low commercialization of energy conversion technologies that are assembled with

PGMs-based electrocatalysts. The new emerging type of alloy enables the mixing of PMGs elements into solid solution, which enhances the properties of and the performance of PGMs HEAs; however, shortcomings associated with HEAs formation is the fabrication method which requires the specialized synthesis method. It was observed during the review of this book chapter that, the new facile method, low temperature and pressure method is required for the development of HEAs.

5. Energy storage: MOFs

MOFs are known as coordination polymers developed from solid-state/zeolite chemistry and coordination chemistry [87, 88]. The use of MOFs in gas is extremely important because they are porous materials [89]. Ever since the establishment of MOFs by Hoskins and Robson, MOFs have grabbed great attention in porous materials because of their ability to store gases. Yaghi et al. went on to popularize research on MOFs after that, particularly after MOF-5 was reported [90]. It is theoretically possible to construct a MOF that is well-suited for a desired application (such as sensors, catalysis or separation) by carefully choosing nodes and linkers (including flaws, such as missing nodes and/or linkers). It's noteworthy that MOFs have an ad hoc naming scheme that often uses numbers (e.g. MOF-5) or names derived from the universities from which they originated (e.g. NU-1000) [91].

5.1 Properties of MOFs

Metal-organic frameworks (MOFs) are crystalline substances with a high surface area, high porosity, and the ability to efficiently adsorb hydrogen. At 77 K, MOFs exhibit good gravimetric hydrogen capabilities, and some of them have exceeded the US Department of Energy (DOE) objective. Numerous MOFs have been reported to display permanent porosity with pore windows between 5 and 25 Å. Through ligand extension, MOFs with large interior surface areas, extending 10,000 m^2/g, and extremely high porosity (up to 90% free volume) have been created. The measurement of gas isotherms has proven crucial in determining if permanent porosity has been achieved after guests have left. MOFs can be created with the right tailoring to function as extremely selective molecular sieves, sensors or catalysts. Among other capabilities of MOFs, gas storage is one of the most promising uses for metal-organic frameworks. As shown in **Figure 1**, the isotherm forms, which are typically Type I with little to no hysteresis, show that durable microporous structures exist under reversible gas physisorption of tiny molecules (**Figure** 7).

5.2 Metal hydrides for hydrogen storage

Due to its high gravimetric energy density and environmental benefits, hydrogen has been proposed as a promising alternative to the widely used fossil fuels as an energy source [92]. Hydrogen can be generated and separated from a variety of sources including water, fossil fuels and biomass [93]. The use of hydrogen as an energy carrier can however be impeded by the lack of safe, energy-efficient and cost-effective storage systems [94]. The most common storage modalities of hydrogen include (i) pressurized gas, (ii) cryogenic liquid and (iii) solid fuel via adsorption onto porous materials [95]. Since the storage of hydrogen in MOFs was innovatively proposed by Rosi et al. [96], a plethora of research has been conducted on the

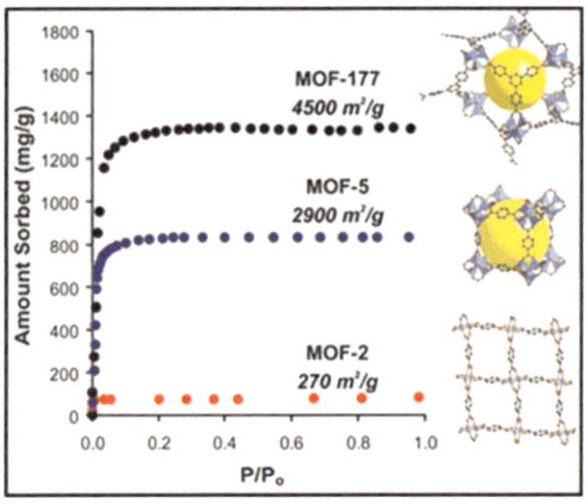

Figure 7.
Schematic representation of type I isotherms of MOFs (MOF 177, MOF-5, MOF-2) N2 sorption measured at −196,15°C showing the conformal microspores. Reprinted with permission from literature.

modification and application of MOFs as hydrogen storage materials [97–99]. The tunability, topological structure and nanoconfined environments of MOFs provide ideal conditions for hydrogen capture, storage and release with considerable safety, convenience and efficiency [100]. Other properties of MOFs that make them attractive for hydrogen storage include their porous structures, high specific surface areas, exposed metal nodes, facile fabrication procedures, controllable chemical functionality and amenability to scale-up [101]. Although MOFs are favourable for hydrogen storage, high storage capacities (up to 4.5–7.5 wt%) are normally achieved at cryogenic temperatures (77 K) and high pressures [87, 94, 102]. The strong pressure and temperature dependence as well as storage capacity requirements of hydrogen physisorption on MOFs therefore limit their practical application [103]. Several approaches have thus been considered to improve the hydrogen storage capacities of MOFs, including the formation of composites by adding dopants and substituting the metal nodes within the MOFs [104]. Some researchers have even proposed nanoconfinement of other materials inside MOFs as an alternative approach to enhancing hydrogen storage [105]. Alternatively, several other materials have emerged as feasible candidates for efficient hydrogen storage such as in metal hydrides. Hydrogen forms metal hydrides with some metals and alloys leading to solid-state storage under moderate temperature and pressure [106]. The reaction of hydrogen with a metal to form a metal hydride results in the generation of heat, i.e. an exothermic reaction. When hydrogen is then required, the stored heat (ΔH) is utilized to release hydrogen from the hydride in an endothermic reaction [107]. Since hydrogen becomes part of the chemical structure of the metal, cryogenic temperature or high pressure are not required to break these chemical bonds; hence, most metal hydrides absorb and desorb hydrogen at ambient temperature and close to atmospheric pressure [108]. Metal hydrides have a higher hydrogen storage density than gaseous or liquid hydrogen; hence, they are volume-efficient storage materials [108]. The use of metal hydrides for hydrogen storage is also favourable over-pressurized gas and other hydrogen storage methods because of their gravimetric and volumetric storage capacities and safe operating pressures [21]. Nanostructured metal hydrides have particularly gained attention due to their improved reversibility, altered heats of hydrogen absorption/desorption and nano interfacial reaction pathways with fast rates [109].

Examples of commonly used metal hydrides as solid-state hydrogen storage materials are binary metal hydrides which include MgH_2, TiH_2 and AlH_3 [110, 111]. The low cost and good reversibility of MgH_2 have made it particularly popular, as well as the fact that it holds the highest energy density (9 MJ/kg) among all the reversible hydrides that apply to hydrogen storage [112]. The sorption kinetics involved in hydrogen generation with MgH_2 are however quite sluggish and it has a high thermodynamic stability requiring temperatures that exceed 300°C for the desorption of hydrogen [113]. Upon alloying, metal hydrides exist either as intermetallic or complex hydrides (alanates, borohydrides and nitrides) which have also been more extensively studied due to their high hydrogen storage capacities [109].

Alternative metals have also been proposed as additives to aid the shortcomings of these metal hydrides. Palladium, for instance, acts not only as a catalyst to facilitate the uptake and dissociation of hydrogen in other metal hydrides but it can also protect the surface from corrosion [28]. Although palladium can absorb large volumetric quantities of hydrogen at room temperature and atmospheric pressure to form palladium hydride, it (as well as other PGMS) is not often considered as a sole hydrogen storage material since it is somewhat expensive and has a low gravimetric hydrogen density [113]. Extensive research has been conducted on the use and optimization of various metal hydrides to optimize the conditions and increase the hydrogen storage capacity of these materials as shown in **Table 4**.

5.3 Application of MOFs and metal hydrides in batteries

The power output of these renewable energy resources such as solar, hydro and wind power is highly fluctuating and intermittent which invites the parallel implementation of electrochemical energy conversion and storage technologies, such as rechargeable batteries [122]. Such storage technologies make sustainable energy utilization easy and efficient [123]. Rechargeable lithium-ion batteries (LIBs) with zero emissions, now particularly dominate the energy storage and conversion devices market, which reduces our reliance on conventional energy resources [124]. The development of high-capacity electrode materials for LIBs, however, is still necessary to meet the sustained growing demand for energy. Thus, research centred on the optimization of efficient conducting materials in energy storage devices such as batteries has soared.

Hydride type	Hydride	T_{abs} (°C)	P_{abs} (bar)	Hydrogen storage capacity (wt%)	Ref
Binary hydrides	MgH_2	30	30	6.7	[114]
Complex hydrides	$LiBH_4$	300	100	9.20	[115]
	$Mg(BH_4)_2$	290–350	1	14.82	[116]
	$Ca(BH_4)_2$	350–400	30–50	9.60	[117]
	$NaBH_4$	350	40	10.80	[118]
Intermetallic hydrides	$LaNi_{4.5}Sn_{0.5}$	25	7.50	0.95	[119]
	TiFe + X wt.% Zr	25	—	1.20	[120]
	$Ti_{1.1}CrMn$	23	330	1.80	[121]

Table 4.
Summary of hydrogen storage parameters and efficiencies of different metal hydrides.

The high porosity, versatile functionalities, diverse structures and controllable chemical compositions of MOFs offer various possibilities for generating adequate electrode materials for rechargeable batteries [125]. The porous structure of MOFs enables a facile electrolyte penetration and ion transportation, while the designable components promise the incorporation of electroactive sites, offering infinite possibilities for the search for candidate electrode materials for different battery systems [126]. Despite these attractive features, MOFs (and their derivatives) as electrode materials face various challenging issues, which impede their practical applications. They suffer from poor electrical conductivity, low tap density, and irreversible structural degradation upon the charge/discharge processes [125].

Metal hydrides on the other hand have been widely investigated not only in LIBs but also in nickel-metal hydrides (Ni-MHy) batteries [127, 128]. Due to their attractive properties, Ni-MHy batteries have often been used for both electric and hybrid vehicles because they provide several advantages compared to lead-acid batteries [129]. In the conventional Ni-MH battery configuration, the charge-discharge processes occur as depicted in **Figure 8**.

Metal hydrides have received increasing interest as materials in these kinds of batteries, both as electrodes and ion conductors [130]. Metal hydride-based materials have the potential to be negative materials for LIBs, owing to their high theoretical Li storage capacity, relatively low volume expansion, and suitable working potential with very small polarization [131]. They also owe their efficiency to their large specific capacity and low voltage hysteresis compared to other conversion materials used for LIBs [124]. Among various metal hydrides, MgH_2 has particularly gained popularity as an anode material for LIBs [132]. Various modification techniques including the addition of TiH_2, as well as other materials (catalyst/carbon or suitable binders) and nanocrystallization, have been implemented to remedy the drawbacks associated with the electrochemical performances of metal hydrides [133]. These include their kinetics

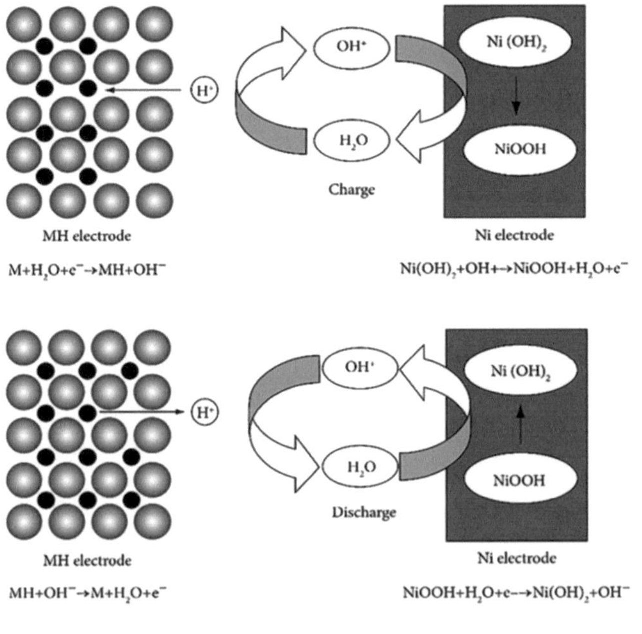

Figure 8.
Ni-MHy batteries. Reprinted with permission from [129].

limitations, structural reorganization, capacity fading and volumetric change during the discharge/charge process [124].

For instance, Yang et al. fabricated MgH_2-based composites with expanded graphite (EG) and TiO_2 using a plasma-assisted milling process to improve the electrochemical performance of MgH_2. The resulting MgH_2–TiO_2–EG composites showed an increase in the initial discharge capacity and cycling capacity compared with a pure MgH_2 electrode. A stable discharge capacity of 305.5 mAh·g^{-1} could be achieved after 100 cycles for the 20 h-milled MgH_2–TiO_2–EG-20 h composite electrode and the reversibility of the conversion reaction of MgH_2 could be greatly enhanced [134].

Additionally, Mo et al. designed a three-dimensional hierarchical metal hydride/graphene composite ($LiNa_2AlH_6$/3DG) that showed outstanding cycling stability with $LiBH_4$ as a solid electrolyte. An ultra-high capacity of 861 mAh·g^{-1} at the current density of 5 A g^{-1} and a long cycle life of 500 cycles with capacity retention of 97% was achieved [135]. Moreover, Weeks et al. analysed and compared the physical and electrochemical properties of an all-solid-state cell utilizing $LiBH_4$ as the electrolyte and aluminium as the active anode material. An initial capacity of 895 mAh·g^{-1} was observed and is close to the theoretical capacity of aluminium due to the formation of a LiAl (1:1) alloy. This demonstrated the possibility of utilizing other high-capacity anode materials with a $LiBH_4$-based solid electrolyte in all-solid-state batteries [136].

5.4 Challenges and advantages of MOFs and metal hydride

Many beneficial features have evolved into a wide variety of MOFs that is high degree of porosity, high surface area, flexible architecture, multifunctional chemical properties and tuneable structure. MOFs are widely used in many applications, including catalysts, supercapacitor, adsorbents, sensors, environmental protection and drug delivery, due to their simplicity of design and homogeneous and fine-tuneable pore architectures. However, a number of disadvantages have also limited the use of these innovative materials in practical applications, including high production costs, poor selectivity, low capacity and challenges with recycling and regeneration. Furthermore, poor electrical conductivity and stability of conventional MOFs inhibit development and application. With many short coming associated with MOFs, synthetic challenges in the discipline emerge from understanding and regulating both structural and compositional complexity because of the enormous array of conceivable topologies and compositions. In addition, despite the enormous variety of known structural types, there have been no definitive findings of shape-selective catalysis in MOFs although zeolite community has a highly established understanding of shape-selective catalysis. Zhang et al. reported the attempts to improve MOFs shapes and widen their application through the introduction of graphene as a template to grow the MOFs which avoids agglomeration, and diminish poor electrical conductivity and stability [137].

6. A computational approach for conversion: Density functional theory and artificial neural network

Machine learning is one of the most powerful too of artificial intelligence and has been utilized for numerical prediction, classification, efficiency and pattern recognition. Among ML tools, artificial neural network (ANN) has become the popular nonlinear algorithm, adaptive structure, tunable and easy to train for various

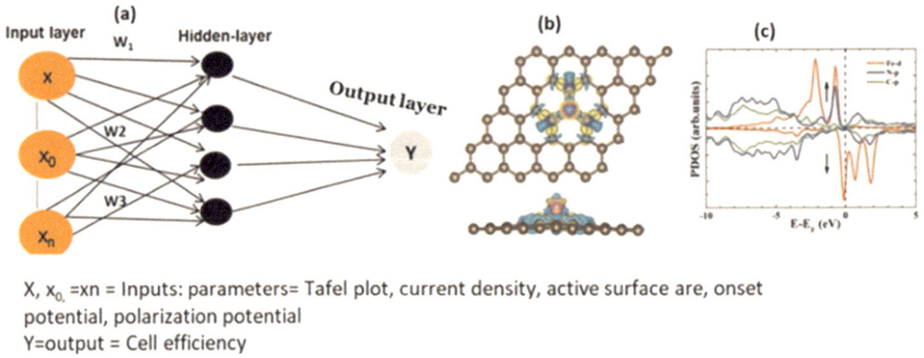

Figure 9.
Typical example of feed-forward approach using artificial neural network architecture for input/output variables, output representing system efficiency, (b) schematic cartoon illustrating the possible adsorption site (charge depletion, and charge accumulation), (c) partial PDOS showing ORR. Reprinted from the literature [139].

applications such as catalyst, biology and energy [138]. In general, ANN architecture is comprised of at least three (3) layers, i.e. input layer, hidden layer and output layer as depicted in **Figure 8**. Each layer contains numerous neurons that connect to the next layer, where they connect it represented by weights. As depicted in **Figure 9**, ANN is composed of single layers (Hidden layers) that represent the parameters such as current and cell potential. The output is represented by the efficiency of the system. ML utilizing the artificial neural network has attracted considerable interest in energy conversion and electrocatalytic reaction to predict the throughput using the algorithms. ML has attracted many researchers in the field of catalysis. Lu et al. leverage the use of neural networks and density functional theory (DFT) for predicting the surface defects for oxygen reduction reaction. Mehiritz et al. report the first model of electrooxidation of ethanol using an artificial neural network (ANN) utilizing the differential evolution (DE) algorithm. The best results (Model) were obtained with a single hidden layer [140].

In addition, theoretical techniques, such as quantum-chemical modeling employing density functional theory (DFT, Vienna ab initio simulation package (VASP) methodologies, are used to expedite the process of finding the ideal material. This route enables us to evaluate different catalytic materials without using experimental analysis, as well as determine the catalyst's adsorption qualities and the impact of the material's composition and structure on the kinetics of the catalytic process. Nrskov and co-authors first use DFT-based techniques to compute the adsorption energy of reactants and intermediates. In this section, the efforts to identify the highly active catalysts and relevant electrocatalysis reaction mechanisms have been studied, particularly focusing on the d-band centre location, the electronegativity of the central atom to the neighboring, along with density of state (DOS). These theoretical properties are utilized in electrochemical systems to determine the appropriate descriptor for catalytic activity, specifically towards the ORR and OER. As illustrated in **Figure 9**, this is the typical schematic diagram depicting the DOS pattern and d-band centre level. There has been extensive research on the theoretical modeling of energy conversion and storage to screen the catalyst process. For example, Sunday et al. reported reactive molecular dynamics on surface PtNiFe heteroatoms to model the catalyst mechanism. Based on the energy barrier, it was found that NiO efficient than monometallic Ni and Pt, the HOads and Hads are likely easier on NiO during water oxidation (dissociation) [141]. Mkhohlakali et al. used the DFT and found the appropriate descriptor for the EOR catalyst process and the geometric and electronic effect has been found as the best descriptor for PdTe towards ethanol oxidation in alkaline media [15]. Furthermore, DFT has also been used for energy storage, particularly hydrogen storage. Kabelo and co-workers have described the hydrogen hybridization between boron and yttrium in yttrium-doped borophene adsorption on borophene materials using DFT indicating the potential use of borophene for hydrogen storage [142]. In addition, adsorption energies, diffusion barriers and reaction activation barriers for reactants, intermediates and products on the catalyst surface are very important microscopic characteristics that can be obtained from first principles calculations based on the density functional theory (DFT) in heterogeneous catalysis. In many energy conversion processes, DFT is used to estimate the rates of the fundamental processes and identify the rate-limiting phases. Shen et al. employed DFT to predict the mechanism for CO_2 reduction on Cobalt porphyrin and it was found that the key intermediates formed when Co is in CO1 and the results were agreed with the experimental results with more details [143]. There is an increasing search for prediction of CO_2 reduction using DFT, and Gao et al. reported the theoretical calculation for CO_2 to CO on Co-quaterypyridine complex surface [144]. In the case of metal hydrides, there is yet no material that can fully satisfy the requirements for the practical onboard application, despite the fact that solid-interstitial and non-interstitial (state) hydrides have made notable advancements in terms of materials synthesis, mechanistic understanding, and performance enhancement.

7. Conclusion and future perspective

Based on the reported research trends, new emerging materials such as HEAs and metal hydrides show the potential and future technology in energy conversion and storage respectively. The second generation of HEAs illustrate the higher electrocatalytic efficiency and stability for both anode and cathode materials because of increased active sites, enhanced reaction kinetics and CO- poison tolerance dues to single-solid solution as compared to conversional PGMs alloys. Both the electrochemical and physicochemical priorities affect the efficiency of energy conversion and storage. The future work following this trend is the consideration of "giant" HEAs compounds. Furthermore, the overall findings therefore not only depict the use of metal hydrides as anode materials but also as solid-state electrolytes. Beyond the conventional Ni-MHy battery, the potential of these hydrides has been demonstrated at a lab-scale upon optimization of a range of innovative energy storage concepts, including the MgH_2-Li chemistry. Extensive research however still needs to be conducted to promote the development of new batteries of higher energy density and lower cost. Computer modeling demonstrates a better approach to understanding the interfacial electrochemical properties and physico-chemical properties. In addition, the efficiency of the electrochemical systems was well predicted machine learning through artificial neural networks, which shows the efficient potential approach to predict the efficiency of fuel cell, hydrogen storage and batteries. Among the algorithm approach, the differential evolution utilizing the single hidden layer shows the efficient and electrocatalytic reaction kinetics prediction. The modeling of structural formation and electrochemical properties of metal hydrides, high entropy and MOFs should be mandatory to strengthen the synthesis and better understanding of electrocatalysts. Although advances centred on solving energy problems have soared with the discovery and implementation of various emerging catalysts, further work still needs to be conducted to ensure sustainable energy systems and technologies. In addition to high efficiency, catalysts for energy generation and storage should result in technologies that are affordable, abundant and generate clean energy.

Author details

Andile Mkhohlakali[1]*, Nonhlahla Ramashala[2], Sivuyisiwe Mapukata[3], Sanele Nyembe[4] and Lebohang Hlatshwayo[1]

1 Analytical Chemistry Division, Mintek, Randburg, Johannesburg, South Africa

2 Council for Scientific and Industrial Research (CSIR), Energy Center, Pretoria, South Africa

3 DSI/Mintek Nanotechnology Innovation Centre, Advanced Materials Division, Mintek, Randburg, Johannesburg, South Africa

4 Molecular Science Institute, School of Chemistry, University of the Witwatersrand, South Africa

*Address all correspondence to: andilem@mintek.co.za

References

[1] Yang M, et al. Zero→Two-Dimensional metal nanostructures: An overview on methods of preparation, characterization, properties, and applications. Nanomaterials. 2021

[2] Wang Y, Chen KS, Mishler J, Cho SC, Adroher XC. A review of polymer electrolyte membrane fuel cells: Technology, applications, and needs on fundamental research. Applied Energy. 2011;**88**(4):981-1007. DOI: 10.1016/j.apenergy.2010.09.030

[3] Mudd GM. Key trends in the resource sustainability of platinum group elements. Ore Geology Reviews. 2012; **46**:106-117. DOI: 10.1016/j.oregeorev.2012.02.005

[4] Hughes AE, Haque N, Northey SA, Giddey S. Platinum group metals: A review of resources, production and usage with a focus on catalysts. Resources. 2021;**10**(9):1-40. DOI: 10.3390/resources10090093

[5] Grandell L, Lehtilä A, Kivinen M, Koljonen T, Kihlman S, Lauri LS. Role of critical metals in the future markets of clean energy technologies. Renewable Energy. 2016;**95**:53-62. DOI: 10.1016/j.renene.2016.03.102

[6] Sudarsono W et al. From catalyst structure design to electrode fabrication of platinum-free electrocatalysts in proton exchange membrane fuel cells: A review. Journal of Industrial and Engineering Chemistry. 2023;**122**:1-26. DOI: 10.1016/j.jiec.2023.03.004

[7] Wang YJ, Fang B, Li H, Bi XT, Wang H. Progress in modified carbon support materials for Pt and Pt-alloy cathode catalysts in polymer electrolyte membrane fuel cells. Progress in Materials Science. 2016;**82**: 445-498. DOI: 10.1016/j.pmatsci.2016.06.002

[8] Kim Y et al. Fabrication of platinum group metal-free catalyst layer with enhanced mass transport characteristics via an electrospraying technique. Materials Today Energy. 2021;**20**: 100641. DOI: 10.1016/j.mtener.2021.100641

[9] Abid N et al. Synthesis of nanomaterials using various top-down and bottom-up approaches, influencing factors, advantages, and disadvantages: A review. Advances in Colloid and Interface Science. 2021, 2022;**300**: 102597. DOI: 10.1016/j.cis.2021.102597

[10] Nyembe S, Mpelane S, Shumbula P, Harris R, Moloto N, Sikhwivhilu L. The effects of gold seeds stabilizing agent on gold nanostructures morphologies. Materials Today Proceedings. 2015;**2**(7): 4149-4157. DOI: 10.1016/j.matpr.2015.08.045

[11] Khan MAR, Al Mamun MS, Ara MH. Review on platinum nanoparticles: Synthesis, characterization, and applications. Microchemical Journal. 2021;**171**:106840. DOI: 10.1016/j.microc.2021.106840

[12] Rompa E. An in Depth Look at Hungary, an Depth Look Hungary. 2019. pp. 1-134

[13] Antolini E. Carbon supports for low-temperature fuel cell catalysts. Applied Catalysis B: Environmental. 2009;**88**(1–2):1-24. DOI: 10.1016/j.apcatb.2008.09.030

[14] Zhang W et al. Electrodeposited platinum with various morphologies on carbon paper as efficient and durable

self-supporting electrode for methanol and ammonia oxidation reactions. International Journal of Hydrogen Energy. 2022;**48**(7):2617-2627. DOI: 10.1016/j.ijhydene.2022.10.157

[15] Mkhohlakali A, Fuku X, Seo MH, Modibedi M, Khotseng L, Mathe M. Electro-design of bimetallic PdTe electrocatalyst for ethanol oxidation: Combined experimental approach and ab initio density functional theory (DFT)—Based study. Nanomaterials. 2022; **12**(20):3607. DOI: 10.3390/nano12203607

[16] Pavlenko V et al. A comprehensive review of template-assisted porous carbons: Modern preparation methods and advanced applications. Materials Science & Engineering R. 2022;**149**. DOI: 10.1016/j.mser.2022.100682

[17] Saod WM, Oliver IW, Thompson DF, Holborn S, Contini A, Zholobenko V. Magnesium oxide loaded mesoporous silica: Synthesis, characterisation and use in removing lead and cadmium from water supplies. Environmental Nanotechnology, Monitoring & Management. 2023;**20**: 100817. DOI: 10.1016/j.enmm.2023.100817

[18] Gong S, Zhang Y, Niu Z. Recent advances in earth-abundant core/noble-metal shell nanoparticles for electrocatalysis. 2020. DOI: 10.1021/acscatal.0c02587

[19] Ren X, Lv Q, Liu L. Sustainable energy & fuels current progress of Pt and Pt-based electrocatalysts used for fuel cells. 2020. pp. 15–30. DOI: 10.1039/c9se00460b

[20] Habibi B, Mohammadyari S. Facile synthesis of Pd nanoparticles on nano carbon supports and their application as an electrocatalyst for oxidation of ethanol in alkaline media: The effect of support. International Journal of Hydrogen Energy. 2015;**40**(34): 10833-10846. DOI: 10.1016/j.ijhydene.2015.07.021

[21] Cui P, Zhao L, Long Y, Dai L, Hu C. Carbon-based electrocatalysts for acidic oxygen reduction. Angewandte Chemie. 2023;**62**:202218269. DOI: 10.1002/anie.202218269

[22] Zhang L, Lee K, Zhang J. The effect of heat treatment on nanoparticle size and ORR activity for carbon-supported Pd –Co alloy electrocatalysts. Electrochimica Acta. 2007;**52**: 3088-3094. DOI: 10.1016/j.electacta.2006.09.051

[23] Zhang L, Chang Q, Chen H, Shao M. Nano energy recent advances in palladium-based electrocatalysts for fuel cell reactions and hydrogen evolution reaction. Nano Energy. 2016;**29**:198-219. DOI: 10.1016/j.nanoen.2016.02.044

[24] Wang Y, Wang D, Li Y. Rational design of single-atom site electrocatalysts: from theoretical understandings to practical applications. 2021:1-38. DOI: 10.1002/adma.202008151

[25] Cao X, Huo J, Li L, Qu J, Zhao Y, Chen W, et al. Recent advances in engineered Ru-based Electrocatalysts for the hydrogen/oxygen conversion reactions. Advanced Energy Materials. 2022;**12**:2202119. DOI: 10.1002/aenm.202202119

[26] Bligaard T, Logadottir A, Kitchin JR, Chen JG, Pandelov S, Stimming U. 2005;(3). DOI: 10.1149/1.1856988

[27] Jung N, Young D, Ryu J, Jong S, Sung Y. Pt-based nanoarchitecture and catalyst design for fuel cell applications.

Nano Today. 2014;**9**(4):433-456. DOI: 10.1016/j.nantod.2014.06.006

[28] Lu Z, Shi Y, Shen L, Tan H. The acidic OER activation-decay process of highly active Ir e Ni mixed oxide modified by capping agent for both particle fining and Ir e OH formation. International Journal of Hydrogen Energy. 2022;**48**(21):7549-7558. DOI: 10.1016/j.ijhydene.2022.11.112

[29] Cherevko S et al. Oxygen and hydrogen evolution reactions on Ru, RuO_2, Ir, and IrO_2 thin film electrodes in acidic and alkaline electrolytes: A comparative study on activity and stability. Catalysis Today. 2016;**262**: 170-180

[30] Cruz JC et al. Nanosized IrO_2 electrocatalysts for oxygen evolution reaction in an SPE electrolyzer. 2011: 1639-1646. DOI: 10.1007/s11051-010-9917-2

[31] Ortel E, Reier T, Strasser P, Kraehnert R. Mesoporous IrO_2 films Templated by PEO-PB-PEO block-copolymers: Self-assembly, crystallization behavior, and Electrocatalytic performance. Chemistry of Materials. 2011;**23**:3201–3209

[32] Nong HN et al. Oxide-supported IrNiO x Core – Shell particles as efficient, cost- effective, and stable catalysts for electrochemical water splitting. Angewandte Chemie. 2015: 2975-2979. DOI: 10.1002/anie.201411072

[33] Lee Y, Suntivich J, May KJ, Perry EE, Shao-Horn Y. Synthesis and activities of rutile IrO_2 and RuO_2 nanoparticles for oxygen evolution in acid and alkaline solutions. The Journal of Physical Chemistry Letters. 2012;**3**:399-404

[34] Gyeom I, Wook I, Oh I, Park S. Crumpled rGO-supported Pt-Ir bifunctional catalyst prepared by spray pyrolysis for unitized regenerative fuel cells. Journal of Power Sources. 2017; **364**:215-225. DOI: 10.1016/j.jpowsour.2017.08.015

[35] Du L, Xing L, Zhang G, Dubois M, Sun S. Strategies for engineering high-performance PGM-free catalysts toward oxygen reduction and evolution reactions. Small Methods. 2020:1-28. DOI: 10.1002/smtd.202000016

[36] Hong S et al. Active motif change of Ni-Fe spinel oxide by Ir doping for highly durable and facile oxygen evolution reaction. DOI: 10.1002/adfm.202209543

[37] Avc N, Sementa L, Fortunelli A. Mechanisms of the oxygen evolution reaction on $NiFe_2O_4$ and $CoFe_2O_4$ inverse-spinel oxides. ACS Catalysis. 2022;**12**:9058-9073. DOI: 10.1021/acscatal.2c01534

[38] Huang Y, Zhou T, Hu Y, Yang Y, Yang F. Ordered derivatives on Ti surface enhance the OER activity and stability of Ru-based film electrode. International Journal of Hydrogen Energy. 2023;**xxxx**. DOI: 10.1016/j.ijhydene.2023.05.108

[39] Woo S, Baik C, Kim T, Pak C. Three-dimensional mesoporous Ir – Ru binary oxides with improved activity and stability for water electrolysis. Catalysis Today. 2020;**352**:39-46. DOI: 10.1016/j.cattod.2019.10.004

[40] Lim J et al. Amorphous Ir atomic clusters anchored on crystalline IrO2 nanoneedles for proton exchange membrane water oxidation. Journal of Power Sources. 2022;**524**:231069. DOI: 10.1016/j.jpowsour.2022.231069

[41] Liu B, Wang S, Wang C, Chen Y, Ma B, Zhang J. Surface morphology and

electrochemical properties of 2 RuO_2-doped Ti/IrO_2-ZrO_2 anodes for oxygen evolution. Journal of Alloys and Compounds. 2018;**778**:593-602. DOI: 10.1016/j.jallcom.2018.11.191

[42] Niu S et al. Applied catalysis B: Environmental low Ru loading RuO_2/(Co, Mn)$_3O_4$ nanocomposite with modulated electronic structure for efficient oxygen evolution reaction in acid. Applied Catalysis B: Environmental. 2021;**297**:120442. DOI: 10.1016/j.apcatb.2021.120442

[43] Abrham G, Martínez-huerta MV, Jesus M. Recent progress on bimetallic NiCo and CoFe based electrocatalysts for alkaline oxygen evolution reaction: A review. Journal of Energy Chemistry. 2022;**67**:101-137. DOI: 10.1016/j.jechem.2021.10.009

[44] Vazhayil A, Vazhayal L, Thomas J, Shyamli Ashok C, Thomas N. A comprehensive review on the recent developments in transition metal-based electrocatalysts for oxygen evolution reaction. Applied Surface Science Advances. 2021;**6**:100184. DOI: 10.1016/j.apsadv.2021.100184

[45] Xu J, Liu G, Li J, Wang X. The electrocatalytic properties of an IrO_2/SnO_2 catalyst using SnO_2 as a support and an assisting reagent for the oxygen evolution reaction. Electrochimica Acta. 2012;**59**:105-112. DOI: 10.1016/j.electacta.2011.10.044

[46] Ai L et al. Robust interfacial Ru-RuO_2 heterostructures for highly efficient and ultrastable oxygen evolution reaction and overall water splitting in acidic media. Journal of Alloys and Compounds. 2022;**902**:163787. DOI: 10.1016/j.jallcom.2022.163787

[47] Xie Y, Su Y, Qin H, Cao Z, Wei H. Ir-doped Co_3O_4 as efficient electrocatalyst for acidic oxygen evolution reaction. International Journal of Hydrogen Energy. 2023;**48**(39):14642-14649. DOI: 10.1016/j.ijhydene.2022.12.292

[48] Yao Z, Tang T, Jiang Z, Wang L, Hu J, Wan L. Electrocatalytic hydrogen oxidation in alkaline media: From mechanistic insight to catalyst design. ACS Nano. 2022;**16**:5153-5183. DOI: 10.1021/acsnano.2c00641

[49] Munonde TS, Zheng H, Nomngongo PN. Ultrasonics - Sonochemistry ultrasonic exfoliation of NiFe LDH/CB nanosheets for enhanced oxygen evolution catalysis. Ultrasonics Sonochemistry. 2019;**59**:104716. DOI: 10.1016/j.ultsonch.2019.104716

[50] AM. Design. Towards the computational design of solid catalysts. Nature Chemistry. 2009;**1**:37-46. DOI: 10.1038/nchem.121

[51] Kaya D et al. Electrocatalytic hydrogen evolution on metallic and bimetallic Pd e Co alloy nanoparticles. International Journal of Hydrogen Energy. 2023;**48**(39):14633-14641. DOI: 10.1016/j.ijhydene.2023.01.049

[52] Capozzoli L et al. Ruthenium-loaded titania nanotube arrays as catalysts for the hydrogen evolution reaction in alkaline membrane electrolysis. Journal of Power Sources. 2022, 2023;**562**:232747. DOI: 10.1016/j.jpowsour.2023.232747

[53] Zhan G, Yao Y, Quan F, Gu H, Liu X, Zhang L. D-band frontier: A new hydrogen evolution reaction activity descriptor of Pt single-atom catalysts. Journal of Energy Chemistry. 2022;**72**:203-209. DOI: 10.1016/j.jechem.2022.05.012

[54] Dubouis N, Grimaud A. The hydrogen evolution reaction: From

material to interfacial descriptors. Chemical Science. 2019:9165-9181. DOI: 10.1039/c9sc03831k

[55] Zhong W et al. Ultralow-temperature assisted synthesis of single platinum atoms anchored on carbon nanotubes for efficiently electrocatalytic acidic hydrogen evolution. Journal of Energy Chemistry. 2020;**51**: 280-284. DOI: 10.1016/j.jechem.2020.04.035

[56] Liu G, Wang C, Wang J. Recent advances in nanostructured electrocatalysts for hydrogen evolution reaction. Rare Metals. 2021;**40**(12): 3375-3405. DOI: 10.1007/s12598-021-01735-y

[57] Wu H, Huang Q, Shi Y, Chang J. Electrocatalytic Water Splitting: Mechanism and Electrocatalyst. Nano Research. 2023;**16**:9142-9157. DOI: 10.1007/s12274-023-5502-8

[58] Ekspong J, Gracia-espino E, Wa T. Hydrogen evolution reaction activity of heterogeneous materials: A theoretical model. The Journal of Physical Chemistry C. 2020;**124**:20911-20921. DOI: 10.1021/acs.jpcc.0c05243

[59] Quaino P, Juarez F, Santos E, Schmickler W. Volcano plots in hydrogen electrocatalysis – Uses and abuses. 2014:846-854. DOI: 10.3762/bjnano.5.96

[60] Rahman ST, Rhee KY, Park S-J. Nanostructured multifunctional electrocatalysts for efficient energy conversion systems: Recent perspectives. Nanotechnology Reviews. 2021;**10**: 137-157

[61] You J, Yao R, Ji W, Zhao Y, Wang Z. Research of high entropy alloys as electrocatalyst for oxygen evolution reaction. Journal of Alloys and Compounds. 2022;**908**:164669. DOI: 10.1016/j.jallcom.2022.164669

[62] Wu D et al. On the electronic structure and hydrogen evolution reaction activity of platinum group metal-based high-entropy-alloy nanoparticles. Chemical Science. 2020; **11**(47):12731-12736. DOI: 10.1039/d0sc02351e

[63] Li D et al. High-entropy Al0.3CoCrFeNi alloy fibers with high tensile strength and ductility at ambient and cryogenic temperatures. Acta Materialia. 2017;**123**:285-294. DOI: 10.1016/j.actamat.2016.10.038

[64] Haruna AB, Onoh E, Ozoemena KI. Emerging high-entropy materials as electrocatalysts for rechargeable zinc-air batteries. Current Opinion in Electrochemistry. 2023;**39**:101264. DOI: 10.1016/j.coelec.2023.101264

[65] High R et al. Microstructural evolution and phase formation in 2nd-generation refractory-based High entropy alloys. Materials (Basel). 2018; **11**(75):1-13. DOI: 10.3390/ma11020175

[66] Pradeep KG, Deng Y, Li Z, Raabe D, Tasan CC. Metastable high-entropy dual-phase alloys overcome the strength–ductility trade-off. Nature. 2016:1-8. DOI: 10.1038/nature17981

[67] Wang K et al. Recent progress in high entropy alloys for electrocatalysts. Electrochemical Energy Reviews. 2022;**5** (s1):1-29. DOI: 10.1007/s41918-022-00144-8

[68] Güler S, Alkan ED, Alkan M. Vacuum arc melted and heat treated AlCoCrFeNiTiX based high-entropy alloys: Thermodynamic and microstructural investigations. Journal of Alloys and Compounds. 2022;**903**. DOI: 10.1016/j.jallcom.2022.163901

[69] Kasar AK, Scalaro K, Menezes PL. Tribological properties of high-entropy alloys under dry conditions for a wide temperature range—A review. Materials (Basel). 2021;**14**(19). DOI: 10.3390/ma14195814

[70] Xin Y et al. High-entropy alloys as a platform for catalysis: Progress, challenges, and opportunities. ACS Catalysis. 2020;**10**(19):11280-11306. DOI: 10.1021/acscatal.0c03617

[71] He B, Zu Y, Mei Y. Design of advanced electrocatalysts for the high-entropy alloys: Principle, progress, and perspective. Journal of Alloys and Compounds. 2023;**958**:170479. DOI: 10.1016/j.jallcom.2023.170479

[72] Yao Y, Huang Z, Xie P, Lacey SD, Jacob RJ, Xie H, et al. Carbothermal shock synthesis of high-entropy-alloy nanoparticles. Science. 2018;**359**: 1489-1494

[73] Song J, Kim C, Kim M, Cho KM, Gereige I. Generation of high-density nanoparticles in the carbothermal shock method. 2021;**2984**:16-19

[74] Li S et al. Nanoporous high-entropy alloys with low Pt loadings for high-performance electrochemical oxygen reduction. Journal of Catalysis. 2020;**383**: 164-171. DOI: 10.1016/j.jcat.2020.01.024

[75] Feng G et al. Sub-2 nm ultrasmall high-entropy alloy nanoparticles for extremely superior Electrocatalytic hydrogen evolution. Journal of the American Chemical Society. 2021; **143**(41):17117-17127. DOI: 10.1021/jacs.1c07643

[76] Abdelhafiz A, Wang B, Harutyunyan AR, Li J. Carbothermal shock synthesis of High entropy oxide catalysts: Dynamic structural and chemical reconstruction boosting the catalytic activity and stability toward oxygen evolution reaction. 2022:10-15. DOI: 10.1002/aenm.202200742

[77] Mishra K, Devi N, Siwal SS, Thakur VK. Insight perspective on the synthesis and morphological role of the noble and non-noble metal-based electrocatalyst in fuel cell application. Applied Catalysis B: Environmental. 2023;**334**:122820. DOI: 10.1016/j.apcatb.2023.122820

[78] Wang D et al. Low-temperature synthesis of small-sized high-entropy oxides for water oxidation. Journal of Materials Chemistry A. 2019: 24211-24216. DOI: 10.1039/c9ta08740k

[79] Chen X et al. Multi-component nanoporous platinum e ruthenium e copper e osmium e iridium alloy with enhanced electrocatalytic activity towards methanol oxidation and oxygen reduction. Journal of Power Sources. 2015;**273**:324-332. DOI: 10.1016/j.jpowsour.2014.09.076

[80] Li SY et al. Sputter-deposited High entropy alloy thin film Electrocatalyst for enhanced oxygen evolution reaction performance. Nano Micro Small. 2022; **469**(39):144015. DOI: 10.1002/smll.202106127

[81] Zhang G et al. High entropy alloy as a highly active and stable electrocatalyst for hydrogen evolution reaction. Electrochimica Acta. 2018;**279**:19-23. DOI: 10.1016/j.electacta.2018.05.035

[82] Kante MV et al. A high-entropy oxide as high-activity electrocatalyst for water oxidation. ACS Nano. 2022;**17**: 5329-5339. DOI: 10.1021/acsnano.2c08096

[83] Zuo X et al. A hollow PdCuMoNiCo high-entropy alloy as an efficient bi-functional electrocatalyst for oxygen reduction and formic acid oxidation.

Journal of Materials Chemistry A. 2022; **10**(28):14857-14865. DOI: 10.1039/d2ta02597c

[84] Chen T et al. PtFeCoNiCu high-entropy solid solution alloy as highly efficient electrocatalyst for the oxygen reduction reaction. iScience. 2023;**26**(1): 105890. DOI: 10.1016/j.isci.2022.105890

[85] Wu D et al. Platinum-group-metal high-entropy-alloy nanoparticles. Journal of the American Chemical Society. 2020;**142**(32):13833-13838. DOI: 10.1021/jacs.0c04807

[86] Amiri A, Shahbazian-Yassar R. Recent progress of high-entropy materials for energy storage and conversion. Journal of Materials Chemistry A. 2021;**9**(2):782-823. DOI: 10.1039/d0ta09578h

[87] Shet SP, Shanmuga Priya S, Sudhakar K, Tahir M. A review on current trends in potential use of metal-organic framework for hydrogen storage. International Journal of Hydrogen Energy. 2021;**46**(21): 11782-11803. DOI: 10.1016/j.ijhydene.2021.01.020

[88] Rowsell JLC, Yaghi OM. Metal – Organic frameworks: A new class of porous materials. Microporous and Mesoporous Materials. 2004;**73**: 3-14. DOI: 10.1016/j.micromeso.2004.03.034

[89] Bucior BJ et al. Identification schemes for metal – organic frameworks to enable rapid Search and cheminformatics analysis. Crystal Growth & Design. 2019;**19**:6682-6697. DOI: 10.1021/acs.cgd.9b01050

[90] Reviews C. Introduction to metal – organic frameworks. Chemical Reviews. 2012;**112**:673-674

[91] Öhrström L, Kemiteknik IK, Högskola CT, Gothenburg S. Let's talk about MOFs—Topology and terminology of metal-organic frameworks and why we need them. Crystals. 2015;**5**:154-162. DOI: 10.3390/cryst5010154

[92] Tarhan C, Çil MA. A study on hydrogen, the clean energy of the future: Hydrogen storage methods. Journal of Energy Storage. 2021;**40**:102676. DOI: 10.1016/j.est.2021.102676

[93] Megia PJ, Vizcaino AJ, Calles JA, Carrero A. Hydrogen production technologies: From fossil fuels toward renewable sources. A mini review. Energy and Fuels. 2021;**35**(20): 16403-16415. DOI: 10.1021/acs.energyfuels.1c02501

[94] Ibarra IA et al. Highly porous and robust scandium-based metal-organic frameworks for hydrogen storage. Chemical Communications. 2011;**47**(29): 8304-8306. DOI: 10.1039/c1cc11168j

[95] Madden DG et al. Densified HKUST-1 monoliths as a route to high volumetric and gravimetric hydrogen storage capacity. Journal of the American Chemical Society. 2022;**144**:13729-13739. DOI: 10.1021/jacs.2c04608

[96] Rosi NL. Hydrogen storage in microporous metal-organic frameworks. Science. 2003;**1127**:10-14. DOI: 10.1126/science.1083440

[97] Yang SJ, Jung H, Kim T, Im JH, Park CR. Effects of structural modifications on the hydrogen storage capacity of MOF-5. International Journal of Hydrogen Energy. 2012;**37**(7): 5777-5783. DOI: 10.1016/j.ijhydene.2011.12.163

[98] Peedikakkal AMP, Aljund IH. Upgrading the hydrogen storage of mof-

5 by post-synthetic exchange with divalent metal ions. Applied Sciences. 2021;**11**(24). DOI: 10.3390/app112411687

[99] Suresh K, Aulakh D, Purewal J, Siegel DJ, Veenstra M, Matzger AJ. Optimizing hydrogen storage in MOFs through engineering of crystal morphology and control of crystal size. Journal of the American Chemical Society. 2021;**143**(28):10727-10734. DOI: 10.1021/jacs.1c04926

[100] Gómez-Gualdrón DA et al. Evaluating topologically diverse metal-organic frameworks for cryo-adsorbed hydrogen storage. Energy & Environmental Science. 2016; **9**(10):3279-3289. DOI: 10.1039/c6ee02104b

[101] Zhang Z, Wang Y, Wang H, Xue X, Lin Q. Metal-organic frameworks promoted hydrogen storage properties of magnesium hydride for in-situ resource utilization (ISRU) on Mars. Frontiers in Materials. 2021;**8**:1-6. DOI: 10.3389/fmats.2021.766288

[102] Yan Y et al. High volumetric hydrogen adsorption in a porous anthracene-decorated metal-organic framework. Inorganic Chemistry. 2018; **57**(19):12050-12055. DOI: 10.1021/acs.inorgchem.8b01607

[103] Thomas KM. Hydrogen adsorption and storage on porous materials. Catalysis Today. 2007;**120**(3–4 SPEC. ISS):389-398. DOI: 10.1016/j.cattod.2006.09.015

[104] El Kassaoui M, Lakhal M, Benyoussef A, El Kenz A, Loulidi M. Enhancement of hydrogen storage properties of metal-organic framework-5 by substitution (Zn, Cd and Mg) and decoration (Li, Be and Na). International Journal of Hydrogen Energy. 2021; **46**(52):26426-26436. DOI: 10.1016/j.ijhydene.2021.05.107

[105] Zeleňák V, Saldan I. Factors affecting hydrogen adsorption in metal–organic frameworks: A short review. Nanomaterials. 2021;**11**(7). DOI: 10.3390/nano11071638

[106] Sakintuna B, Lamari-Darkrim F, Hirscher M. Metal hydride materials for solid hydrogen storage: A review. International Journal of Hydrogen Energy. 2007;**32**(9):1121-1140. DOI: 10.1016/j.ijhydene.2006.11.022

[107] Desai FJ, Uddin MN, Rahman MM, Asmatulu R. A critical review on improving hydrogen storage properties of metal hydride via nanostructuring and integrating carbonaceous materials. International Journal of Hydrogen Energy. 2023. DOI: 10.1016/j.ijhydene.2023.04.029

[108] Züttel A. Materials for hydrogen storage. Materials Today. 2003;**6**(9): 24-33. DOI: 10.1016/S1369-7021(03)00922-2

[109] Schneemann A et al. Nanostructured metal hydrides for hydrogen storage. Chemical Reviews. 2018;**118**(22):10775-10839. DOI: 10.1021/acs.chemrev.8b00313

[110] Rusman NAA, Dahari M. A review on the current progress of metal hydrides material for solid-state hydrogen storage applications. International Journal of Hydrogen Energy. 2016;**41**(28):12108-12126. DOI: 10.1016/j.ijhydene.2016.05.244

[111] Ley MB et al. Complex hydrides for hydrogen storage - new perspectives. Materials Today. 2014;**17**(3):122-128. DOI: 10.1016/j.mattod.2014.02.013

[112] Ali NA, Ismail M. Advanced hydrogen storage of the Mg–Na–Al system: A review. Journal of Magnesium and Alloys. 2021;**9**(4):1111-1122. DOI: 10.1016/j.jma.2021.03.031

[113] Adams BD, Chen A. The role of palladium in a hydrogen economy. Materials Today. 2011;**14**(6):282-289. DOI: 10.1016/S1369-7021(11)70143-2

[114] Zhang X et al. Realizing 6.7 wt% reversible storage of hydrogen at ambient temperature with non-confined ultrafine magnesium hydrides. Energy & Environmental Science. 2021;**14**(4): 2302-2313. DOI: 10.1039/d0ee03160g

[115] Zhang X et al. Nano-synergy enables highly reversible storage of 9.2 wt% hydrogen at mild conditions with lithium borohydride. Nano Energy. 2021; **83**:105839. DOI: 10.1016/j.nanoen.2021.105839

[116] Clémençon D, Davoisne C, Chotard JN, Janot R. Enhancement of the hydrogen release of Mg(BH 4) 2 by concomitant effects of nano-confinement and catalysis. International Journal of Hydrogen Energy. 2019; **44**(8):4253-4262. DOI: 10.1016/j.ijhydene.2018.12.143

[117] Yan Y, Rentsch D, Remhof A. Controllable decomposition of Ca(BH4) 2 for reversible hydrogen storage. Physical Chemistry Chemical Physics. 2017;**19**(11):7788-7792. DOI: 10.1039/c7cp00448f

[118] Leading SS, Reversible H, Storage H. Core–shell strategy leading to high reversible hydrogen storage capacity for NaBH4. ACS Nano. 2012;**6**: 7739-7751

[119] Borzone EM, Baruj A, Blanco MV, Meyer GO. Dynamic measurements of hydrogen reaction with LaNi5-xSn x alloys. International Journal of Hydrogen Energy. 2013;**38**(18):7335-7343. DOI: 10.1016/j.ijhydene.2013.04.035

[120] Search H et al. First hydrogenation enhancement in TiFe alloys for hydrogen storage. Journal of Physics D: Applied Physics. 2017;**50**

[121] Kojima Y. Hydrogen storage materials for hydrogen and energy carriers. International Journal of Hydrogen Energy. 2019;**44**:18179-18192

[122] Mehek R, Iqbal N, Noor T, Bin Amjad MZ. Metal – Organic framework based electrode materials for lithium-ion batteries: A review. RSC Advances. 2021: 29247-29266. DOI: 10.1039/d1ra05073g

[123] Kim T, Song W, Son D-Y, Ono LK, Qi Y. Lithium–ion batteries: Outlook on present, future, and hybridized technologies. Journal of Materials Chemistry A. 2019;**7**:2942-2964. DOI: 10.1039/C8TA10513H

[124] Cheng Q, Sun D, Yu X. Metal hydrides for lithium-ion battery application: A review. Journal of Alloys and Compounds. 2018;**769**:167-185. DOI: 10.1016/j.jallcom.2018.07.320

[125] Zhao R, Liang Z, Zou R, Xu Q. Metal-organic frameworks for batteries. Joule. 2018;**2**(11):2235-2259. DOI: 10.1016/j.joule.2018.09.019

[126] Pettinari C, Tombesi A. MOFs for electrochemical energy conversion and storage. Inorganics. 2023;**11**(2). DOI: 10.3390/inorganics11020065

[127] Chem JM, Liu Y, Pan H, Gao M, Wang Q. Advanced hydrogen storage alloys for Ni/MH rechargeable batteries. Journal of Materials Chemistry. 2011;**21**:4743-4755. DOI: 10.1039/c0jm01921f

[128] Oumellal Y, Rougier A, Nazri GA, Tarascon JM, Aymard L. Metal hydrides for lithium-ion batteries. Nature Materials. 2008;7(11):916-921. DOI: 10.1038/nmat2288

[129] Arun V et al. Review on Li-ion battery vs nickel metal hydride battery in EV. Advances in Materials Science and Engineering. 2022;**2022**. DOI: 10.1155/2022/7910072

[130] Li HW, Zhu M, Buckley C, Jensen TR. Functional materials based on metal hydrides. Inorganics. 2018;**6**(3):1-5. DOI: 10.3390/inorganics6030091

[131] Zeng L, Kawahito K, Ichikawa T. Metal hydride-based materials as negative electrode for all- solid-state lithium-ion batteries. Alkali-ion Batteries. 2016. DOI: 10.5772/62866

[132] Zhang B, Xia G, Sun D, Fang F, Yu X. Magnesium hydride nanoparticles self-assembled on graphene as anode material for high-performance lithium-ion batteries. ACS Nano. 2018;**12**(4):3816-3824. DOI: 10.1021/acsnano.8b01033

[133] Sun Z et al. Enhancing hydrogen storage properties of MgH2 by transition metals and carbon materials: A brief review. Frontiers in Chemistry. 2020;**8**:1-14. DOI: 10.3389/fchem.2020.00552

[134] Yang S, Wang H, Ouyang L, Liu J, Zhu M. Improvement in the electrochemical lithium storage performance of MgH_2. Inorganics. 2018;**6**(1):1-9. DOI: 10.3390/inorganics6010002

[135] Mo F et al. Stable three-dimensional metal hydride anodes for solid-state lithium storage. Energy Storage Materials. 2019;**18**:423-428. DOI: 10.1016/j.ensm.2019.01.014

[136] Weeks JA, Tinkey SC, Ward PA, Lascola R, Zidan R, Teprovich JA. Investigation of the reversible lithiation of an oxide free aluminum anode by a LiBH4 solid state electrolyte. Inorganics. 2017;**5**(4). DOI: 10.3390/inorganics5040083

[137] Xinyu Zhang HP, Zhang S, Tang Y, Huang X. Recent advances and challenges of metal – organic framework/graphene-based composites. Composites Part B, Engineering. 2022;**230**

[138] Li H, Zhang Z, Liu Z. Application of artificial neural networks for catalysis: A review. DOI: 10.3390/catal7100306

[139] Bai X et al. A direct four-electron process on Fe-N3 doped graphene for the oxygen reduction reaction: A theoretical perspective. RSC Advances. 2017;7(38):23812-23819. DOI: 10.1039/c7ra03157b

[140] Mehrizi AA et al. Artificial neural networks modeling ethanol oxidation reaction kinetics catalyzed by polyaniline-manganese ferrite supported platinum-ruthenium nanohybrid electrocatalyst. Chemical Engineering Research and Design. 2022;**184**:72-78. DOI: 10.1016/j.cherd.2022.05.046

[141] Oyinbo ST, Jen T. Hydrogen evolution reaction in an alkaline environment through nanoscale Ni, Pt, NiO, Fe/Ni and Pt/Ni surfaces: Reactive molecular dynamics simulation. Materials Chemistry and Physics. 2021;**271**:124886. DOI: 10.1016/j.matchemphys.2021.124886

[142] Ledwaba K, Karimzadeh S, Jen T. Emerging borophene two-dimensional nanomaterials for hydrogen storage. Materials Today Sustainability. 2023;**22**:100412. DOI: 10.1016/j.mtsust.2023.100412

[143] Shen J, Kolb MJ, Göttle AJ, Koper MTM. DFT Study on the Mechanism of the Electrochemical Reduction of CO_2 Catalyzed by Cobalt Porphyrins. 2016. DOI: 10.1021/acs.jpcc.5b10763

[144] Jingfeng Gao GD. DFT study on the mechanism of the CO 2 -to- CO conversion by Co-quaterpyridine complexes. Computational & Theoretical Chemistry. 2022;**1214**: 113794. DOI: 10.1016/j.comptc.2022.113794

Chapter 7

New Trends in Industrial Equipment for the Improvement of Asphalt Roofing Process

Alberto Barragan-Garcia, Miguel Fernandez-Munoz, Alberto Vidal-Sanchez and Efren Diez-Jimenez

Abstract

Roofing techniques are the key for weather resistance and energy efficiency of buildings. Installing asphalt roofing rolls is one of the most popular roof protection methods. This is usually carried out manually. Workers apply heat to the rolls by means of burners. Operators must follow quite a few steps to roof: place the rolls on the ground, unroll, apply energy, and secure. From 20 to 25 rolls per day can be installed by an operator using this manual procedure (200–250 m^2). So, the fact that manual installation means such a reduced work capacity has pushed the development of industrial equipment. Besides, requirement for reduction of CO_2 emissions is forcing to develop systems that optimize the fuel consumptions or even to replace fuel burners by other types of electric heating devices. In this chapter, a review of the state of the art and market of equipment for accelerating asphalt roofing process is presented. A detailed description of some systems and patents is given. The main geometric, physical, and performance parameters will be described and compared. Two new systems based on torches and infrared heating show double installation speed that is the usual manual roofing rate.

Keywords: asphalt roofing, asphalt coating, bituminous sheet, lightweight trolley mechanism, infrared heaters, fuel burners

1. Introduction

Historically, construction has been a wasteful process that sometimes could be automated or anyhow upgraded. Particularly, the manual installation of asphalt roofing rolls has several efficiency issues and its susceptible of being developed such as: safety and workers' health, installation speed, emissions, and quality of installation.

Currently, as experts in the field can ensure, installation of asphalt roofing rolls is usually carried out manually. The operators must accomplish several tasks: choosing the position of the rolls, calculating how many rolls are needed, getting them unrolled on the ground, activating rolls bonding with the surface by heating them, and ensuring everything is glued tight. In addition, during manual installation, there is not any heat control or measuring, since heat torches are hand-controlled by the workers themselves, which sometimes leads to an excessive application of heat and energy loss. Using the explained current method, an average of 20–25 installed rolls per worker a day is only achievable. The poor work capacity of manual installation has uncovered the necessity to look for new systems or equipment that makes the process easier and more efficient.

Most of the devices designed to assist in these tasks are based on one or more torches. SO_2 and CO_2 emissions generated by gas burners must be diminished as much as possible. Environmental administrations are heeding roofing processes in order to reduce emissions. Emission factors in the manufacturing of asphalt [1] were devised by the Asphalt Roofing Manufacturing Association (ARMA). A measurement of 75.2 kg CO_2 per roll was done [2] and determined that the responsible for the greater number of those emissions was the installation stage. Six million tons of asphalt per year are used in this installation process, which produces a large amount of CO_2, SO_2, and other non-eco-friendly emissions.

An alternative option to gas burners [3] could be infrared heaters, but a high electrical power source must be available, so its usability is limited. Therefore, the most polyvalent solution is to optimize heat transfer from torches, burners to asphalt rolls and, thus, gas expense.

Roofing workers' health must also be secured besides installation speed and gas consumption. Workers do not usually employ safety prevention equipment at the expense to improve their speed while performing manual installation. This practice also increases the risk of accidents. A proper way to prevent injuries is to design systems and equipment for efficient operation [4]; fatalities and occupational illnesses also can come from a non-designed bad posture during manual roofing.

In this book chapter, a state of the art is made of the main patents and products for the improvement and assistance in the installation of asphalt roofing rolls, based on both torches and infrared heating. Then, two new systems based on torches and infrared heating are described and compared with previous systems. This novel equipment permits the operators to work in more secure conditions avoiding occupational injuries. It must be lightweight and usable on any class of roof, will increase installation speed, and optimize gas and electricity consumption. Consequently, this new equipment even supposes a human factor improvement because while working, humans are also an emission source (eating, creating waste, using electricity, etc.), so, if the working time is reduced, so will be worker's impact.

In contrast, infrared heaters can supply steady heating power, free from the need of fuel burners. Infrared heaters are electrically powered and driven. That power could have been obtained from green energy sources, and also it is cleaner and safer for operators in their workplace. Enhancement of infrared heaters can also imply large heat transfer rates [5].

The designed new equipment is based on a trolley, which ensures the installation of asphalt in a cleaner, faster, and safer way. One version of the devices incorporates eight infrared heaters located radially and longitudinally around the asphalt roll, while the second version has a set of five parallel torches located in front of the roll to heat the asphalt up. Operators do not have to lift or unroll asphalt rolls, since they are directly placed on the floor and unrolled by auxiliary wheels as the same operator controls heat application. This conciliates a uniform and continuous roll heating to properly stick onto the ground, at the same time it keeps unrolling. Moreover, to guarantee the adherence of the asphalt in those places where the rolls overlap, the equipment counts with two small compaction drum rollers in both lateral sides.

The mechanical and thermal design of both models is presented, comparing heat transfer efficiency with computer fluid dynamics (CFD) simulations and calculations. Furthermore, some photos of the prototypes and operational tests are provided, demonstrating the advantages of its use rather than using manual installation. A comparison of these new systems with previous patents and commercial equipment is given in conclusions section.

2. State of the art

There can be found several inventions to assist in the installation of asphalt coating, most of them based on one or more torches. In 1993, the patent ES 2041192 – entitled "Method and apparatus for applying a bituminous sheet to a substrate" [6] relates a preheated constant temperature roller (close to 300°C), wherein bituminous sheets slide toward their installation (**Figure 1**).

The patent DE-A 1652399-Device for heating and gluing a rolled material on a substrate [7] was published in 1971 (**Figure 2**). This patent claimed for a mechanism that allows the installation of a sheet by the application of heat to it. It includes a series of torches placed along all the bituminous sheet width. It has some disadvantages, for example, the direction of operator walking is backward, and this may cause falls from a certain dangerous height. Also, it prevents the operator from seeing what is happening behind his back, which could cause injury to other operators or apply the sheet on material remains causing fissures. Also, the heat application on the sheet is done from a great distance with respect to the installation point, and the bitumen primer that must be put on the substrate where the sheet is installed does not heat up. Moreover, the linear heating of the bituminous sheet causes that heat must be applied for a longer time, since it heats up from room temperature. Besides, there is only one general gas-regulating valve, which causes a perceptible pressure drop between the first torch and the last one. Last, the device could be too long, making it difficult to transport and place it on the construction site.

The patent US 4725328 [8]—a single ply roofing applicator, from 1988, presents the same characteristics commented for the previous mechanism, differing only in

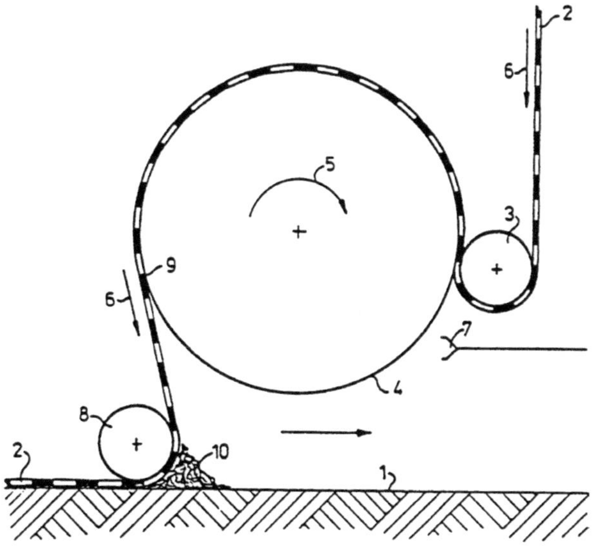

Figure 1.
ES 2041192 patent design [6].

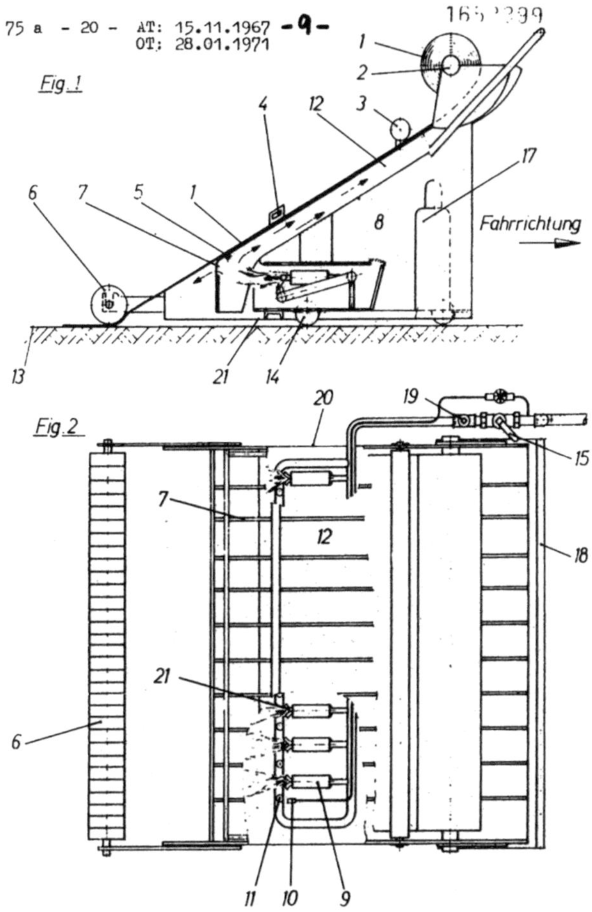

Figure 2.
DE-A 1652399 patent design [7].

two aspects. First, two torches are incorporated, placed perpendicular to the substrate, which are placed at both ends to promote adherence in the overlaps. Secondly, two rollers are added to each side of the mechanism to improve adherence to the overlaps by means of pressure against the substrate. It also shares some disadvantages with the previous mechanism, such as backward direction of operator walking, the linear heating of the bituminous sheet, and that there is only one gas-regulating valve.

In 1982, the patent US 4354893 [9]—combination roofing material unrolling and heat applying apparatus, was published. This patent describes a lighter design than the described above, since it dispenses with the roller system and the base to support the gas cylinder. It incorporates an extra handle, which permits the torches to get toward the roll as it unwinds, therefore enhancing heat application. As the previous ones, the direction of operator travel is backward, the warming is done from room temperature, that there is only one gas-regulating valve and, in addition, the unrolling system of the bituminous sheet is carried out by means of a bar incorporated into the axis of the roll. This is not efficient because the bituminous sheet is flexible and tends to deform. The appearance of folds that can cause problems for water filtration is favored.

Other significant patent is US2006037710 [10]—a membrane applicator. It varies in two aspects with respect to the previous patents: a gas cylinder is supported by a base, and it includes the incorporation of a roller on one flank of the mechanism to guarantee the adherence of the overlap. Their downsides, again, are the reverse direction of travel, but also heat must be applied for a longer time since it heats up from room temperature, and to have only one gas-regulating valve, which is also a pitfall.

As well as patents, it is interesting to analyze what commercialized mechanisms exist. Among them, Seal-Master 1030 (Schäfer Technic GmbH) [11] is very relevant. It is a machine that allows the continuous installation of bituminous sheets in civil engineering works. It is mainly characterized by the large roller on which the bituminous sheet roll is placed for unwinding. This mechanism is not suitable for use on roofs due to its great weight and the use of sheets of greater length than those marketed for waterproofing roofs. It also incorporates a base to place the gas cylinder that is used to supply the torches that are located in the lower part of the mechanism. Finally, the roller used to ensure adherence of the bituminous sheet throughout its width also stands out. On the Schäfer Technic GmbH website, it is indicated that the weight of the set is 660 kg, which makes this mechanism not easy to transport. A site crane would be needed to raise it to the roof, and overload problems on punctual spots of the roof could seriously damage the structure of the construction.

Unify-ER (RES Automatisation Contrôle) [12] is characterized by its remote control (controlled by an operator) and by a greater number of mobile mechanisms than the rest of the systems, which makes this mechanism considerably increase in weight and makes it difficult to use on roofs. It also has a frame on which to place the

Feature to improve	ES 2041192	DE-A 1652399	US 4725328	US 4354893	US 2006/0037710
Reverse travel direction		x	x	x	x
Far away heat source		x			
Heating from room temperature		x	x	x	x
A single gas regulation valve		x	x	x	x
Feature to improve	**Unify-ER**	**Seal-Master 1030**	**FORSTHOFF-P2**	**BITUMAT B2**	**Bitumenbrenner**
Reverse travel direction		x			
Heating from room temperature					x
Excessive weight	x	x			
Very long mechanism	x	x			
Electronic components and batteries	x	x			
Does not seal overlaps					x
Exclusive use for overlaps			x	x	

Table 1.
Details of the characteristics to be improved in the different patents and commercialized products.

gas cylinder, a system of rollers to improve the unrolling of the bituminous sheet, and some side stops that guide the roll and prevent folds from appearing when the sheet is placed on the substrate. It is based on the patent US 8887782—membrane applying apparatus, from 2014. The company claims for an installation speed of 1 m/min, which means a rate of 18 rolls per hour employing two workers. This leads to an installation of 480 m^2 a day per person.

The principal drawbacks of these two systems are that they are too large to operate on small roofs, too heavy, and challenging to transport. In addition, they have an electronic and battery system, which increases the possibility of having failures that prevent the continued use of the device.

On the other hand, the devices FORSTHOFF-P2 (Forsthoff GmbH) [13] and BITUMAT B2 (Leister Technologies AG) [14] can be found. They only serve to heat the sheets in the overlaps, which is an essential requirement for a good waterproofing of the roofs, but they do not heat the rest of the sheet.

And finally, there is Bitumenbrenner (Bamert Spenglerei GmbH) [15], which allows the entire surface of the bituminous sheet to be heated as it uses four individualized torches equally spaced between them, spreading across the width of the sheet. This system also needs the heat to be applied for a longer time, since it heats up from room temperature, at least two operators are required (one of them unrolls the sheet while the other applies heat to it), and it does not allow a correct overlap between sheets, this being the most important process in the installation of waterproofing.

Table 1 presents main characteristics of previous state-of-the-art semiautomatic roofing systems. **Table 1** includes those features that should be improved or changed.

3. Recent developments

As seen in the state of the art, most systems are based on torches. The devices described above present a series of disadvantages. The main ones are an excessive weight, and the fact that they are used for very specific tasks (some only serve to fix the overlaps, and others precisely do not seal the overlaps well).

To overcome all previous limitations from **Table 1**, two novel prototypes have been designed, simulated, and tested with torches and radiators as heating devices. Heaters are fixed to a lightweight structure, which can be pushed forward by an operator, and which unfolds the roll as it heats up, with emphasis on the overlaps. The efficiency of both systems is compared, and the speed of installation is determined.

3.1 Infrared heating

The patent protecting these novel equipment is P201830702 [16]—entitled "Lightweight mechanism for the rapid laying of bituminous sheets in flat roof waterproofing." A more in-depth analysis can be read in [17]. The mechanism is based on a trolley that allows a quicker installation of asphalt roofing rolls, in a cleaner and safer way than manual installation, by using infrared heaters. The worker does not need to lift the rolls because they are laid on the ground. As the worker advances, he pushes the trolley and the roll is unrolled helped by an auxiliary wheel. Eight infrared heaters are located longitudinally and radially around the roll, as can be seen in **Figure 3**. This layout provides a continuous and uniform heating of the roll, improving its adherence to the ground. Moreover, two drum rollers have been placed in the laterals of the trolley to ensure the adhesion on the overlaps.

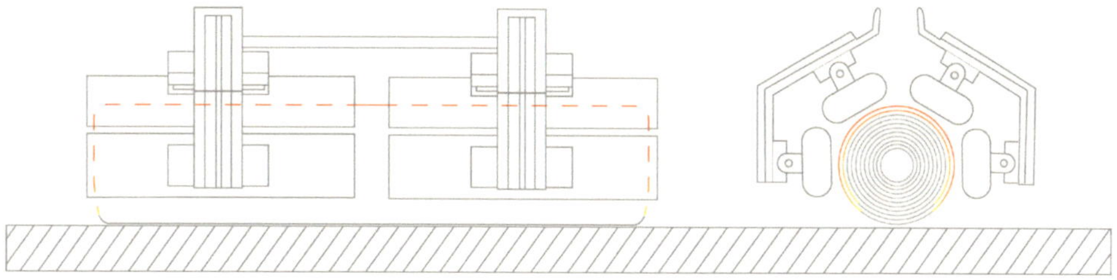

Figure 3.
Design of the structure for the heaters [17].

Figure 4 shows the front, lateral, and top views of the design. The trolley (2) is pushed forward from the handle (1), whose height can be modified to suit the operator. The mechanism rests on four wheels (5) that allow the movement of the trolley. The rear-mounted wheels are locked in the movement direction to help maintain the installation direction. The asphalt roofing roll (7) is located inside the structure. The trolley keeps the roll in the desired position with two locating stops (6). Besides, there is an auxiliary wheel (8), placed in the longitudinally center of the roll, and at an optimum height, which transmits the pushing force to the roll. This causes the rotation movement of the roll, facilitating its unwinding. Two small drum rollers (3) improve the overlapping.

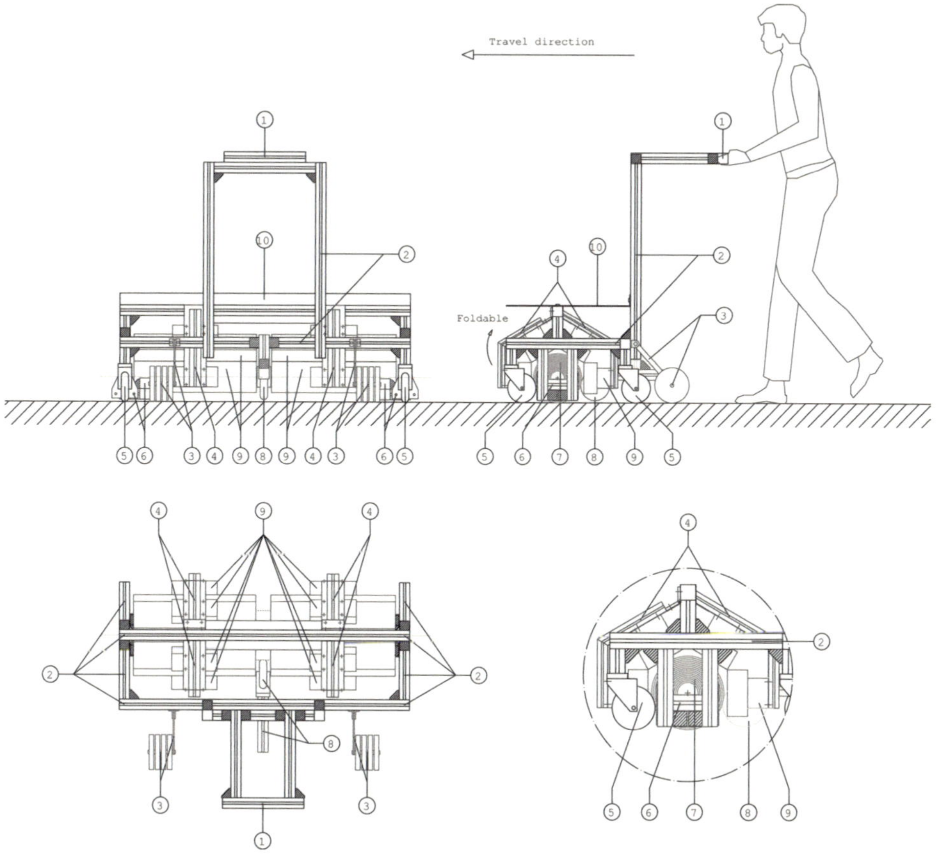

Figure 4.
Front, lateral, and top views of the infrared heaters design [17].

Al last, there are the eight infrared heaters (9) and the enclosure removable cover (10). A detail of the distribution of the heaters is also shown.

1200 W infrared heaters have been selected. The roll received a maximum power flux of 11,794 W/m^2. To determine the adequate heating up time for installing the first layer, a temperature selection choice was defined. On the one hand, the temperatures must be high enough along the external layer to bring most of the asphalt to a viscosity behavior point (140°C). But on the other hand, lower temperatures at the interlayer point are needed to avoid interlayer adhesion of the roll. Then, the top surface temperature must be much greater than 140°C, temperature must be at least 110°C at the middle section, and lower than 70°C the interlayer point. After 35 seconds of warming, the temperature is reached. Important to remark that when the installation begins, the operator has to pre-warm the roll for 32.5 seconds. **Figure 5** depicts the desired temperature profiles after heating for 35 seconds.

The time necessary to install the second and subsequent layers can be also calculated. After heating up of initial layer, the first–second interlayer temperature is 69.18°C. So, the heating up time for the second layer will be shorter because the layer is already preheated. As can be seen in **Figure 5**, after 2.5 seconds of warming, the outer surface of the roll is 69°C, the same temperature as the outer surface of the second layer at the end of the heating up time of the first layer. Thus, it will be needed to hold on for 32.5 seconds for raising the temperature of the following layer from 69 to 169°C. An average installation speed of 1 m/min, is demonstrated for 3 kg/m^2 rolls. This means more than the current manual roofing rate, around 400–420 m^2 per person a day.

Following the design and simulations, a prototype including infrared heater was assembled (**Figure 6**). The temperature criteria selected during thermal design were validated, and also the maximum temperature to prevent interlayer adhesions (below 60–80°C).

Finally, the equipment was tested outdoors. It was powered by a 14 kW diesel portable generator and pre-warmed for 2 minutes. Then, the installation began. One asphalt roofing roll was completely adhered to the ground, as shown in **Figure 7**. Approximately 8 m

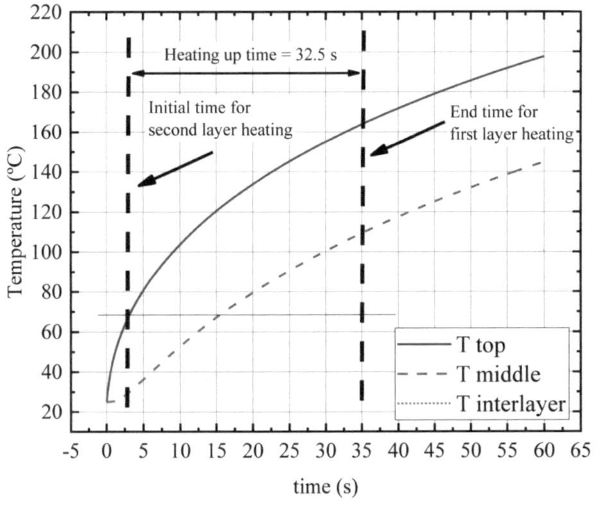

Figure 5.
Temperature versus time of the top surface, middle section, and interlayer point of the roll [17].

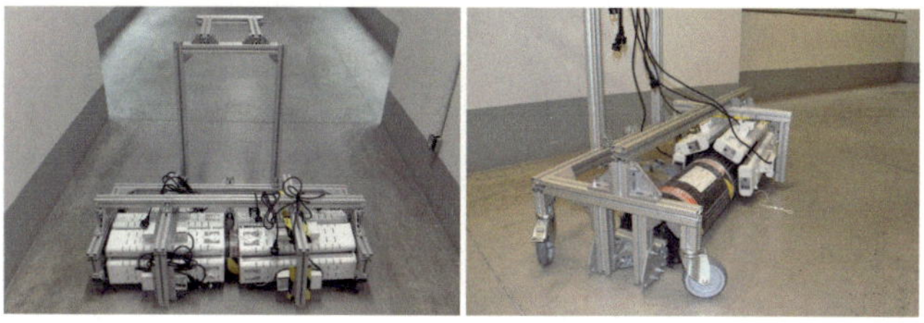

Figure 6.
Infrared heating prototype.

Figure 7.
Outdoor installation test [17].

of the asphalt roll were adhered after 8 minutes of warming, demonstrating the claimed installation speed.

3.2 Fuel burners

A second similar to the previous one model is presented, but in this prototype, gas torches are used instead of infrared heaters. A more in-depth analysis can be read in [18]. The structural design, as well as the analysis of the auxiliary wheel, is the same as in the previous model, but presents some slight differences, which can be seen in **Figure 8**. In this case, the trolley has two auxiliary wheels (5) instead of one at the optimum determined height. Moreover, a thermal insulation cover, preferably made of sheet metal and filled with insulating material (6), has been included, to improve the heat transfer.

The five torches (2) are aligned in parallel in front of the roll. The torches are connected in series with a flexible pipeline (8). Each burner can be controlled separately (7), but they are also controlled by a general regulator (10). An on/off manual valve (9) commands the activation/deactivation of the fuel burners. The gas comes from a pressurized propane bottle through a 10-meter-long flexible pipeline (11). There is very common that a worker just used a single torch, and thus, the heat flux is just applied in localized area. This leads to a nonuniform heating, inducing adherence problems in the overlapping areas. Using a layout with several torches, the heating uniformity can be improved. Gas pressure and fuel burner selection were calculated through a transient CFD simulation iterative process. They also determined the heat transfer to the roll.

In the proposed design, a thermal insulation is included to increase the efficiency of the heat transfer and to reduce the thermal losses by keeping the hot air close to the roll. A CFD analysis has been done to measure the effect of including this thermal insulation cover, and the results are shown in **Figure 9**.

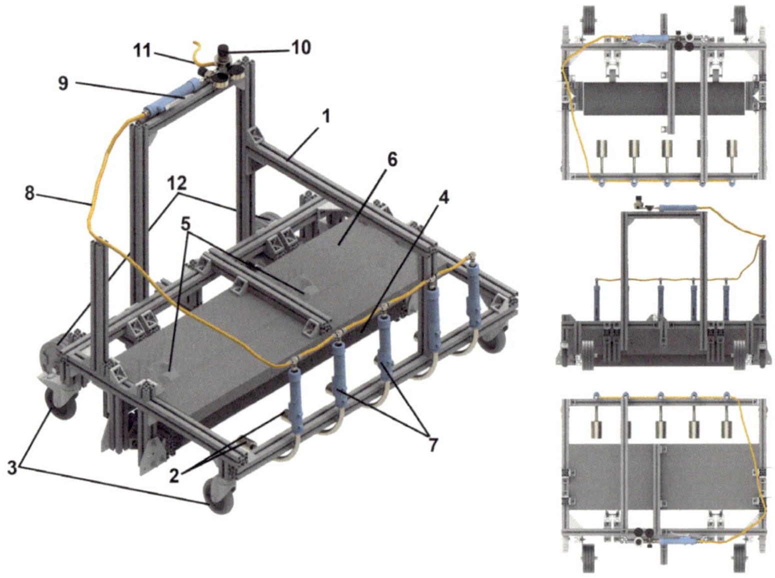

Figure 8.
3D model of the fuel burners design [18].

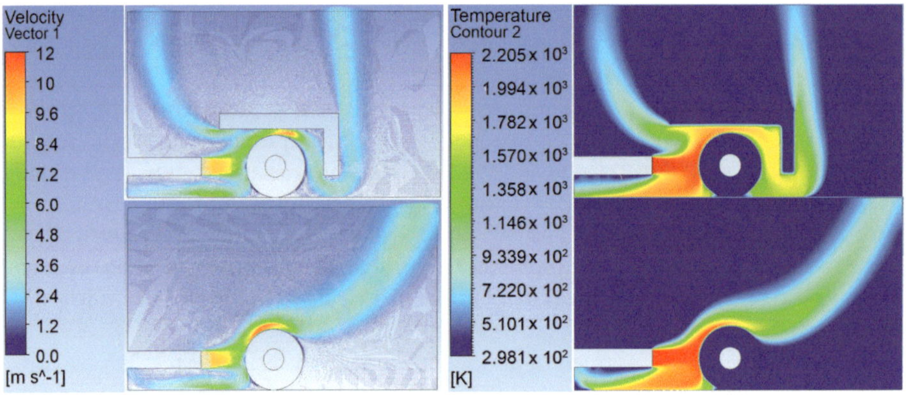

Figure 9.
CFD results: Velocity (left) and temperature maps (right) with or without cover, time = 15 s [18].

Based on the design and the simulations, a prototype was assembled (**Figure 10**). The hot air flow speed and temperature were determined in laboratory preliminary tests. The propane consumption was also measured, being 2800 g/h of propane per torch.

The flame turns blue at the torch outlet, which means that air is about 1800–1980°C for propane [19]. In the yellow zone of the flame, the temperatures are lower, between 500 and 750°C. This can be seen in **Figure 10(left)**. In **Figure 10(right)**, the fuel burners operating at a pressure of 2.8 bar also can be seen, and the reached temperature after 10 second of warming on the roll surface. Almost half of the surface of the roll is above 200°C, which is high enough for a properly adherence.

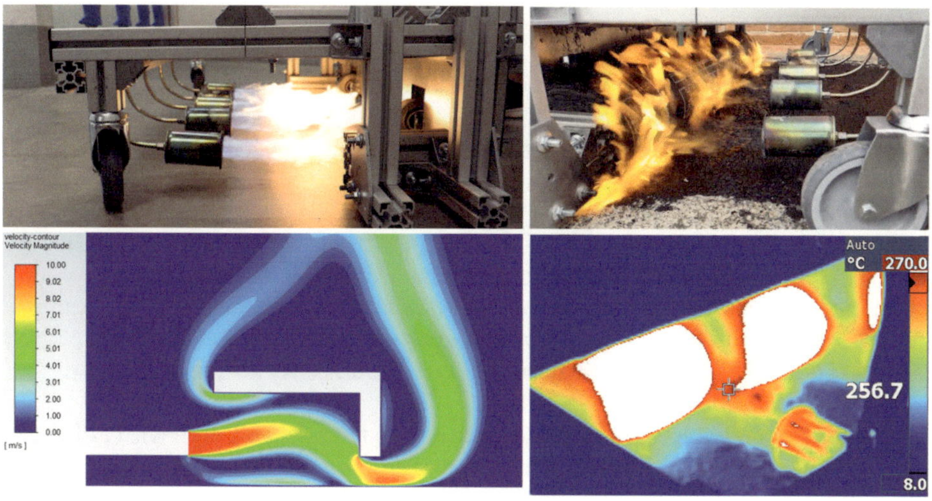

Figure 10.
Flow and consumption test of the five torches (left) and temperature of the roll after 10 seconds of heating (right) [18].

Figure 11.
Prototype and results of the asphalt rolls installation.

Figure 12.
Overlaps quality test and bond check.

Once the preliminary test was done, the outdoor tests of the installation of several adjacent rolls were performed (**Figure 11**). An installation speed of about 1.75 m^2/min was reached, with a 2.8 bar torch pressure and a consumption per torch of 2800 g/h of propane. A good adhesion was obtained on most of the surface of the bituminous sheet. In addition, the overlaps were adhered perfectly one over the other, ensuring isolation (**Figure 12**).

4. Conclusions

A state of the art of the main patents and commercialized products for the improvement and assistance in the installation of asphalt roofing rolls has been made, based on both torches and infrared heating inventions. Their main characteristics have been detailed and collected, as well as the aspects that would require improvement.

Later, two novel systems based on torches and infrared heating have been analyzed and described. The one with infrared heaters has demonstrated an installation speed of 1 m/min, on average, for 3 kg/m2 rolls, which leads to approximately 400–420 m^2 per person a day, and the one with fuel burners has reached an installation speed of 1.75 m^2/min for 3 kg/m^2 rolls, approximately 700–735 m^2 per person per day, more than double the usual manual roofing rate. A summary and comparison of the installation speeds are provided in **Table 2**.

This shows greater efficiency in the system with torches, although with radiators it also achieves more than the usual manual roofing rate. Using higher-power radiators, it could be equal to, and even to surpass, the torch model. But, there appears the requirement of electric generators, which in certain circumstances could be more complicated. Both have advantages and disadvantages and may be preferable for depending on what applications. Although the torch system may seem very polluting, by replacing the traditional manual method by the new system, work times are greatly reduced and larger efficiency is achieved, reducing total final emissions. In conclusion, the development of this type of systems can be very useful to improve the quality, safety, efficiency, and work time of asphalt roofing.

Device	Installation speed (m/min)	Installation speed (m^2 per person a day)	Installation speed (rolls per hour)
Unify-ER	1	480	18 (two workers)
Seal-Master 1030	2.5	1200	45
FORSTHOFF-P2	11	—	—
BITUMAT B2	0.8–12	—	—
Bitumenbrenner	?	?	?
Infrared heating prototype	1	400–420	15–16
Fuel burners prototype	1.75	700–735	26–28

Table 2.
Installation speeds of the prototypes and commercialized products.

Acknowledgements

The research leading to these results was partially funded by Ingeniería Industrial Ibérica S.A. and Cubiertas ALMADI 2017 S.L.

The authors wish to recognize the work of Alba Martínez Pérez during preparation of the figures and 3D pictures.

Conflict of interest

The authors declare no conflict of interest.

Appendices and nomenclature

ARMA	Asphalt Roofing Manufacturing Association
CFD	Computer Fluid Dynamics
LPG	Liquefied Petroleum Gas

Author details

Alberto Barragan-Garcia[1,2], Miguel Fernandez-Munoz[1], Alberto Vidal-Sanchez[1] and Efren Diez-Jimenez[1*]

1 Mechanical Engineering Area, Universidad de Alcalá, Alcalá de Henares, Spain

2 Industrial Engineering Department, Ingeniería Industrial Ibérica S.A., Madrid, Spain

*Address all correspondence to: efren.diez@uah.es

References

[1] Trumbore D et al. Emission factors for asphalt-related emissions in roofing manufacturing. Environmental Progress. 2005;**24**(3):268-278. DOI: 10.1002/ep.10071

[2] Vaz W, Sheffield J. Preliminary assessment of greenhouse gas emissions for atactic polypropylene (APP) modified asphalt membrane roofs. Building and Environment. 2014;**78**:95-102. DOI: 10.1016/j.buildenv.2014.04.019

[3] EL-Mesery HS, Abomohra AE-F, Kang C-U, Cheon J-K, Basak B, Jeon B-H. Evaluation of Infrared Radiation Combined with Hot Air Convection for Energy-Efficient Drying of Biomass. Energies. 2019;**12**(14):2818. DOI: 10.3390/en12142818

[4] Young-Corbett DE. Prevention through Design: Health Hazards in Asphalt Roofing. Journal of Construction Engineering and Management. 2014;**140**(9):06014007. DOI: 10.1061/(ASCE)CO.1943-7862.0000892

[5] Erchiqui F. Application of genetic and simulated annealing algorithms for optimization of infrared heating stage in thermoforming process. Applied Thermal Engineering. 2018;**128**:1263-1272. DOI: 10.1016/j.applthermaleng.2017.09.102

[6] Van Toor AC, Bax NXC. Method and apparatus for applying a bituminous sheet to a substrate, EP0466249. EP0466249. 1993

[7] Wobbermin H, Kugler R, Frueh E, Bucher H, Vorrichtung zum Erwaermen und Aufkleben von Materialbahnen auf Unterlagen, DE1652399. DE1652399. 1971

[8] Warren A. Single ply roofing applicator, US4725328. US4725328. 1988

[9] Kugler WE, Pacello JM. Combination roofing material unrolling and heat applying apparatus, US4354893. US4354893. 1982

[10] Vaillancourt P. Membrane applicator, US2006037710. US2006037710. 2006

[11] "Seal-Master 1030"

[12] Unify-er. 2019. www.modbitapplicator.com/produits.php. [Accessed: November 04, 2019]

[13] "FORSTHOFF-P2"

[14] "BITUMAT B2"

[15] "Bitumenbrenner"

[16] Díez-Jiménez E, Vidal-Sánchez A, Corral-Abad E, Gómez-García MJ, Mecanismo ligero para la puesta rápida de láminas bituminosas en impermeabilizaciones de cubiertas planas, P201830702. P201830702. 2018

[17] Díez-Jiménez E, Vidal-Sánchez A, Barragán-García A, Fernández-Muñoz M, Mallol-Poyato R. Lightweight equipment for the fast installation of asphalt roofing based on infrared heaters. Energies. 2019;**12**(22):4253, 1-20. DOI: 10.3390/en12224253

[18] Barragán-García A, Fernández-Muñoz M, Díez-Jiménez E. Lightweight Equipment Using Multiple Torches for Fast Speed Asphalt Roofing. Energies. 2020;**13, 2020**(9):2216. DOI: 10.3390/EN13092216

[19] Haynes WM. CRC Handbook of Chemistry and Physics. Boca Raton, Florida, USA: Apple Academic Press; 2014

Section 3

Energy Conversion and Storage

Chapter 8

Energy Storage Efficiency

Patricia Scholczova

Abstract

Renewable energy sources with their growing importance represent the key element in the whole transformation process worldwide as well as in the national/global restructuring of the energy system. It is important for a sufficient energy system is to find a solution and key element to complete energy supply, that is, energy storage. Reasons and background, which make the energy storage so crucial, imply that exact, enduring development of energy storage is an indispensable part of the full energy supply. There are some necessary components for further development and implementation of renewable energy sources, and these components involve not only a flexible generation system but also network expansion, demand-side integration, and storage. As the energy storage is a much needed component that can facilitate a low carbon energy system, energy storage technologies find their applications in two major areas, and these are electricity network energy storage and transport/mobility. Interest toward energy storage has also grown due to technical and innovative progress in the field of energy storage technologies. Additionally, energy storage can be considered from different perspectives, which always give corresponding benefits, emphasizing the importance and attractiveness of energy storage.

Keywords: energy storage system, renewable energy, energy supply, storage technologies, energy storage efficiency

1. Introduction

A strong deployment of flexibility solutions for energy storage is necessary to provide the system with the capacity to adapt the dynamics of the load from the frequency response to interannual flexibility. The main candidate for such solutions to offer flexibility networks, response to demand, and dispatchable and flexible energy production is energy storage.

There are five major subsystems in energy power systems, namely, generation, transmission, substations, distribution, and final consumers, where energy storage can help balance client demand as well as the generation itself.

Energy storage is a making a lot of possibilities for technology for various applications, such as power top shaving, renewable energy utilization, boosted structure energy systems, and advanced transporting within multiple areas. Important elements regarding application of energy storage necessary to explain in the introduction are:

- Peak charge or discharge rate of storage—as the maximum power or rate of energy transition to or from storage,

- Peak storage – the largest possible size or capacity available for storing energy. We can simply calculate and get more information about this peak storage capacity if we take into account two factors - namely the discharge speed and the number of peak storage hours that the device where the energy is stored, can actually provide for us,

- Energy density—the sum of energy that can be contained in each mass of a substance or system [1].

In the process of productive use of energy storage, a balance needs to be struck between implementing frameworks that facilitate cost monitoring and establishing a secure climate for investment in the field.

Energy storage is also one of the leading forces in the implementation of renewable energies and plays a key role in sustaining a strong and efficient modern electricity grid, with minimizing the power volatility, increasing the reliability of the electrical grid, and allowing the storage and delivery of electricity produced by intermittent renewable energy sources.

In terms of how to ensure transparency and predictability for market participants, we have learned that adaptation policies that incorporate cost-tracking capabilities of technology are the perfect way to do this.

To reduce costs, which is one of the most contentious topics relating to storage, one of the most successful approaches was to tightly track costs as the share of the use of renewables and energy mix in general expanded.

If renewable energy prices continue to fall and grid parity is reached in various countries, a new period of policies will be required to ensure further growth of renewables in the energy mix.

2. Importance of energy storage

Energy storage is important for developing electricity, since storage technology allows us to 'reserve' electricity, which is of tremendous advantage not only in terms of technical growth but also in economic terms. As energy storage is being used more often than ever before, diversification and protection of supply have improved, which means that the energy market will be balanced. Another energy storage aspect is the use of green energy sources. Ancillary networks for grid convergence are probable options for the advantage of clean energy sources. Asset utilization is supported as well as voltage control and device stability due to the likelihood of a long-term reserve, which is not necessarily feasible for any form of storage technology [2].

The second application is transport and mobility, but this form of application is not as large as for the usage of electricity, but even energy storage in this area provides the possibility of minimizing fuel consumption by providing a kinetic force against a gasoline-fueled internal combustion engine. A particular interest can be defined by the lengths of the charge/discharge, which will be deemed to be the most relevant features that have contributed to a great deal of concern on the end-user side.

Energy storage in transport and mobility has benefits on a broader scale, as increasing system reliability and reducing greenhouse gas emissions and technology related to this form of storage are flywheels or supercapacitors [2].

Prioritizing the use of energy from the on-site renewable energy source over the utility grid (based on fossil fuel) also aims to minimize the consumption of fossil fuels and, as a result, to advance the priorities of the clean environment [3]. In 2015, the European Commission released two Communications: "*Providing a New Deal for Energy Consumers*" and "*On a New Energy Market Layout*" [4]. Their message was that there are three columns of future customer energy plan, which include:

a. Consumer empowerment,

b. Clever houses, work environment,

c. Information monitoring and security.

The importance of this was clearly emphasized from the beginning: minimizing energy expenses with self-generation and consumption [4] as well as increasing the customer's role via intermediation and cumulative engagement systems [5].

To fulfill the environmental and energy policy goals of the European Union, the energy market will have to experience systemic improvements in the coming decades, where electricity will play a key role in the transition phase and increase its share of final energy use.

3. Application of energy storage technologies

Energy storage offers a variety of useful services and cost benefits to electrical systems, and companies are adopting storage technology for a variety of reasons. Large-scale energy storage also allows today's electrical systems to operate more efficiently. This efficiency gain means lower costs, less pollution, and more stable power.

Traditional energy sources such as coal and natural gas power plants must be cycled on and off in response to changing demand and rarely operate at peak efficiency. Not only does this make energy more expensive but it also creates more pollution than is necessary to meet our energy needs. In addition, due to long start-up times, these large-scale power generation facilities cannot keep up with real-time demand spikes, which can lead to blackouts.

It is possible to use the technology for a variety of applications, ranging from offering ancillary services to grid operators to reducing end-user costs behind the meter. There are a diversity of uses for pumped hydropower, flywheels, and thermal storage systems; however, battery energy storage systems have seen the most use across a wide range of applications.

- Energy arbitrage

Using energy arbitrage, prices can be offset in markets characterized by significant variations in locational marginal prices (LMPs) over time. Low LMP levels are used to purchase and store wholesale electricity for resell when high LMP levels are reached.

- Black start

In the event of a network-wide power outage, the black start tool is used to restore power. Since it has to work without being connected to an energy source, it can cause many problems. Energy storage systems are ideal for black start applications because they can operate in standby mode and restart other grid systems on their own.

Hardware-in-loop (HIL) simulations, optimization algorithms, and other intelligent methods and techniques are essential and appropriate for advanced energy management technologies to boost storage life and operability. The HIL energy storage testing enables the optimum calculation of battery capacity and energy density along with the algorithm management tuning. In general, HIL gives insight into the optimum storage configuration. HIL simulations are also used as part of the validation process of either machine models running in real time or laboratory testing of components outside their traditional system [6].

The operation of an energy storage device may pose a problem of optimization where the cost function is defined by a financial metric, a grid gain metric, or a combination of the two. Restrictions are placed on the model and on the features of the energy storage system. There are many optimization methods that are mostly applicable to decision issues, which are mathematical programming, stochastic programming, dynamic programming, and optimal control. Predictive strategies are a model-free monitoring technique that uses weather forecasts without model or historical evidence. Control of temperature settings is also suggested as their control at the component level is simple to be applied by traditional controllers. In the sense of the choice of parameters, it can be defined as: weather forecast only and/or weather forecast and building characteristics. If we regard the weather forecast only, the controller calculated the setting of the wall temperature as a function of the cloud forecast and the actual electrical price situation relative to the optimum price. Compared to unpredictable performance, thermal comfort was vastly improved, and the price and energy savings were recorded at 41% and 30%, respectively. Thus, this form of predictive control could provide good results in the case of an active storage device [7].

4. Key attributes of energy storage

- Reducing imbalances between energy demand and production.
- Managing the amount of power required to supply customer when it is needed.
- Improving power efficiency and secure supply of electricity to customers.
- Enhancing the stability and reliability of transmission and delivery systems.
- Increasing the use of current facilities, deferring or removing expensive upgrades.
- Strengthening availability and improved market demand of distributed generation sources.
- Improving efficiency of green energy generation.
- Cost savings by capacity deferral and transmission of payment [8].

4.1 Advantages and disadvantages of energy storage

Without storage, the production and use of electricity must still be balanced. And since generation and consumption are physically different, transport capability is also a limiting factor. Owing to these limitations, the value of energy on wholesale markets will change rapidly and considerably over time, often even adversely. Electricity spot market rates reflect this abrupt shift in value [9].

Various storage devices are found in energy systems. They can be cataloged as: chemical or electrochemical, mechanical, electromagnetic, or thermal storage. Generally, the energy storage plant consists of a storage medium, a power conversion device, and a balance of plant [10]. Obviously, these forms of energy storage have many promising features, but on the other hand, certain important elements are still lacking.

Electricity storage systems have major beneficial energy, economic, and environmental impacts:

a. Reduction of backup power plants by satisfying fluctuations and peaks in demand,

b. Mitigation of electricity-loss costs by bridging the difference during power outages and by easing short-term volatility,

c. Support for green energy by smoothing out the uncertainty of renewable energy sources and allowing electricity to be dispatched as needed,

d. Steadily upgrading production and costs.

Despite the apparent benefits for the use of storage systems, it is evident that more measures are needed to address the major obstacles. The key one is the initial capital cost of storage per kW of all storage technology. It should be remembered that the cost of using Pumped Heat Electrical Storage (PHES) and Compressed air energy storage (CAES) systems is comparatively low. Battery technology is less efficient than PHES technology but can be found to be comparatively cost-effective.

While there have been significant technological improvements in energy storage systems, in many countries, it is still not cost-competitive for electricity consumers (whether at a residential, commercial, or utility scale) to store their energy. It means that when a consumer demands electricity, supply across the transmission and distribution networks must be carried out in real-time. One of the biggest disadvantages of energy storage is the fact that energy storage usually uses electricity and stores it but afterward distributes it back to the grid, which is called "round-trip" as a proportion of energy put in to energy returned, measured in %. This is inefficient, because the energy lost in the process of this round trip could be stored better, and no wastage of energy would be possible in this cycle. It depends on the use of storage technology; the higher the round-effectiveness, the lesser will be the amount of energy lost [11]. The last disadvantage is certainly the necessity of a significant amount of additional energy, which represents a reserve used for energy storage [12]. Reserve or in other words "back-up" can have two types, i.e. capacity and operational backup. The optimum proportion of variable renewable energy (VRE) sources in the composition of energy sources depends on different factors [13]. Backup power, network flexibility, transmission-system quality and capacity [14], as well as load efficiency characteristics [15] and real local weather models will determine the amount and variety of

VREs required to backup and that can be safely fed into the system. By adding storage space to the energy grid, it is possible to gain greater resilience of VRE by supplying contingency ability for peak load shaving or valley loading [16]. There is considerable variation in national markets because of the different endowments of indigenous fossil fuels and renewable energy potential, levels of technological development, and environmental and energy security risks. As a result, the relative importance of each of the above characteristics defining electricity as a 'mixed good' varies between countries [17].

5. Types of energy sources

5.1 Hydroelectric (hydropower) energy

Hydroelectricity (hydropower) is generated by the gravitational flow of water through the turbine attached to a generator. Most of the hydropower is created from water stored in a reservoir with a wide dam. The largest hydropower plant in China is currently in operation with 22.5 gigawatt (GW). The rising portion of hydroelectricity is created by water flowing down the river straight through the turbine or by water flowing via the pipes and the turbine near the spur of the river before returning to the river, and this form of hydroelectricity is called run-of-the-river hydropower. Advantage of it is that vast areas of land are not submerged behind a dam, but on the other hand, a smaller volume is valuable for storage [18].

Environmental impact relies in some cases on eliminating the likelihood of carbonic acid gas release during power generation. It has been found that areas with hydropower potential that value more highly to utilize the sources of power that depend on fossil fuels emit up to four times more the quantity of greenhouse emission than necessary [19]. This suggests that hydropower production does indeed have the power to cut back greenhouse emissions.

Increased opportunities for hydropower would also result from changes in how electricity markets are run. Sub-hourly scheduling tries to encourage greater participation and compensation for flexibility. Hydropower may be compensated for forward market scheduling. Independent System Operators (ISOs) arranging hydropower resources across several hours or days would also enable hydropower optimization in the context of other resources. This "fixed-schedule" method might boost plant earnings by 63–77%. Why? Because the existing market structure advantages fossil fuel producers with time-independent output. They may burn fuel to create power at any time of day or night. Pumped hydropower or energy-limited hydropower does not have this benefit and must instead guess the lowest cost time to "refuel" and the greatest price time to sell.

5.2 Wind energy

Wind turbines convert the mechanical energy of the wind right into electricity [20]. Wind speed is normally measured at a height of 10 m. The suspensions in the elevation will adjust the wind speed at just a few hundred meters. Hills or mountains have a major impact on wind direction. The technical device, such as the wind turbine, should draw as much input out of the wind as possible with the use of wind power. Wind turbines slow down the wind by transferring energy from wind to electricity. However, the pre-and post-wind turbine mass flow remains steady [21].

DOI: http://dx.doi.org/10.5772/intechopen.109851

The maximum power in the wind extracted from the wind turbine is 59% and with connecting the principle of preserving mass and momentum through a theoretical disk that extracts energy from the air result is called Betz's law [18].

Wind turbines are often found in the coastal region of mountain passes, ridges, and offshore. Onshore wind farms typically consist of some thousands of midsize 1–8 MW generators to power portions of communities. Offshore wind farms typically consist of several to a lot of medium- to large-sized 3–15 megawatt (MW) turbines; for example, one of the 12 W generators specifically built for overseas use has an elevation of about 150 m above the sea level, which means that the rotor capacity is 220 m, which is the height of the Eiffel Tower and can provide electrical energy for approximately 16,000 homes.

Collecting and storing energy from wind turbines is possible using battery storage with electrical batteries, which is a very common type of storage for wind and solar energy. Other solutions could be compressed air storage and hydrogen fuel cells to store energy surplus [22].

To get a perfect score of 100% on the efficiency table, the wind turbine must capture the full kinetic energy of the wind. Using all of the kinetic energy, however, will result in zero velocity or no wind on the other side of the turbine. At the same time, we must keep in mind that these wind turbines are intended to operate at a specific speed that maximizes production. If the wind turbine is designed to produce the most energy at a speed of 20 mph, the highest amount of energy produced will be at that speed, while the amount will be less at lower speeds. For comparison, traditional power plants have a theoretical maximum output or load factor of roughly 50% on average. Although the same is true for wind power, the fact that it is an environmentally benign and renewable source of energy gives it the necessary push. The entire concept that promotes wind power as the best future power source is based on the idea of maximizing output while keeping expenses to a minimum. When the initial cost of infrastructure development and payback time are considered, it is safe to say that wind power is cost-effective in the long run.

5.3 Solar PV

Solar power may be concentrated, better known as concentrated solar power (CSP), through mirrors or reflective lenses that focus sunlight onto a collector containing a fluid to heat the fluid to an extreme temperature; then, the heated fluid flows from the collector to an engine, where a little of it is converted to electricity, and a few kinds of CSP allow the warmth to be stored for several hours. One kind of collector is termed "*long parabolic trough mirror reflectors,* "and therefore, the second type is "*central tower receiver* "with a field of mirrors surrounding it. Within the central power, the focused light heats a circulating storage fluid known as concentrated solar power (CSP). CSP is a dispatchable form of solar power, whereas PV is not to be dispatched. PV only works while the sun is shining. CSP can be dispatched on demand, much like flipping on a solar switch. So, CSP is not competing with PV [23]. During days without sun, a storage plant additionally produces electricity but only for few hours, so basically without storage, the CSP ability to store is around 25%. With storage availability, the capability element raises to around 65%. CSP is great for helping to satisfy the power need and power requirements, and importantly, CSP collectors can ramp their power manufacturing up and down quicker than coal or nuclear plants [24].

Photovoltaic systems demand more maintenance than other forms of energy generation. Solar systems has a tendency to become the greatest source of power generation by boosting the efficiency of solar cells, which is now around 43%, although manufacturing these cells has not yet been accomplished. A battery will never drain below 50%, preventing the cells from deteriorating and thereby increasing the battery's life expectancy. It is evident that a battery with double the amperage must be utilized instead of a battery with the same power as required. All of these elements influence solar system efficiency; therefore, by boosting solar system efficiency, power system stability may be enhanced. As a result, we might claim that this strategy is still one of the most expensive.

5.4 Bioenergy

Bioenergy (from conventional Greek bios, life) is extracted from organic materials such as wood, agricultural products, or organic waste and is originated from a recently produced organic material, known as biomass, as an anti-fossilized biomass fuel. It is also used in electricity, heating, cooling, and transport. It can be used in liquid forms, including biofuels; in gaseous forms, like biogas; or in solid forms. Bioenergy is the oldest form of energy used by humans, but it is also at the forefront of Europe's new attempts to step away from fossil fuels and decarbonize our economy. As a result of European climate and energy policies, the use of bioenergy is increasing exponentially [25]. European policies consider all bioenergy to be renewable energy and the foremost important measure from the EU within the fight against temperature change, so much hope currently lies on the performance of bioenergy. The sustainable use of biomass for heating/cooling and the generation of electricity will result in a variety of energy, economic, employment, and environmental benefits. Biomass can be processed at times of low demand and can be used to provide energy as required. Depending on the type of conversion facility, biomass will also play a role in managing the growing share of intermittent renewable energy from wind and solar energy in the electricity system. The possibility to store biomass enables the production of heat to satisfy seasonal demand. In addition, biomass enables the production of high-temperature heat that cannot easily be provided by other low-carbon sources [26]. In 2011, 95% of bioenergy was consumed as heat, 4.7% as transport fuel, and 72,700 kWh as electricity [27]. As far as bioenergy storage is concerned, the sufficient way to store biomass is to accommodate seasonal production and to ensure daily supply to the biomass utilization facility [28]. Wet storage systems may be used for higher yield intended for wet use, such as in brewing and anaerobic digestion systems, with tight monitoring of storage times to prevent unnecessary depletion of feedstock. Storage structures usually used for dry agricultural residues should be secured against spontaneous combustion and excessive decomposition, and the actual storage moisture depends on the type of storage used [28].

The effectiveness of various biomass-to-energy conversion processes varies greatly. For example, producing electricity from pure biomass is only about 30–35% effective; however, producing heat from the same material is frequently more than 85% efficient. In general, utilizing bioenergy for heat and electricity is a far more efficient approach of decreasing greenhouse gas emissions than using bioenergy for transportation fuel. Organic waste and agricultural or forestry residues are more resource-efficient than many other types of feedstock because they do not place additional strain on land and water resources and offer significant greenhouse gas savings.

5.5 Ocean energy

Oceans cover about 70% of earth's surface area, making them the globe's biggest solar batteries. The installed capacity of ocean energy in 2019 was 530 MW [29]. Sunlight warms the surface area of water by a lot compared to the deep ocean water, and also this temperature-level difference develops thermal power. Simply, a tiny low part of the heat trapped within the ocean might power the planet [30]. Underwater compressed air energy storage (UWCAES) and underwater pumped hydro storage (UWPHS) are the second type of ocean energy storage available at the moment [31]. Air is contained in foldable bags on the seafloor in the UWCAES system. This underwater storage system has a great advantage over its ground equivalent. In standard CAES, air is contained in a tank of a set capacity, while the compressed air releases the pressure within the pipe, which reduces the flow to the turbine. UWCAES systems operate well at depths of 400–700 meters below sea level; this water depth provides the pressure required for most turbine compressors where compressed air energy storage is usually used. Seawater is used as a working solvent instead of air in the UWPHS system. This system uses solid steel or concrete spheres. To 'release' as an energy storage unit, the mechanism allows high-pressure seawater to penetrate the sphere through an opening by means of a turbine attached to the generator. Such energy storage systems are best used in peak-saving grid applications. Round-trip efficiencies for UWCAES and UWPHS are in the range of 70–85%. Concerns on how to create the most efficient energy storage of seawater are still on the way and raise the most important question-how to install them and how to do it cost-effectively [31]?

5.6 Geothermal energy

Geothermal energy is derived from hot water or steam that, in both cases, emanates from hot rocks or soil and exists below the earth's surface. Some of the high-temperature rocks are discovered around volcanic activities as well as low-temperature rocks and soil that we can find everywhere, including in the ice field [18]. Geothermal energy is used to supply heat directly or at high temperature it is able to supply electricity. Lower temperature (0C–120C) is employed to warm buildings, and hot-temperature warmth (120C–400C) is normally used to generate electrical power. For electricity production, there are three types of geothermal plants: flash steam, dry steam, and binary geothermal plant [18]. Energy storage for geothermal energy works by technology that transfers heat energy from underground water to electricity, and after that, extra energy is stored into the underground water. Geothermal storage also has disadvantages such as fluctuation in the binary geothermal plant, mostly used in the US as a heating agent, that can cause more damage than we imagined. If it is used for other purposes, environmental impact is pretty huge, especially when geothermal fluid is not stored and recycled in a pipe, which can then absorb toxic compounds such as arsenic, boron, and fluoride. These poisonous compounds may be taken to the surface and leak as the water evaporates [32].

In the framework of the effectiveness of geothermal energy, we are mainly talking about the fact that geothermal systems can provide any combination of forced-air heating, radiant in-floor heating, domestic hot water, and air conditioning all from the same unit.

The heating efficiency of a geothermal heat pump is assessed by the coefficient of performance (COP), while the cooling efficiency is reflected by the energy efficiency ratio (EER). These metrics relate the number of units of heat given or withdrawn to the number of units of power consumed to complete the task. Ground source heat pumps generally provide 4 units of heat energy for every unit of electricity consumed.

So, we are talking about 400% effectiveness. In comparison, the most efficient conventional systems on the market today are just 98% efficient (**Figures 1** and **2**).

6. Energy storage technologies

Energy storage technology is often classified on the basis of its use in large- or small-scale applications. Flywheels, pumped energy storage, and compressed air storage are among the forms of storage typically used in large scales. Battery systems are expected to have a significant impact on the small-scale installation of clean energy resources in commercial and residential buildings [35]. Some technologies provide short-term energy storage; others can endure for much longer. In short-term response energy storage devices as a short-term duration applications whose big advantage is quick discharging, we identify flywheels or supercapacitors. Moreover, their application prevents the collapse of power systems, and use of short-term response energy system is more practical in terms of renewable energy like wind [36]. Use of long-term energy storage devices is expected to be functional even more in the upcoming years, and these devices are compressed air technology, batteries, or pumping hydro storage. Energy storage contains technologies such as flywheels or flow batteries that are part of a so-called "distributed" storage, and technologies such as PHES or CAES are very important in "bulk" storage [37]. **Figure 3** shows these technologies divided into categories:

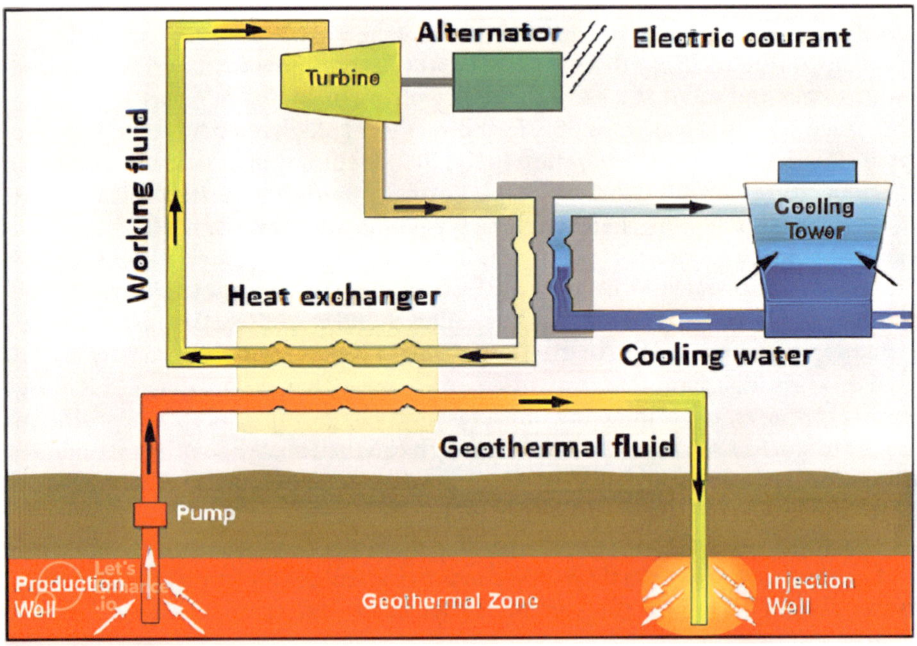

Figure 1.
Schematic of a typical binary cycle geothermal power plant [33].

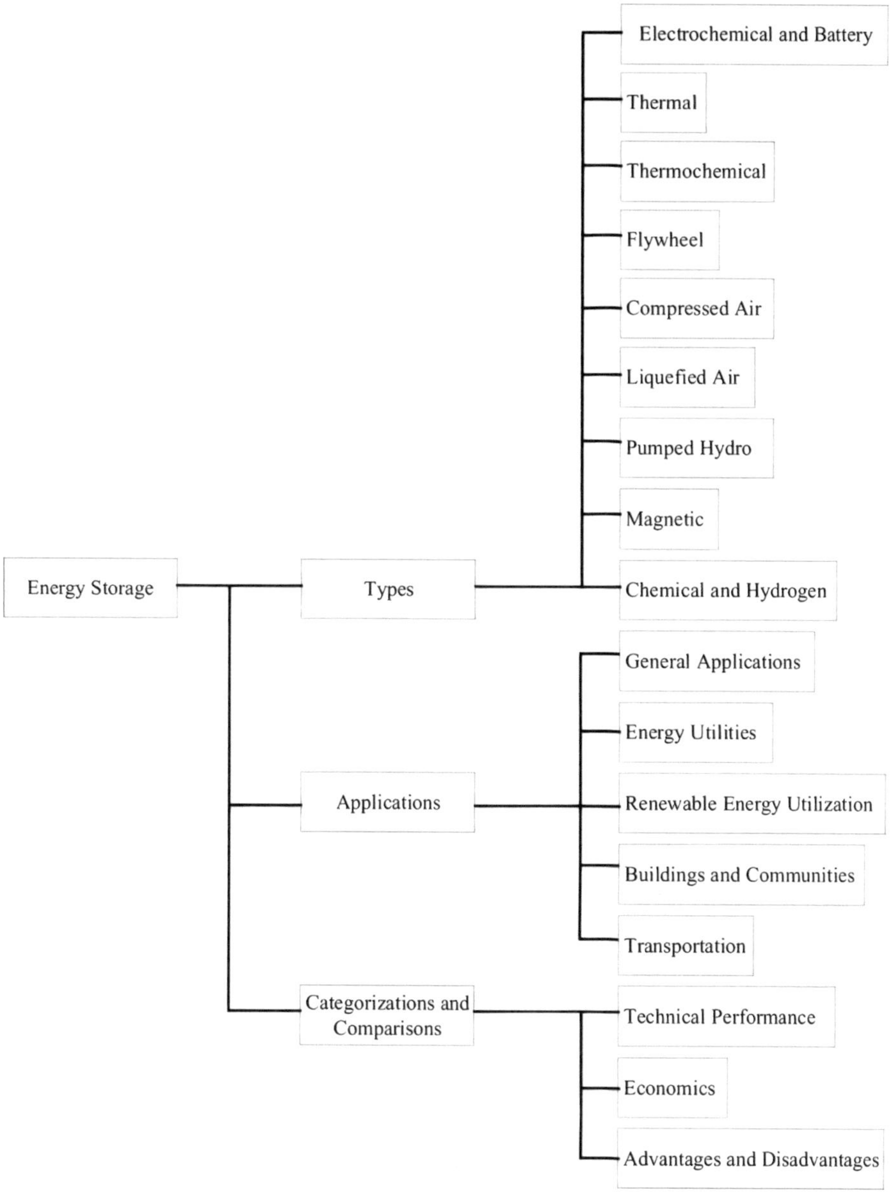

Figure 2.
Storage types and applications [34].

6.1 Battery energy storage system (BESS)

A battery is an electrochemical cell that transforms chemical energy right into electricity. While the battery is filled, the direct current is transformed into chemical energy; when the chemical energy is discharged, it is converted back into the flow of electrons in a direct current form [36]. Batteries are the most common storage devices for electricity. However, the term battery consists of a variety of technologies applying various operating concepts and materials. It is necessary to differentiate between two essential principles of battery: electrochemical and redox flow [36].

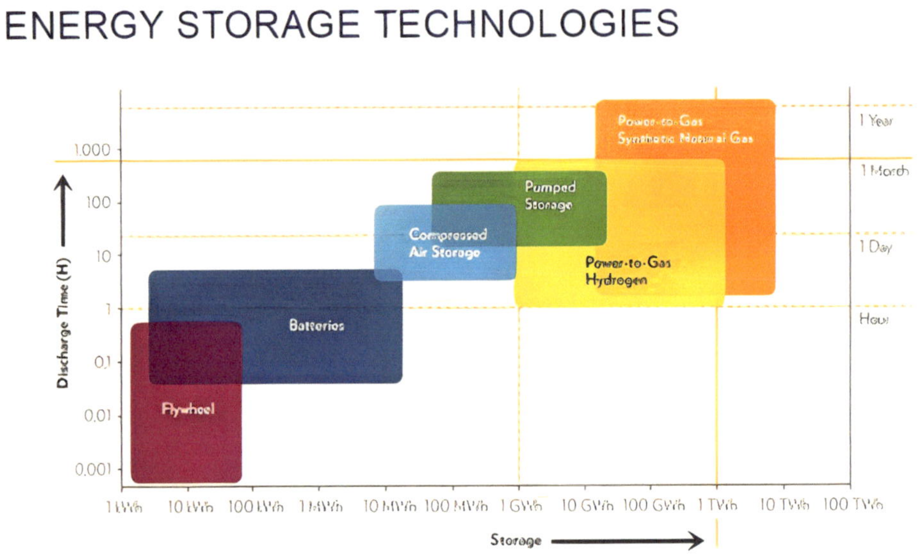

Figure 3.
Energy storage types and their discharge time [38].

Battery-storage growths have mainly focused on transportation systems as well as smaller systems for portable power or periodic backup power, although system dimension and volume are less crucial for grid storage than portable or transport applications [39]. Future utility applications of batteries could be focused on providing peak distribution capability deferral as well as top shaving at the substation along with reliability improvement [40]. Study into battery storage at the grid range is concentrated on longevity for multitudes of charge/discharge cycles and life time, high round-trip-efficiency, capability to react rapidly to adjustments in lots or input, and practical capital costs [41]. Utility packages of batteries in future might be targeted on supplying distribution ability deferral and peak shaving on the substation in addition to reliability enhancement [40]. Batteries are regularly compared to supercapacitors for various energy applications, and it is predicted that exploiting their features (i.e., frequent electricity storage functionality without sacrificing their cycle) by means of integration should help cope with future electric-storage demanding situations. For big-scale electrical storage (e.g., strength from renewable strength sources), the use of flow batteries seems to be the most appropriate option, even though charges and development continue to be a challenge [42]. Attributes of battery storage system are:

Rated power capacity—the overall viable rapid discharge functionality (in kilowatts [kW] or megawatts [MW]) of the BESS, or the maximum rate of discharge that the BESS can obtain, starting from a completely charged position [43].

- Storage capacity—the overall volume of energy stored (in kilowatt-hours [kWh] or megawatt-hours [MWh]).

- Storage length—the amount of time for storage to discharge to its energy capacity earlier than the exhaustion of its energy capacity.

Cycle lifestyle/lifetime—the period of time or cycles that a battery storage unit can deliver daily charging and unloading earlier than loss or full-size deterioration. Self-discharge occurs while the accumulated charge (or energy) of the battery is depleted by internal chemical reactions or when it is not discharged to do work for the grid or the consumer. Self-discharge, calculated as a percentage of the rate misplaced for a given period of time, decreases the amount of energy needed for discharge and is a significant criterion that is not to be ignored in batteries designed for longer-lasting programs [43].

BESSs account for around 5% of worldwide energy storage capacity, far less than pumped-storage hydropower. According to Fortune Business Insights, the global battery energy storage market is estimated to reach €19.74 billion by 2027, growing at a 20.4% compound annual growth rate (CAGR). Given its availability, efficiency, and recent developments in electrochemical storage technology, a BESS is expected to be a leader in energy storage in the next years. Alternatively, on the other hand, it can compete with battery power storage systems, gaining the upper hand in some situations.

6.2 Pumped hydroelectric storage (PHES)

Pumped hydroelectric energy storage (PHES) is the most established storage innovation in the world today. The International Energy Agency (IEA) estimates that PHES installments make an increase in capacity of 26 GW, and energy storage capability of PHES will overtake battery storage globally by 2050. The current storage volume of PHES plants is measured at 9000 GWh, although the battery capacity is just 7 GWh [44]. By 2023, electricity production from PHES has expanded by one quarter to 146 TWh; however, the estimated operating hours of the PHES were considerably unpredictable, and the spectrum of capacity factors was broad due to the uncertainty of market conditions. Storage consist of a reservoir, specifically upper and lower [18], Besides the power stored behind hydropower dams, 97% of all energy stored for electrical usage is currently in the form of pumped hydropower storage [18]. Where current surplus energy or costs are low at the moment, water is pumped through pipelines from the pumping station in the lower reservoir to the upper reservoir, as seen in **Figure 4**. At the same time, as pumped hydro storage is introduced into the grid or the capability of the water reservoirs is more advantageous, the hydropower facility can provide most of the load necessities, obviating the need to build huge top-load fuel mills [46].

Pumped storage hydropower (PSH) systems account for more than 94% of the global energy storage capacity. Water-spinning turbine, as water flows down from a higher tank to a lower reservoir, generates electricity in PSH. This energy storage system (ESS) may provide large storage capacity at a low cost, fulfilling the needs of bigger electrical networks. The difficulty with pumped hydro storage systems is that they take years to create and need significant investments.

6.3 Flywheels

Flywheel energy storage systems use a rapidly rotating flywheel to help convert mechanical energy to electrical energy and back to electrical energy. This system consists of four main parts as follows:

Solid cylinders, bearings, motor/generators, and vacuum-tight housings. To generate kinetic energy, the motor draws energy from the electrical grid to rotate a

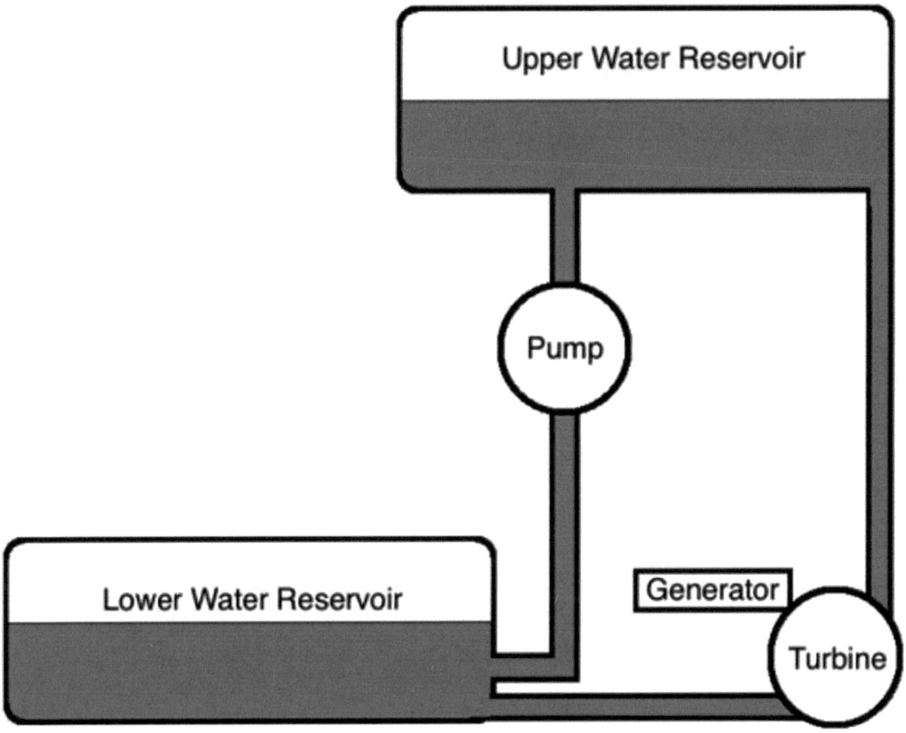

Figure 4.
Pumped hydro storage [45].

cylinder or disc at speeds up to 60,000 rpm. Flywheels are considered "dynamic" energy storage systems because they must be accelerated by an external force before they can store energy. A flywheel is a rotating wheel or disk, mostly made from steel or carbon fiber, that turns around an axis and is usually used in short-term energy storage devices for motion applications such as powertrain engines and road cars. In these applications, the flywheel soothes the power load during deceleration by dynamic braking operation and provides a lift during acceleration [47]. A flywheel generates power as a rotational kinetic energy and later converts the power into electrical energy. The flywheel is an electric motor, a holding system for energy, and a generator at the same time. As excess power is available, the power and electric motor spins the flywheel roughly at a high speed. For the energy applied to the flywheel, a small amount is used to hold the flywheel rotated; the remainder is storage energy [18]. Flywheels are suitable for storing surplus electricity from intermittent solar and wind power on the electrical grid; on the other hand, flywheels cannot conserve an immense amount of power due to the fact that they have a high loss rate, so the generated electrical energy can be consumed very rapidly and has a high loss rate.

Because of these losses, the cost per kWh of energy stored is high, and power-specific costs are relatively low [48].

Simply put, the advantages and disadvantages of flywheels can be summarized as systems notable for their longevity (up to decades), easy maintenance, and fast response time. But they can only operate for a short period of time (**Figure 5**).

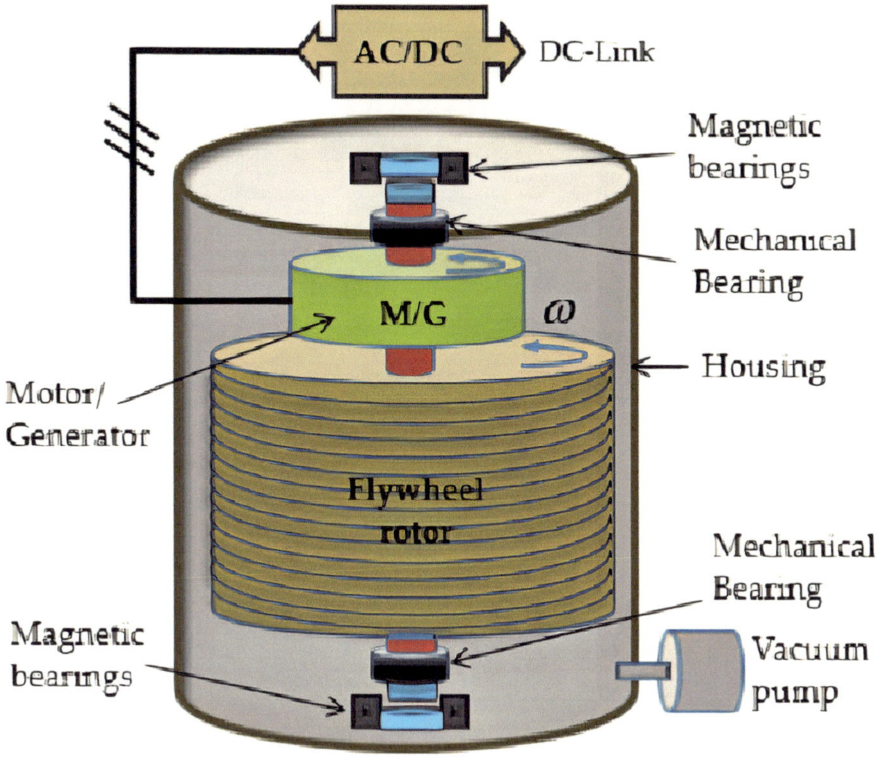

Figure 5.
Flywheel energy storage [49].

6.4 Compressed air energy storage (CAES)

The method to gather periodic renewable power is fixed by utilizing compressed air power storage space. This manner of storage opens up possibilities to keep power for a long-time period and then to resupply that electricity to the grid. Excess recurring electricity is made use of to compress air. Using this form of storage ensures that intermittent renewable energy may be stored for a long period of time, in contrast to flywheels or supercapacitors. The place of CAES is mainly underground, and globally, there are just a few buildings, just one in Europe-Germany and two in the U.S. CAES is normally attached to a power-producing system, such as a wind turbine, and when energy is required, compressed air is expanded and transferred back to a power-extending engine. For better perception, **Figure 6** contains how the CAES system looks exactly.

CAES systems are widely employed in the manufacturing and mining industries. However, adopting this technology in some applications, particularly residential solutions, and their installation itself might be difficult and expensive.

7. Economic aspects of energy storage

Market integration means the process of step-by-step harmonizing of the rules of the various power markets, culminating in the harmonization of all cross-border

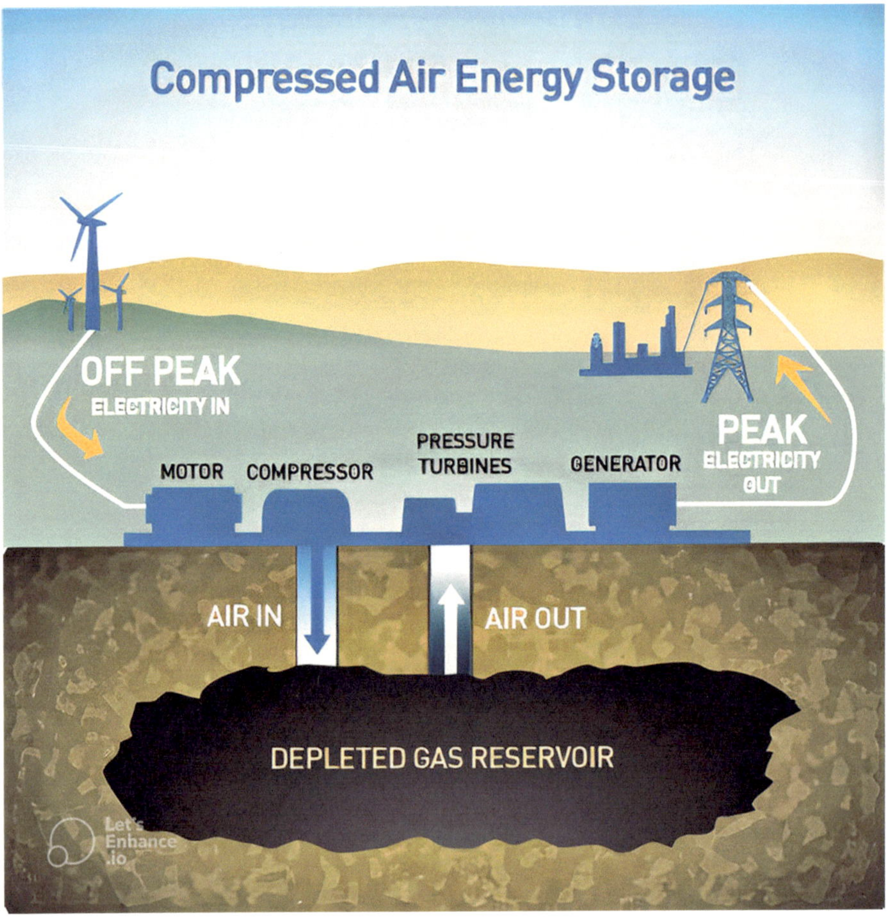

Figure 6.
CAES [50].

market rules that allows electricity to respond to price signals and flow freely across borders (as do goods and services in the internal market).

As several aspects of the evolving modern energy economy, the greatest results come from the cooperation of bringing components mentioned above together in an actively and carefully built system that consists of: dispersed renewable energy generation, scalable baseload infrastructure, smart grid, demand response, and rapid energy storage response [51]. Markets integrate into regional markets for cross border trade of electricity and its continuously increasing.

A reinforced, interconnected European network requires coupled markets, versatile production, increased backup and storage capability, demand response measures, clear worth signals, responsiveness of support, and cost-effectiveness to balance the fluctuation of energy sources across Europe [52]. These are the aspects that help to improve the functioning of the electricity system and financial market [53].

7.1 Energy storage systems costs

The cost of large-scale mechanical storage devices is influenced by location. Pumped hydro systems need locations that can accommodate both a storage reservoir and a sufficient elevation variation to produce potential energy. Compressed air energy storage (CAES), like pumped hydro systems, has been constrained by the availability of natural resources to supply low-cost air storage [54]. Calculation of related costs and operating prices for energy storage devices is a problem because of not only a wide variety of innovations but also a multitude of external factors.

In contrast to pumped hydro and compressed air systems, where the storage medium is virtually free, finding low-cost heat-retaining materials is crucial for thermal storage systems. Although heat may be stored directly as steam, molten salts are the most popular choice since they can reach greater temperatures [55].

Although most studies employ such a meter, there is no globally accepted standard or method for calculating the costs of energy storage, due to the fact that different metrics emphasize different aspects of storage cost and operation.

One way to make apple-to-apple comparisons between storage technologies is through the use of the Leveled Cost of Energy (in this case, the Leveled Cost of Storage or "LCoS"), where the technology per kWh is calculated as a function of the total project life cost divided by the expected lifetime power output. The cost of electricity in this calculation includes any capital expenses associated with electricity generation for direct consumption (ccap, gen), capital expenses for electricity generation that goes to storage (ccap, gen2stor), capital expenses for storage technologies (ccap,stor), fuel (pfuel), or purchased electricity (pelec) costs (accounting for generator-efficiency losses, gen, and round-trip-efficiency losses of storage charge and discharge, RTE), and to compute LCOE, costs may be discounted (using discount rate r) to find the net present value, which is then divided by the discounted quantity of energy provided during the system lifespan (**Figures** 7 and **8**).

Whatever calculation is chosen, one important point to make when calculating the LCOE or LCOS is that the cost is also affected by the demand for electricity or stored energy. Although a storage system may be technically capable of cycling continuously for 24 hours a day, demand is determined by use patterns. This implies that even if a system is built to provide longer-duration storage, actual cycling behavior might mean that the charging and discharging cycles are frequently just a fraction of the installed capacity, limiting the power produced by the system and raising the levelized cost [54]. For example, where a compressed air energy storage (CAES) system may have a higher initial capital cost than a Li-ion battery system, the CAES system's lifetime power output is far higher than the Li-ion battery system (which normally lasts only 10 years), which reduces the LCoS [57].

However, the LCoS formula does not adequately represent other crucial points, including spatial limitations (vital for CAES and pumped hydro systems), safety issues about battery explosions, and technological features that are best suited for various applications. Seems that we are still in the process of creating or implementing a

$$LCOS = \frac{\Sigma(Capital_t + O\&M_t + Fuel_t) \cdot (1+r)^{-t}}{\Sigma MWh_t \cdot (1+r)^{-t}}$$

Figure 7.
LCoS calculation [56].

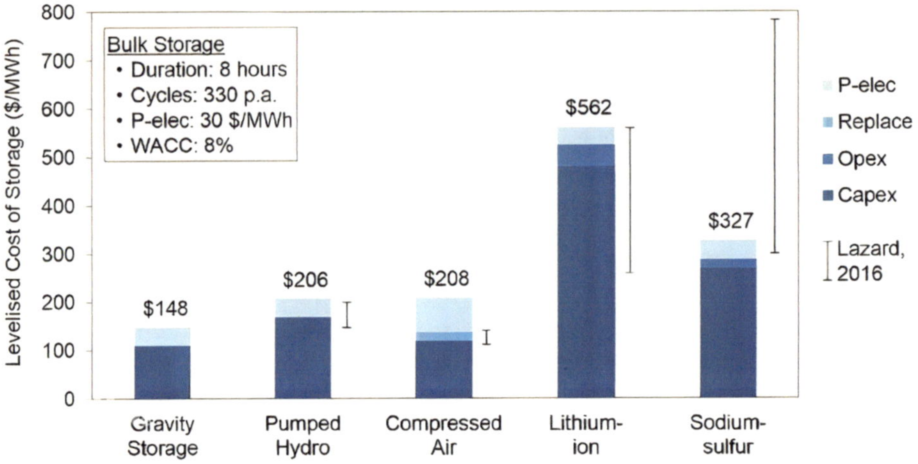

Figure 8.
LCos calculation based on different storage types [57].

formula for calculating energy storage efficiency that would take into account all the abovementioned parameters.

Other costs to consider are:

- Construction and commissioning

C&C expenses, also known as engineering, procurement, and construction (EPC) costs, include site design costs, equipment purchase/transportation costs, and labor/parts for installation [58]. Cost reductions for C&C are not likely to be as significant since these expenses are more mature than those that are more directly related to each technology. The cost of grid integration is primarily determined by the system footprint and weight (with discrete steps in costs), the degree of factory assembly versus on-site assembly (the total cost may be the same regardless of where the assembly occurs), and the architecture (open racks vs. containerized systems) [59]. The literature consensus C&C expenses were raised by 15% for the technology with the lowest energy density indicated as the highest liters per watt-hour (L/Wh). This figure was multiplied by the normalized volume per watt-hour multiplied by 0.33 to obtain a lithium-ion C&C cost of €100/kWh, which is somewhat higher than the €80/kWh reported by McLaren et al. [60]. While improvements have been achieved in recent years, the anticipated C&C cost of €100/kWh is on the low end of current forecasts, with minimal room for future cost reduction owing to "learning."

- Operations and maintenance

Expenses that are necessary to maintain the functionality of the storage system during the period of its economic lifespan are linked to the demand for energy. According to the available literature, fixed O&M costs for all battery chemistries range between 6 and €20 per kW-year, with the majority falling between €6 and €14 per kW-year [58]. While lithium-ion batteries may have higher costs for safety and battery management systems (BMSs), the larger size of other battery technologies can result in higher O&M costs, and their relatively safe operational characteristics contribute to lower O&M costs.

7.2 Financial instruments and support

A number of public or private funding instruments are currently in operation in European countries to promote renewable energy and energy storage itself. The choice of instruments depends on the point at which technology or projects are created. The bulk of funding instruments come into three primary categories:

- Energy market tools (feed-in tariffs, premium, green bonds, tenders, fiscal incentives);
- Equity funding mechanisms (venture capital, equity, R&D grants, capital/project grants, contingent grants);
- Debt financing mechanisms (mezzanine debt, senior debt, guarantees).

The criterion used to test the funding systems, the financial instruments, and the support are as follows:

Efficiency-applies, on the one hand, to extra generation costs and, on the other, to regulation costs. Although the additional generation costs reflect the welfare effects in general, the policy costs additionally consider the distributional effects or the issue of which stakeholder pays for the additional costs.

Effectiveness- analyzes the effect of funding systems on the business diffusion of clean energy technology.

Certainty for investors-the degree to which policy instruments are capable of minimizing the uncertainties of energy and renewable energy ventures, which could be of a fiscal, technical, or political type.

Long-term competitiveness.

Market compatibility (only applicable to help schemes, not applicable) [61].

7.3 Financial benefits of energy storage

Energy management is a critical issue for businesses seeking to maintain and reduce operational costs. Energy storage systems give businesses the control over distributed energy resources, allowing them to save money on demand charges, provide critical continuous power to protect against grid variability, and better integrate renewable energy sources to foster more sustainable and financially sound business practices. Every planning and execution approach should be connected to the real-time control and organizational functionality of the ESS in conjunction with Distributed Energy Resources (DER) in order to achieve a rapid integration process [62].

The bulk of C&I-scale(commercial and industrial) facilities must pay demand charges based on peak power use. This expense often accounts for 30–70% of the total energy expenditures on a commercial electric bill. Energy arbitrage can result in significant cost savings by discharging energy during peak usage and cost periods, reducing load during those peak periods, and resulting in lower demand charges.

Load shifting is a critical component of this method for lowering energy expenses. BESS (described in Chapter 5 (5.1)) and related software assess consumption patterns and storage to efficiently identify the ideal time to charge and discharge stored energy, moving peak loads to off-peak hours.

Finally, BESS may be utilized to "smooth out" grid fluctuations, effectively becoming the major source of site power and relegating the grid to a secondary energy source. This allows the site to not only disregard grid power outages (and the possible expenses associated with them), which are limited only by the site's storage capacity, but also maintain a constant power factor and eliminate any grid oscillations that may compromise sensitive equipment.

Moreover, the energy storage could be used to adjust the amount of electricity produced from renewable energy sources. Energy is retained when demand and energy prices are low so that it can be used when.

a. demand and energy prices are high and

b. output from intermittent renewable energy production is low [62].

- Cost avoid or revenue gain of ancillary services: it is well-recognized that energy storage can provide many forms of ancillary services. In short, they are what could be considered support facilities that are used to keep the municipal grid running. Two of the more common ones are: the spinning reserve and the accompanying load [63].

- Cost saving or revenue improvement by bulk energy arbitration: Arbitration includes the procurement of low-cost power available during low-demand storage times so that low-priced energy can be used or sold at a later time when the price of electricity is high.

At present, in many parts of Europe, energy storage projects have to pay for both extracting electricity from the grid and pumping power into it, and this legacy policy has long been seen as both a major obstacle to making an economic argument for energy storage and one that could be overcome reasonably quickly. The Committee of Members of the European Parliament (MEPs) has recently pointed out that this is one of a number of 'shortcomings' in network codes across Europe. Further changes will be made to the European Energy Taxation Directive in 2023 to 'ensure a harmonized taxation on all storage and hydrogen production.' In the meantime, the EU also responded to the fact that the share of energy costs charged as tax is much higher than energy consumption itself [64], i.e. one of the appropriate and important steps and solutions should be to tax reduction.

8. Environmental impacts and future promising technologies for renewable storage

Key environmental impacts include: lifetime energy efficiency, lifecycle greenhouse gas emissions, supply chain criticality, material intensity, recyclability, and environmental health and social impacts as and safety and human rights. Energy efficiency of the life cycle is important since high performance sustained over a long planned lifetime minimizes the criteria for technological uptake and the related impacts. Supply chain criticality recognizes not only the geological availability of essential commodities but also the possible supply chain vulnerabilities and threats associated with fiscal, technical, social, or geopolitical influences. Owing to the high usage of nonrenewable materials in main energy storage systems, content intensity is an important parameter. Battery storage systems typically have a higher material density relative to other technologies.

The recyclability of battery storage technologies has the capacity to minimize high material intensity by recycle, reuse, or remanufacturing. Poor recyclability highlights the need to implement innovative approaches to infrastructure and technologies [65].

Environmental health is significant as an adverse effect on habitats or human health. The supply chain will negate the benefits of moving into a green energy system. Since batteries are material-intensive technologies, they have the most important impacts. The effect varies depending on the location of extraction, manufacturing, and end-of-life due to variations in technologies, production pathways, and local environmental and social norms. The most important mining impacts in China include pollution and water and soil emissions from lead, graphite, and phosphate mining, both of which have severe health impacts. There are important human rights impacts associated with the resource market for lithium-ion batteries, in particular lithium and cobalt. Cobalt mining is mostly undertaken by artisanal and small-scale miners who work in precarious environments in handheld mines without adequate protective equipment and widespread child labor.

8.1 Storage and future

Targets aimed at zero emissions are more difficult and costly than net-zero goals, which utilize negative emissions technology to achieve a 100% reduction. Pursuing a zero, rather than net-zero, aim for the energy system may result in high power costs, making the achievement of economy-wide net-zero emissions by 2050 more difficult.

Storage can help poor countries lower their power costs while also delivering local and global environmental advantages. Lower storage costs enhance both the savings in electricity and the environmental advantages.

For example, E-mobility as one of the possible solutions for energy storage in coming years seems realistic. E-mobility is expanding, and now we are seeing hybrid vehicles, e-bikes, scooters, and kick bikes on the streets. They will shortly be followed by more fuel cell vehicles running on hydrogen. The use of all these vehicles will be expanded beyond their planned use as means of transport to also provide energy storage: they will charge when renewable energy is available in the system and feed back into the micro-grid battery as required. Such vehicle-to-grid and vehicle-to-building systems will become more prevalent as regulatory barriers are eliminated. With compact storage with more secure solid-state batteries and hydrogen bottles, our phones will never run out of batteries again [66]. Emerging battery storage systems would reduce energy storage costs and boost new opportunities in the energy market. In the coming decades, we are anticipating new types of batteries, such as solid-state batteries, to increase the efficiency of airplanes, vehicles, or medical devices. The other alternatives are sulfur-based chemistries, as their continued development would ensure the appropriate use of renewable-grid installations or magnesium batteries, which may meet their maximum capacity and be ready for commercialization within the next 5 years [67].

We can utilize energy more efficiently and reduce carbon emissions by storing it. Energy storage capacity would need to expand from 140 GW in 2014 to 450 GW in 2050 to keep global warming below 2°. Currently, just 3–4% of the power generated by utilities worldwide is stored.

Conclusion? We still have a long and difficult road ahead of us.

9. Conclusion

Arising from government policies to minimize greenhouse gas emissions and improve economic feasibility combined with storage, clean energy sources and more advanced storage systems in comparison of 20 years ago will inevitably increase dramatically over the next decades. To find out how this 'Grid of the Future' will look and work is a global problem that every nation will have to face in its own way. This continues to be a main theme of future studies, as developed and emerging countries face a wide variety of problems in terms of energy and power systems. While the prospects for decarbonizing the energy market are facing significant challenges in terms of technical breakthroughs, stationary energy storage has already achieved a competitive edge, partially due to cost declines in the production of batteries. The increase of incorporation of storage would lead to closer relations between supply and demand in the transition to a decarbonized energy environment. As a consequence, the value of the device would not depend on the volume and size of the hardware used, as is the case for current storage modes (hydro- or fossil fuel-based). Storage should also be known as an additional type of investment in grids (copper grid lines), an aspect of demand for service responsiveness, and a stabilizing factor for power grids. The energy storage market is increasingly shifting from an equipment-based industry to a service-based industry where various device players (generators, grid managers, aggregators, and final customers) will be given access to storage-related facilities. This summary of energy storage shows the future role of emerging services in the transition of energy systems. However, regulation and implementation of new market rules across MS in the EU are essential. In conclusion, despite all the above market problems, incomplete regulations, and the lack of legislation, the most important human factor is the ability to cooperate that EU member states must understand and ensure the functioning of the renewables market and their subsequent storage.

Acknowledgements

This paper, which I would like to add as my contribution to the book entitled "Energy Consumption, Conversion, Storage, and Efficiency," was developed independently without the help of other persons, instructors, or external instructors.

Conflict of interest

"The author declares no conflict of interest."

Notes/thanks/other declarations

I would like to thank my professor from the Technical University of Berlin, Mrs. Lydia Scholz, who introduced me to this issue and helped me focus more on this area during my studies. Last but not least, my thanks go to my partner Antreas for his patience and understanding during writing.

Energy Storage Efficiency
DOI: http://dx.doi.org/10.5772/intechopen.109851

Author details

Patricia Scholczova
MBL, Bratislava, Slovakia

*Address all correspondence to: patricia.scholczova@gmail.com

References

[1] Carnegie R, Gotham D, Nderitu D. Preckel Utility Scale Energy Storage Systems Benefits, Applications, and Technologies. State Utility Forecasting Group. Indiana, United States: Purdue University, Energy center, Indiana Utility regulatory commission; 2013

[2] Irina S. Renewable Energy Storage Possibilities. Saarbrucken, Germany: AV Akademikerverlag; 2013

[3] Mishra S. Efficient Power Flow Management and Peak Shaving in a Microgrid-PV System. Tulsa, USA: Transmission Planning Engineer, American Electric Power

[4] European Commision. Communication from the Commission to the European Parliament 339 and 340. Brussels, Belgium: European Commission, European Union; 2015

[5] IRENA, Adapting Renewable Energy Policies to Dynamic Market Conditions. In 2014

[6] Cleary T, Ballew B. Electrical energy storage hardware-in-the-loop simulation of a hybrid electric diesel and electric only locomotive. In: 2015 Joint Rail Conference. San Jose, California, United States: American Society of Mechanical Engineers; 2015

[7] Thieblemont H, Haghighat F, Ooka R, Moreau A. Predictive control strategies based on weather forecast in buildings with energy storage system: A review of the state-of-the art. Energy and Buildings. 2017;**153**:485-500. DOI: 10.1016/j.enbuild.2017.08.010

[8] Energy storage. Energy.Gov. Available from: https://www.energy.gov/oe/energy-storage

[9] Swinkels V. Bottlenecks for Energy Storage in Europe – And how to Address Them. SQ Consult based in Netherlands, written with cooperation of Fraunhofer ISI in Karlsruhe, Germany and Universitat de Catalunya in Barcelona, Spain. Available from: https://www.sqconsult.com/en/news/bottlenecks-for-energy-storage-in-europe-and-how-to-address-them

[10] Ould Amrouche S, Rekioua D, Rekioua T, Bacha S. Overview of energy storage in renewable energy systems. International Journal of Hydrogen Energy. 2016;**41**(45):20914-20927. DOI: 10.1016/j.ijhydene.2016.06.243

[11] Round trip efficiency. Energymag. Cameron Thouati, Energy storage blog, www.energymat.net, Hartland Wisconsin. 2014. Available from: https://Energynet/round-trip-efficiency/

[12] Esteban M, Zhang Q, Utama A. Estimation of the energy storage requirement of a future 100% renewable energy system in Japan. Energy Policy. 2012;**47**:22-31. DOI: 10.1016/j.enpol.2012.03.078

[13] Delucchi MA, Jacobson MZ. Meeting the world's energy needs entirely with wind, water, and solar power. The Bulletin of the Atomic Scientists. 2013;**69**(4):30-40. DOI: 10.1177/0096340213494115

[14] Czisch G, Giebel G. Realisable scenarios for a future electricity supply based 100% on renewable energies. In: Energy Solutions for Sustainable Development. Denmark: Risø National Laboratory; 2007

[15] Widen J. Correlations between large-scale solar and wind power in a future

scenario for Sweden. IEEE Transactions on Sustainable Energy. 2011;**2**(2):177-184. DOI: 10.1109/tste.2010.2101620

[16] Zsiborács H, Baranyai NH, Vincze A, Zentkó L, Birkner Z, Máté K, et al. Intermittent renewable energy sources: The role of energy storage in the European power system of 2040. Electronics (Basel). 2019;**8**(7):729. DOI: 10.3390/electronics8070729

[17] Hall PJ, Bain EJ. Energy-storage technologies and electricity generation. Energy Policy. 2008;**36**(12):4352-4355. DOI: 10.1016/j.enpol.2008.09.037

[18] Jacobson MZ. 100% Clean, Renewable Energy and Storage for Everything. Cambridge, England: Cambridge University Press; 2020

[19] Severnini E. The unintended impact of ecosystem preservation on greenhouse gas emissions: Evidence from environmental constraints on hydropower development in the United States. PLoS One. 2019;**14**(1):e0210483. DOI: 10.1371/journal.pone.0210483

[20] American Geosciences Institute. What are the advantages and disadvantages of offshore wind farms?. Alexandria, Virginia: American Geosciences Institute. Online article. 2016;**1**:1-1. Available from: http://www.americangeosciences.org/critical-issues/faq/what-are-advantages-and-disadvantages-offshore-wind-farms

[21] Quaschning V. Understanding Renewable Energy Systems. London, England: Earthscan; 2004

[22] Arnold B. Collecting and storing energy from wind turbines [Online]. 2014;**1**:2-7. Available from: https://www.azocleantech.com/article.aspx?ArticleID=488

[23] Kraemer S. CSP doesn't compete with PV - It competes with gas. SolarPACES. 2017;**1**:1-1. Available from: https://www.solarpaces.org/csp-competes-with-natural-gas-not-pv/

[24] Concentrating Solar Power. Stanford.edu. Available from: http://large.stanford.edu/courses/2013/ph240/rajavi1/

[25] Eubioenergy.com. Available from: https://www.eubioenergy.com/whatisbioenergy/

[26] European Commision. Commission Staff Working Document State of Play on the Sustainability of Solid and Gaseous Biomass Used for Electricity, Heating and Cooling in the EU. Brussels, Belgium: European Commission, European Union; 2014

[27] Available from: http://chrome-extension://gphandlahdpffmccakmbngmbjnjiiahp or https://usewoodfuel.co.uk/wp-content/uploads/2020/01/aebiom_european_bioenergy_outlook_2013.pdf

[28] Zafar S. Biomass Storage Methods. Salman Zafar: Blogging Hub; 2021. Available from: https://www.cleantechloops.com/strorage-biomass/

[29] Ocean energy. Irena.org. . Available from: https://www.irena.org/ocean. [Accessed: November 30, 2022]

[30] Wikimedia.org. Available from: https://upload.wikimedia.org/wikipedia/commons/c/c7/High_School_Earth_Science_14-26.pdf

[31] Ocean Energy Storage. The Liquid Grid. 2017. Available from: http://www.theliquidgrid.com/marine-clean-technology/marine-energy-storage. [Accessed: November 30, 2022]

[32] White DE. Geothermal energy. Bulletin of Volcanology. 1966;**29**(1):481-483. Available from: https://www.

nationalgeographic.org/encyclopedia/geothermal-energy/. [Accessed: November 30, 2022]

[33] Ouali S, Hazmoune M, Bouzidi K. Low temperature geothermal energy for rural development. Antalya, Turkey: Researchgate.net. 07-10 October 2015:1-8. Available from: https://www.researchgate.net/publication/283499353. [Accessed: 3 Feb, 2023]

[34] Journal of Energy Storage editors. Science Direct-review of Energy Storage Types. Vol. 272020

[35] Hemmati R. Technical and economic analysis of home energy management system incorporating small-scale wind turbine and battery energy storage system. Journal of Cleaner Production. 2017;**159**:106-118. DOI: 10.1016/j.jclepro.2017.04.174

[36] Faias S, Santos P, Sousa J, Castro R. An overview on short and long-term response energy storage devices for power systems applications. Renewable Energy and Power Quality Journal. 2008;**1**(06):442-447. DOI: 10.24084/repqj06.327

[37] Elgqvist E. Energy Storage Economics. U.S.A: National renewable energy laboratory; 2017

[38] Energy Storage – HiSoUR – Hi so you are [Internet, Hisour.com.]. Available from: https://www.hisour.com/energy-storage-42824/

[39] Battery Storage to Drive the Power System Transition. Europa.eu. Available from: https://ec.europa.eu/energy/sites/ener/files/report

[40] Joint Legislative Commission on Energy Policy. Deq.nc.gov. Available from: https://deq.nc.gov/media/17484/download

[41] Mohamad F, Teh J, Lai C-M, Chen L-R. Development of energy storage systems for power network reliability: A review. Energies. 2018;**11**(9):2278. DOI: 10.3390/en11092278

[42] European Comission directorate. The Future Role and Challenges of Energy Storage. Available from: http://chrome-extension://gphandlahdpffmccakmbngmbjnjiiahp or http://www.fze.uni-saarland.de/AKE_Archiv/AKE2014H/Vortraege/AKE2014H_04BemtgenEU2014_EnergyStorageStrategyPaper_36p.pdf

[43] Grid Integration toolkit, Grid- Scale Battery Storage. Nrel.gov. Available from: https://www.nrel.gov/docs/fy19osti/74426.pdf

[44] 2018 hydropower status report. Hydropower.org. Available from: https://www.hydropower.org/publications/2018-hydropower-status-report. [Accessed: November 30, 2022]

[45] Blonbou R, Monjoly S, Bernar J-L. Dynamic Energy Storage Management for Dependable Renewable Electricity Generation. London, UK: Energy Storage - Technologies and Applications. InTech; 2013

[46] Benitez LE, Benitez PC, van Kooten GC. The economics of wind power with energy storage. Energy Economics. 2008;**30**(4):1973-1989. Available from: https://econpapers.repec.org/article/eeeeneeco/v_3a30_3ay_3a2008_3ai_3a4_3ap_3a1973-1989.htm. [Accessed: November 30, 2022]

[47] The renewable electron economy part VII: Stationary energy Storageâ⇆¦Key to the renewable grid. Futurelab. 2007. Available from: https://www.futurelab.net/blog/2007/10/renewable-electron-economy-part-vii-stationary-energy-storage%C3%A2%C2%80%C2%A6key-

renewable-grid. [Accessed: November 30, 2022]

[48] Luo X, Wang J, Dooner M, Clarke J. Overview of current development in electrical energy storage technologies and the application potential in power system operation. Applied Energy. 2015;**137**:511-536. DOI: 10.1016/j.apenergy.2014.09.081

[49] Analysis of Standby Losses and Charging Cycles in Flywheel Energy Storage Systems Scientific Figure on ResearchGate. Available from: https://www.researchgate.net/figure/Structure-and-components-of-flywheel-energy-storage-system-FESS_fig1_343930266 [Accessed: December 1, 2022]

[50] Compressed Air Energy Storage (CAES). Pge.com.. Available from: https://www.pge.com/en_US/about-pge/environment/what-we-are-doing/compressed-air-energy-storage/compressed-air-energy-storage.page

[51] Siegel RP. The pros and cons of energy storage systems. Triple Pundit: People, Planet, Profit. Abgerufen am 04. 2017 von. 2017;**1**:1-1. Available from: http://www.triplepundit.com/special/energy-options-pros-and-cons/energy-storage-systems-pros-cons/

[52] Of the european parliament and of the council of 5. On the Internal Market for Electricity (Recast) (Text with EEA Relevance). Official Journal of the European Union; 2019

[53] Commission staff working document, European Commission Guidance for the Design of Renewables Support Schemes

[54] Hittinger E, Ciez R. Modeling Costs and Benefits of Energy Storage Systems. TA – Annual Review of Environment and Resources. Princeton, New Jersey. 2020;**45**(1):445-469. Available from: https://www.annualreviews.org/doi/abs/10.1146/annurev-environ-012320-082101

[55] Liu M, Tay N, Bell S, Belusko M, Jacob R. Review on Concentrating Solar Power Plants and New Developments in High Temperature Thermal. 2016

[56] Available from: Resource-innovations.com or https://www.resource-innovations.com/resources/lcos-key-metric-cost-energy-storage. [Accessed: November 30, 2022]

[57] Lazard. Lazard's levelized cost of storage analysis—version 7.0. Available from: https://www.lazard.com/media/451882/lazards-levelized-cost-of-storage-version-70-vf.pdf

[58] Energy storage publications. DNV. Available from: https://www.dnv.com/power-renewables/energy-storage/energy-storage-publications.html?. [Accessed: November 30, 2022]

[59] Mongird K, Viswanathan V, Balducci P, Alam J, Fotedar V, Koritarov V, et al. An evaluation of energy storage cost and performance characteristics. Energies. 2020;**13**(13):3307. DOI: 10.3390/en13133307

[60] McLaren J, Gagnon P, Anderson K, Elgqvist E, Fu R, Remo T. Battery Energy Storage Market: Commercial Scale, Lithium-ion Projects in the U.S. United States. 2016:4-13. Available from: https://www.osti.gov/servlets/purl/1337480

[61] Financing Renewable Energy in the European Energy Market. 2011

[62] Ibrahim H, Adrian I. Techno-Economic Analysis of Different Energy Storage Technologies. Energy Storage - Technologies and Applications, January. InTech. 2013:8-10. DOI: 10.5772/52220.2

[63] Mohd A, Ortjohann E, Schmelter A, Hamsic N, Morton D. Challenges in integrating distributed energy storage systems into future smart grid. In: 2008 IEEE International Symposium on Industrial Electronics. Cambridge, United Kingdom: IEEE; 2008

[64] European association for storage of energy. The Way Forward for EnergyStorage Grid Fees. General Overview and Best Practices Across Member States. Brussels, Belgium: Lidia Tamelini, European Association for storage of energy. 2022. Available from: http://chrome-extension://gphandlahdpffmccakmbngmbjnjiiahp or https://ease-storage.eu/wp-content/uploads/2022/07/2022.07.07_The-Way-Forward-for-Energy-Storage-Grid-Fees_EASE.pdf

[65] Florin N, Dominish E. Report Prepared by the Institute of Sustainable Futures for the Australian Council of Learned Academies. 2017

[66] Sweco Group, Maria Xylia, Julia Svyrydonova, Sara Eriksson, Aleksander Korytowski. Urban Energy Report beyond the tipping point: Future energy storage. Urban Insight 2019;**1**:38-44. Available from: https://www.swecourbaninsight.com/urban-energy/beyond-the-tipping-point-future-energy%20storage/ [Accessed: November 30, 2022]

[67] Energy focus, magazine from the energy industries council, Towards the Ion Age. 2019

Chapter 9

Medium-High Temperature Composite Phase Change Materials Based on Porous Ceramics

Jun Qiu and Xibo He

Abstract

Medium-high temperature thermal energy storage usually uses composite phase change materials (CPCMs) composed of inorganic salts and porous skeletons, due to their high energy density, wide phase change temperature range, and stable physical/chemical properties. Inorganic salts provide enough heat storage capacity, and the porous skeleton is a stable packaging carrier that solves the low thermal conductivity and easy leakage of the former. Compared with alloy matrices, porous ceramics have higher corrosion resistance, economic benefits, and more stable thermal performance, which is more suitable for medium-high temperature thermal engineering applications. This chapter focuses on the medium-high temperature CPCMs based on the porous ceramic matrix, including the preparation process and thermal properties of CPCMs, the material types and selection principles of porous ceramic, and the system-level comprehensive study and application. This work aims to provide a coupling relationship between porous ceramics and inorganic salts so that the reader can obtain the ideal CPCMs in a specific application.

Keywords: porous skeleton, medium-high temperature, phase change materials, shape-stabilization, thermal energy storage

1. Introduction

Energy is an indispensable component of human survival and social development. The excessive exploitation and abuse of fossil energy such as coal, oil and natural gas has caused the current global energy crisis and environmental pollution [1, 2]. Therefore, the development of renewable energy (solar energy) and energy conservation are considered the main directions of future energy development. "Green energy" is limited by its constraints showing intermittency and instability [3–5]. Solving the time-space mismatch between the supply and demand of thermal energy is one of the most critical obstacles, which makes thermal energy storage (TES) technology become an important support to promote the energy revolution. Thermal storage materials are a central part of TES [6]. Real-time thermal energy distribution with thermal storage material as the medium for the demand of the energy-using side can significantly improve the energy efficiency, stability, and economic benefits of TES [7, 8].

Solar photo-thermal and industrial waste heat are the main targets in the strategy of practical thermal energy storage conversion and their temperature ranges are shown in **Table 1** [11]. Medium-high temperature latent heat TES technology (>120°C) to store excess thermal energy is the most ideal choice for packed bed latent heat TES technology (PLTES). The high operating temperature range makes the available phase change materials (PCMs) mainly divided into inorganic salts and alloys [12]. For medium-high temperature PCMs, inorganic salts are the most widely accepted objects, including chlorinated salts, nitrates, carbonates, and fluorinated salts, referenced as shown in **Table 2** [14–16]. Molten salts are bound to undergo a solid-liquid phase transition during charging/discharging cycling in practical

Applied environmental objectives	Operating temperature range	Reference
Concentrated solar power	700–900°C	[2]
Industrial waste heat recovery	100–750°C	[3]
Solar energy	500–700°C	[9]
Regenerative combustion	800–1000°C	[10]

Table 1.
Temperature range of practical thermal energy storage-conversion.

Compound (wt.%)	Melting point (°C)	Latent heat (kJ/kg)	Density (kg/m^3)	Thermal conductivity (W/m•K)
NaCl [13]	800	492	2160	2.0
KCl [13]	774	353	1980	1.25
$MgCl_2$ [13]	714	452	2140	0.86
NaCl-$MgCl_2$ (48:52) [14]	450	430	2230	0.95
KCl-NaCl-$MgCl_2$ (33:24:43)	380	400	1800	1.15
$NaNO_3$ [15]	270	180	2200	0.53
$Ca(NO_3)_2$ [15]	561	145	250	0.32
$LiNO_3$ [15]	254	360	2380	0.58
$LiNO_3$-$NaNO_3$ (6:7) [15]	200	232.7	2090	0.84
$Ca(NO_3)_2$-$NaNO_3$ (3:7) [15]	217	135.8	2130	0.58
Na_2CO_3 [13]	854	275.7	2533	2.0
K_2CO_3 [13]	897	235.8	2290	2.0
$CaCO_3$ [9]	1330	142	2390	1.5
K_2CO_3-Na_2CO_3 (51:49) [14]	710	163	2400	1.73
Na_2SO_4 [16]	884	165	2680	1.26
NaF [9]	996	468	2258	1.5
NaF-MgF_2 (72:25) [14]	650	860	2820	1.15

Table 2.
Performance parameters of common molten salts.

applications, which makes the structural requirements of PLTES particularly high and difficult to massively promote. Therefore, there is a need to prepare inorganic salt-based PCMs into shape-stable composite phase change materials (CPCMs) with high energy density, wide operating temperature range, stable physical/chemical properties, and low prices through rational encapsulation methods.

It should be noted that the inorganic salt-based CPCMs have the dilemma of low thermal conductivity and severe corrosion [17]. Mating inorganic salts with porous skeletons to obtain inorganic salt-based CPCMs can effectively solve these problems. Inorganic salt provides sufficient heat capacity, and the porous skeleton acts as a stable carrier and reduces overcooling, so CPCMs have excellent heat storage and release effects [18]. A skeleton matrix generally requires high porosity and a large surface area to provide more PCM loading volume. Ceramic is known to be one of the best choices for porous skeletons with high-temperature corrosion resistance, and high thermal conductivity. Preparation of CPCMs from porous ceramics and inorganic salts can enhance heat transfer and support-encapsulation, with higher corrosion resistance and economic benefits compared to alloy carriers. However, there are fewer comprehensive reviews focusing on medium-high temperature ceramic-based CPCMs, with some content bias: the preparation process is not specific; the materials selection is not comprehensive, and system-level studies and applications are not mentioned.

Our group has been focusing on the study of medium-high temperature CPCMs based porous ceramic (C-CPMs), and thus have a comprehensive understanding of that field. This chapter summarizes the recent contents of C-CPCMs in detail. Firstly, the preparation process and the influence of thermal properties of medium-high temperature CPCMs are described in detail, followed by a summary of the material types and selection principles of porous ceramic skeletons; finally, the CPCMs are described for system-level research and application.

2. Preparation process and thermal properties of ceramic-based CPCMs

Medium-high temperature ceramic-based composite phase change materials are usually composed of inorganic salts with a porous skeleton. Inorganic salts are embedded into porous ceramic skeletons by an efficient method for the preparation of shape-stabled CPCMs. It solves the problems of poor thermal conductivity, substrate corrosion, and material encapsulation in inorganic salt-based PCMs, and its sensible-latent heat gives it a high heat capacity. At present, the preparation of medium-high temperature ceramic-based shaped CPCMs is mainly divided into three categories: mixing-impregnation method and cold pressing-sintering method.

2.1 Mixing impregnation

Mixing impregnation is based on the capillary force of the porous ceramic, which confines the solid-liquid PCM in the pores to prevent leakage and can effectively improve the overall thermal conductivity. The simple preparation process of mixing-impregnation is shown in **Figure 1**. It could be divided into two main steps: (1) preparation of ceramic porous skeleton as a carrier for the inorganic salt, and (2) infiltration of the PCM to provide enough heat capacity. Typically, a ceramic skeleton is prepared by rational pore-making techniques and high-temperature sintering, and the molten PCM penetrates the porous carrier by capillary forces.

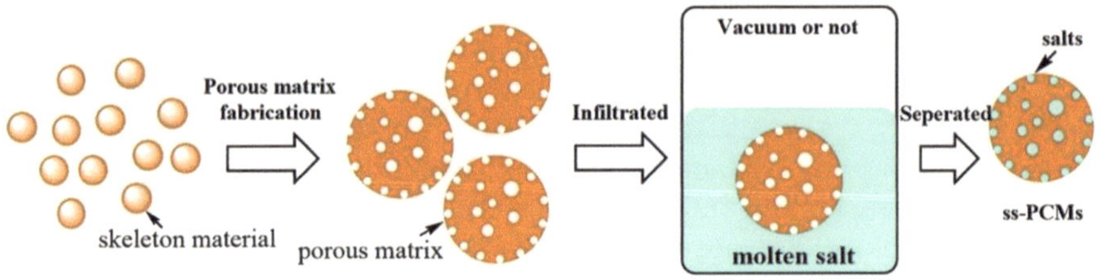

Figure 1.
Process flow for mixed impregnation [19] (Reprinted with permission from Elsevier [OR APPLICABLE SOCIETY COPYRIGHT OWNER]).

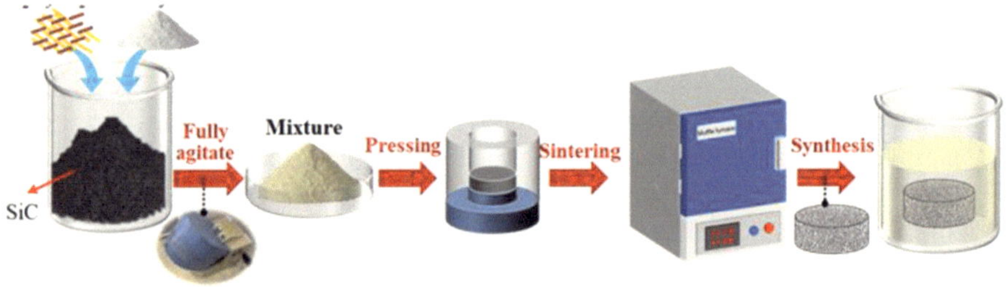

Figure 2.
Process flow of ternary chloride/SiC CPCM [23] (Reprinted with permission from Elsevier [OR APPLICABLE SOCIETY COPYRIGHT OWNER]).

The basic equipment includes a high-speed mill, a high-temperature tube furnace, and a vacuum sintering furnace. Mixing impregnation consists mainly of melt impregnation and vacuum impregnation. Most of the PCMs could be adsorbed into the ceramic skeleton by melt impregnation. However, some pores may still exist inside the carrier, leading to a reduction in the total CPCMs' thermal storage density. Therefore, vacuum impregnation under high-temperature and vacuum environments can introduce more PCM to support the matrix. In general, CPCM units are subjected to continuous thermal cycling charging/discharging tests to ensure their mechanical strength and thermal stability.

Jiang et al. [20] prepared shaped-stabilized CPCMs based on modified diatomite by loading $NaNO_3$ into the porous ceramic skeleton using melt-impregnation. The infiltration process of the sample was carried out in a muffle furnace under atmospheric pressure, followed by sandpapering of the excess cured salt. The ss-PCM has excellent cyclic thermal stability, thermal storage density, and efficiency. Wang et al. [21] used the same method to prepare high-temperature ss-CPCMs with ternary chloride salts as the PCM, porous Si_3N_4 as the skeleton, and heat transfer enhancers. The pore permeability of PCM in the skeleton was 88.14%, and the thermal conductivity of ss-CPCMs was significantly improved. To develop high-performance CPCMs that can be mass-produced at low cost, our group [21, 22] used vacuum-impregnation to combine SiC ceramics with high-enthalpy ternary chloride salts as shown in **Figure 2**. The controlled porosity of the skeleton results in excellent thermal properties (thermal conductivity-heat capacity). PCM relies on capillary force to enter the carrier in a vacuum environment and has a high loading capacity. Other current research about mixing-impregnation is summarized in **Table 3**, including compositions and results.

Authors	Compositions	Main research content
Liu [24]	Bamboo-derived SiC/LiOH-LiF	CPCM has high solar absorptivity and high thermal conductivity;
Song [25]	SiC/LiOH-LiF	Modulation of porosity of ceramics by starch; Excellent thermal conductivity and stability;
Zhang [26]	Solar salt/SiC	Excellent wettability and corrosion resistance; The temperature distribution is more uniform;
Miliozzi [27]	Solar salt/diatomite	Experimental study of high-temperature thermal storage concrete materials; Thermal and mechanical properties increase;
Fang [28]	$CaCl_2{\bullet}6H_2O$-$CO(NH2)_2$-SiO_2	SiO_2 as a porous ceramic carrier for CPCM; The CPCM exhibits excellent thermal stability;

Table 3.
Research related to the melt-impregnation method.

It needs to be clarified that the mixing impregnation requires proper regulation of the pore structure of the porous skeleton. First, the porous ceramics' pore size should be within a certain range. Too small pore size will limit the microflow of the PCM, and too large pores will not provide sufficient capillary. In addition, the carrier's porosity needs to be regulated. Too low porosity reduces the total heat storage capacity, and too high porosity affects the thermal cycling stability of the CPCPMs [29]. Nomura et al. [30] investigated the effect of the ceramic materials' pore size (diatomite) in CPCMs on the melting point. Due to the nano-size effect, the smaller the pore size of diatomite, the lower the melting point of the CPCMs. Liu et al. [31] also demonstrated that increasing the porosity or decreasing the pore size could effectively increase the infiltration ratio of inorganic salt. Both the sintering temperature and impregnation temperature during the preparation process affect the CPCMs' thermal properties [32]. The former is a key factor in regulating the pore structure of the ceramic carrier, while the latter ensures the loading.

2.2 Cold press-sintering

Cold press-sintering can usually be summarized in three steps: mixing grinding, hydrostatic forming, and high-temperature sintering, as shown in **Figure 3**. First, the inorganic salts were mixed and ground with ceramic materials, then the mixture was poured into molds and pressed into desired shapes, and finally, the CPCMs were prepared by high-temperature sintering. The basic equipment includes high-speed mills, tablet presses, and high-temperature sintering furnaces. It needs to be clarified that the stability of CPCMs prepared by the cold pressing-sintering method stems from mechanical locking during pressing and mounding under the external force, and bonding after the melting and solidification of PCM/ceramic particles during the high-temperature sintering. In addition, the structure of CPMs is related to the mounding process. Conventional isostatic pressing results in a more homogeneous and dense material; uniaxial pressing produces a hierarchically arranged structure.

Porous ceramics are inherently capable of adsorbing molten salts to prevent leakage, making them the preferred target for the preparation of ceramic-based CPCMs by cold pressing-sintering. Ye et al. [33] used MgO to encapsulate PCMs.

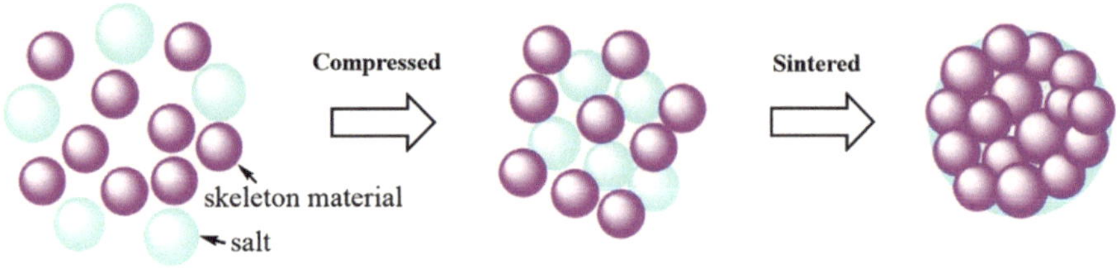

Figure 3.
Preparation process of cold pressed-sintered [19] (Reprinted with permission from Elsevier [OR APPLICABLE SOCIETY COPYRIGHT OWNER]).

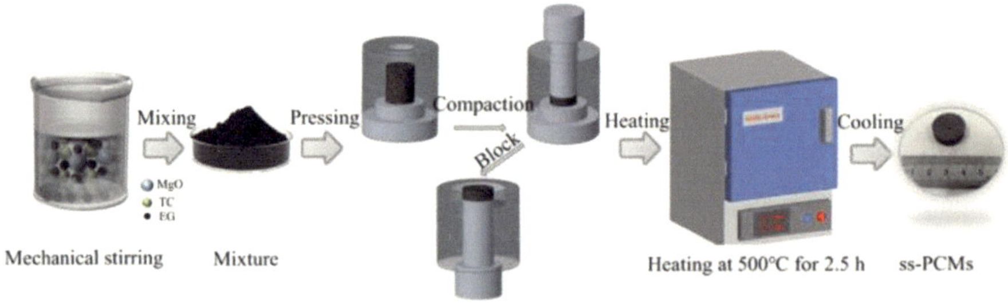

Figure 4.
Process flow of TC/MgO/EG CPCMs [35] (Reprinted with permission from Elsevier [OR APPLICABLE SOCIETY COPYRIGHT OWNER]).

They successfully obtained Na_2CO_3/MgO composites by using cold pressing-sintering and adding carbon nanotubes (MWCNTs) as additives. The CPCM's thermal conductivity increases with the increase in the weight fraction of MWCNTs. Deng et al. [34] prepared KNO_3/diatomite-shaped-stabilized CPCMs by mixing-sintering. Diatomite has high porosity, which acts as a high-strength carrier and effectively limits the leakage of PCMs. The 65% loaded CPCMs have good physical/chemical properties with latent heat of 60.52 J/g and an effective melting point of 330°C. To further enhance the thermal performance of ss-PCM, our group [35] added EG to ternary chloride (TC)/MgO ceramics-shaped CPCMs. The specific preparation process of CPCMs is shown in **Figure 4**. MgO carrier as the supporting skeleton could improve the mechanical strength of CPCMs, and EG as the additive could increase the thermal conductivity and improve the heat transfer process. CPCMs have a promising future in high-temperature solar applications. Some other current research about cold press-sintering is summarized in **Table 4**, including compositions and results.

For the cold pressing-sintering process, the ratio between the materials, the amount of molding pressure, and the sintering temperature all have an important effect on the CPCMs' mechanical and thermophysical properties [40]. Qin et al. [32, 41] investigated the Na_2SO_4/diatomite CPCMs ratio. It was finally determined that CPCMs containing 45% diatomite were optimal in terms of energy density, leakage prevention, and mechanical strength. The effect of sintering temperature (300–500°C) on ceramic-based CPCM was analyzed by Ji et al. [42]. The maximum mass fraction of PCM and compressive lightness decreased with increasing sintering temperature, and the thermal conductivity remained almost constant.

Authors	Compositions	Main research content
Qian [36]	Na_2SO_4/diatomite	The maximum loading of Na_2SO_4 is 65%; Improved subcooling and thermal stability;
Wang [37]	K_2CO_3/Coal ash	Prepare SSPCMs by two-step; The melting point is about 850°C;
Ge [38]	$LiCO_3$/MgO/C	Good wettability makes ceramics denser; Carbon as the thermal conductivity enhancer;
Liu [39]	$NaNO_3$/ anorthite-cordierite	The wettability of $NaNO_3$ on cordierite was better; All have good chemical compatibility;
Jiang [40]	Na_2SO_4-NaCl/Al_2O_3	The stability of CPCM decreases with salt content; 300% increase in thermal conductivity;

Table 4.
Research related to the cold press-sintering method.

3. Selection of ceramic porous skeleton

Porous ceramic skeletons are generally prepared from inorganic non-metallic powder materials by a molding process and high-temperature sintering, possessing a large number of irregular pore structures. For PCMs, especially inorganic salts, their low thermal conductivity and corrosion seriously affect industrial applications in solar thermal utilization, waste heat recovery, and other fields. Porous ceramics encapsulated PCM could effectively solve the above problems. The pore structure of porous ceramics is generally required to be open-pores, and particle stacking or precursor combustion is used to obtain a high porosity.

The selection of ceramic skeletons for medium-high temperature CPCMs generally must meet the following requirements:

1. The basic melting point of the porous ceramics should be higher than 1250°C to ensure a stable shape of the CPCMs;

2. The skeleton material could form a good porous structure or be porous;

3. Required to maintain good, long-period thermal cycle stability;

4. Good chemical compatibility between ceramic materials with PCMs;

5. Ceramic materials with higher specific heat capacity should be selected as the carrier to realize higher sensible and latent heat.

3.1 Porous oxide and non-oxidized ceramic materials

Some porous oxide (MgO, Al_2O_3) and non-oxide materials (SiC, AlN) could be prepared into porous ceramic by suitable processes to provide encapsulation space for inorganic salts. The pore structure of CPCMs prepared from them tends to be uniform, and the leakage of CPCMs could be limited by capillary force and surface tension. In addition, these materials possess excellent high-temperature resistance and physical/chemical stability, making them a reliable choice for the application of medium-high-temperature CPCMs.

MgO has become a widely used carrier material in ceramic-based CPCMs due to its advantages of high thermal conductivity, high specific heat, good wettability with inorganic salts, and wide application temperature range. Li et al. [43] conducted an in-depth study on MgO-based CPCMs and analyzed the effects of particle size and density on CPCMs' microstructure and thermal properties. The results demonstrate that light and small MgO could realize denser structures and better encapsulation. They also found that CPCMs prepared from similar-sized compositions present higher thermal conductivity and mechanical strength. Sang et al. [44] prepared shaped CPCMs using ternary chloride as PCMs and MgO as the carrier (the optimal ratio 5:5). The CPCMs were uniformly embedded in the pore structure of MgO ceramics by cold compression and mixed sintering. Similarly, Ye et al. [33] added MWCNTs as thermal conductivity enhancers to the Na_2CO_3/MgO CPCMs. The results showed that MgO as a porous support matrix ensured shape stabilization. Our group encapsulated ternary chlorides with MgO as a carrier and EG as a thermally conductive additive. The CPCM has good mechanical strength and thermal stability. The microstructure and the thermal properties of the composite are shown in **Figure 5**.

Porous Al_2O_3 is another low-cost ceramic material with excellent thermal stability and corrosion resistance and has also been applied for medium-high temperature CPCMs. Jiang et al. [45] prepared Na_2SO4-NaCl high-temperature CPCMs by cold press sintering. Alumina as the encapsulation skeleton has a microporous coral mesh structure, which can effectively prevent the leakage of PCM. After 100 thermal cycles, the CPCM's latent heat and solidification temperature only decreased by 0.4% and 0.3% without phase separation and chemical reaction. The molten PCM possessed good chemical compatibility with the alumina skeleton. Ji et al. [42] prepared $NaNO_3$-KNO_3/EG/Al_2O_3 shape-stable CPCMs at different sintering temperatures, with Al_2O_3 and EG as skeletal support materials. The results showed that the PCM loading mass fraction and compressive strength decreased with increasing sintering temperature. The thermal conductivity of CPCM is independent of the sintering temperature but increases with EG mass fraction.

Porous SiC has the advantages of high porosity, good permeability, and high thermal conductivity. The porous SiC ceramics obtained through an effective preparation process have high strength, low density, interconnected pore structure, and stable thermal properties. The porous SiC ceramics encapsulated with inorganic salt can solve the low thermal conductivity and leakage of the traditional PCM, and it is

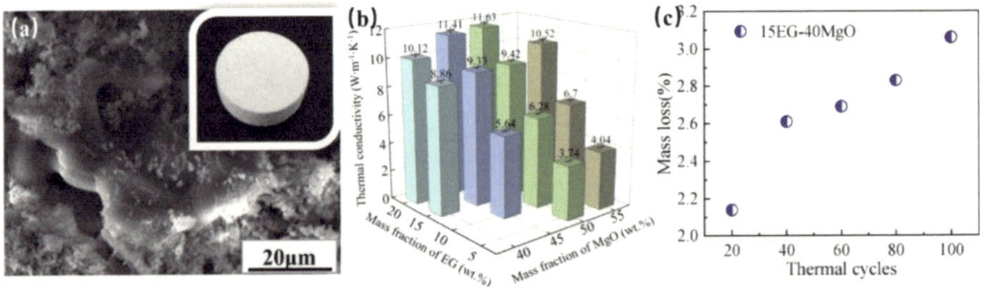

Figure 5.
(a) SEM image after sintering, (b) thermal conductivity with different proportions, and (c) variations of mass loss after 100 cycles [35] (Reprinted with permission from Elsevier [OR APPLICABLE SOCIETY COPYRIGHT OWNER]).

an excellent skeleton material for composite phase change materials. Luo et al. [46] encapsulated PCM through gradient SiC foam as the backbone for fast and stable thermal storage. The fabrication strategy of gradient SiC foams is shown as (a). The strong capillary force generated by the gradient pore structure effectively limits the leakage of PCM. In addition, the interconnected SiC backbone enables the thermal conductivity of CPCMs to be increased to 1.9 W/(m•K) and realizes an efficient solar thermal storage process. Our group regulates the porosity of SiC ceramics based on starch pore-formation, which in turn optimizes the thermal conductivity and mechanical strength of CPCMs [23]. A continuous and stable fast heat transfer channel was constructed by high-temperature sintering and loaded with ternary chloride salt to ensure effective heat storage density. The cell shape, microscopic pore structure, and temperature evolution with time during charging/discharging of CPCM are shown in **Figure 6**. The CPCMs have a thermal conductivity of 19.72 W/(m•K) and an effective heat storage density of 513.46 kJ/kg.

In addition to the ceramic materials mentioned above, other materials could be used as skeleton options for medium-high-temperature ceramic-based CPCM. They are inherently porous or capable of generating porous skeletons to load inorganic salts. They also improve the thermal conductivity of the CPCM, limit PCM leakage, and have good thermal cycle stability. The studies of CPCMs based on other porous oxide and non-oxidized ceramics are summarized in **Table 5**, including compositions and results.

3.2 Porous clay-like mineral materials

Clay minerals such as diatomite, and expanded perlite (EP), inherently have high-quality porous structures and large specific surface area, which can better realize the embedding of PCMs as the core material and effectively reduce the leakage. They serve as a ceramic skeleton matrix for CPCMs and are capable of long cycles and stable operation at medium-high temperatures, maintaining good chemical compatibility and stability, making them an ideal choice for commercial applications. The wide range of raw materials, low cost, and high performance make porous clay mineral skeleton materials an ideal choice for commercial applications.

Diatomite is a natural mineral material whose main component is SiO_2. It has a unique porous structure with a specific surface area of 40–65 m^2/g and a porosity of 80–90%. This gives diatomite excellent adsorption properties, which in turn

Figure 6.
(a) the cell shape, (b) the microscopic pore structure, and (c) the temperature evolution during charging/discharging of CPCM [23] (Reprinted with permission from Elsevier [OR APPLICABLE SOCIETY COPYRIGHT OWNER]).

Authors	Compositions	Main research contents
Liu [47]	NaCl-$LiNO_3$/AlN	Combination of high thermal conductivity and high heat storage density The average solar absorptance is 90%;
Wang [48]	CaF_2-ODE/AlN	A novel thermal enhanced CPCMs with based AlN; Excellent thermal cycling stability;
Guo [49]	$NaNo_3/SiO_2$	The CPCM was prepared by sol-gel process; The latent heat was increased with the increase of the roasting temperature.
Zhao [50]	Na_2SO_4/TiO_2	CPCMs prepared by a facile particle-stabilized emulsion templating method; PTC improved the heat transfer and heat density;
Yu [51]	$NaNO_3/Ca(OH)_2$	The excellent mechanical property was obtained at a pressure of over 220 MPa;

Table 5.
Research related to porous oxide and non-oxidized ceramic materials.

encapsulate PCM with good thermal stability. In addition, diatomite is abundant, readily available, low-cost, and characterized by high temperature and corrosion resistance. Therefore, it has been widely used as a skeleton support material for CPCMs over the years.

Qin et al. [41] prepared high-temperature Na_2SO_4/diatomite-shaped CPCMs by mixing-sintering. The diatomite acts as a shape-stabilized skeleton and possesses excellent chemical-thermal stability with the PCM. They experimentally determined that 45% diatomite was the optimal formulation for CPCMs with an energy density of more than 360 KJ/kg in the range of 700–900°C. Qian et al. [52] used diatomite as matrix and loaded three different PCMs by vacuum impregnation to prepare low, medium, and high CPCMs for solar thermal utilization. They also proved that the melting point, latent heat, and subcooling of CPCMs are constant through their self-designed experimental device. However, since the main component of diatomite is the natural SiO_2, it has the disadvantage of low thermal conductivity (0.2 ~ 0.4 W/(m•K)) compared with metal encapsulation materials. Li et al. [53, 54] synthesized MnO_2-modified diatomite by hydrothermal reaction and then prepared CPCMs by vacuum-impregnation. The preparation process of MnO_2-modified diatomite is shown in **Figure 7**. After characterization, it was proved that the ceramic-based CPCMs possessed a faster heat transfer rate and better photothermal conversion capability. In addition, they also treated the diatomite with microwave acid to obtain higher porosity and added EG to improve thermal conductivity. Jiang et al. [19] used $CaCO_3$-modified diatomite as a porous ceramic skeleton for CPCMs. Due to the dense-continuous skeleton, the CPCMs' thermal conductivity was improved by 129% and exhibited excellent thermal stability in 500 thermal cycles.

Expanded perlite (EP) is a traditional thermal insulation material. Its volume expands rapidly after high-temperature calcination, allowing it to acquire a foamy honeycomb structure with strong adsorption properties, which could be used to prepare high-performance CPCMs. Based on the excellent adsorption capacity of EP, Li et al. [55] prepared $NaNO_3$/EP shaped CPCMs. The results showed that skeletons at different treatment temperatures (300–900°C) exhibited different adsorption strengths, with the highest adsorption after heat treatment at 500°C for 2 h. The CPCMs have good thermal conductivity and thermal cycling stability.

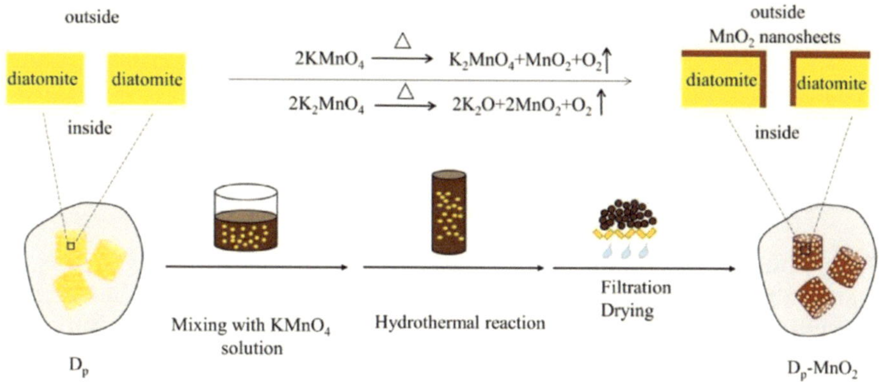

Figure 7.
Preparation process of MnO_2 modified diatomite [53] (Reprinted with permission from Elsevier [OR APPLICABLE SOCIETY COPYRIGHT OWNER]).

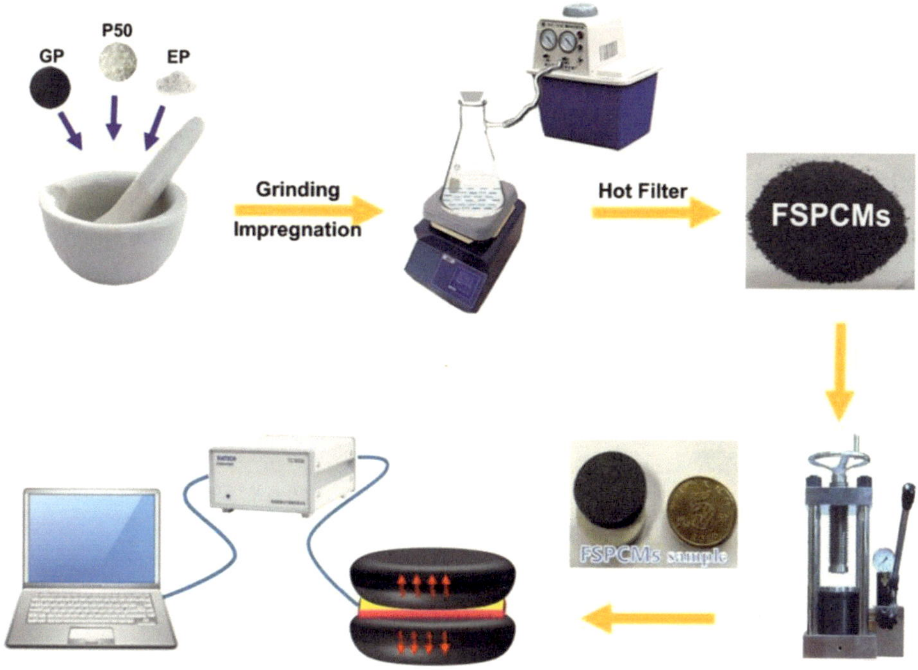

Figure 8.
Preparation of composite FSPCMs [57] (Reprinted with permission from Elsevier [OR APPLICABLE SOCIETY COPYRIGHT OWNER]).

Zhao et al. [56] regulate the porous structure of EP by polyvinyl alcohol (PVA) to increase the nanopore size and specific surface area. Then they proposed novel CPCMs with high excellent loading (73.1%) and heat storage capacity (174.6 J/g). After 500 thermal cycling experiments, the samples possessed good thermal stability and leakage resistance. Zuo et al. [57] optimized the EP-based CPM by adding graphite, and the basic preparation process is shown in **Figure 8**. Characterization results demonstrate that the addition of graphite improves the thermal stability and thermal conductivity of CPCMs. EP/P50/GP80 exhibit the highest heat transfer rates and have good chemical stability.

4. System-level research and applications

4.1 Numerical and experimental study of ceramic-based CPCMs

The optimized preparation process and suitable PCM make the ceramic-based inorganic salt CPCMs have excellent thermal properties, which solves the low thermal conductivity and leakage existing in traditional PCMs. However, a single heat storage unit cannot visualize the heat transfer process at the level of latent heat Thermal energy system (LTES), so system-level research has become one of the hotspots of research at this stage. Our group built a medium-high temperature packed-bed heat storage experimental platform in the laboratory to carry out system-level research [5, 58]. Through system-level simulation calculations and experiments, the thermal properties of the CPCMs within the TES are analyzed, which in turn could provide sufficient guidance for their wide-ranging applications.

Yao et al. [59] investigated the latent heat storage device of a 3D porous skeleton encapsulated PCM based on MCRT and FVM methods to systematically evaluate the heat transfer process under different design parameters. The results showed that the solar thermal efficiency and thermal storage efficiency increased by 71% and 94%, respectively. Li et al. [60] investigated a high-temperature packed-bed LTES with ceramic-based CPCMs and analyzed the heat transfer-flow behavior at the system level. The results show that the system exhibits higher charging/discharging efficiency. The heat transfer efficiency is enhanced when the radiative heat transfer influence is taken into consideration. Finally, the charging/discharging cycle decreases with the increase of the thermal conductivity enhancer.

In addition, Li et al. [61] investigated the thermal performance of MgO-based CPCMs from component to device levels, as shown in **Figure 9**. They analyze the cross-scale problem of CPCMs from the unit-component-device level through numerical calculations and experiments. The CPCMs have MgO as the porous skeleton, $NaLiCO_3$ as the PCM, and graphite as the thermal conductivity enhancer. The results show that the increase in the mass loading of the TCEM in the CPCMs module, and the entrance velocity of the HTF all improve the heat transfer performance at the component level. Zhao et al. [62] investigated the heat transfer process of two types of CPCM components for shell-and-tube TES: single-tube and concentric-tube components. Through

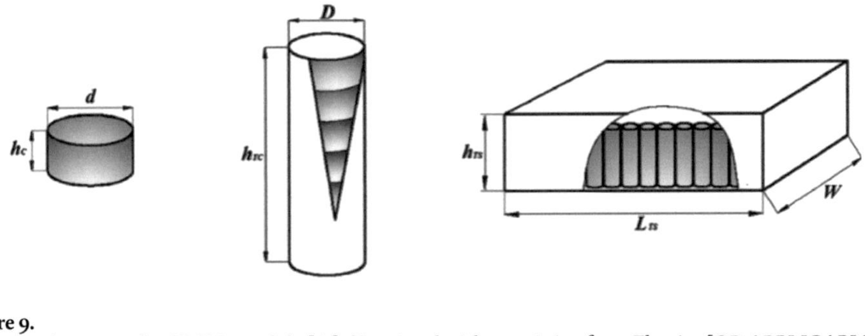

Figure 9.
Schematic diagrams of a CPCMs module [61] (Reprinted with permission from Elsevier [OR APPLICABLE SOCIETY COPYRIGHT OWNER]).

numerical calculations and experiments, they finally concluded that the concentric tube CPCM components have a better heat transfer performance, and the heat storage and release time is 10% and 15% lower than the single-tube one.

4.2 Application of ceramic-based CPCMs

Medium-high temperature ceramic-based CPCMs could charge or discharge a large amount of heat in the form of sensible and latent heat to solve the conflict between the supply and demand of renewable energy. It could be seen that ceramic-based CPCMs have high heat storage capacity, long cycle thermal life, excellent corrosion resistance, and fast heat transfer rate at medium-high temperatures. This ensures the prospect of a wide application in TES systems, mainly including solar thermal utilization, and waste heat recovery.

Solar thermal technology is severely limited by its instability and discontinuity, so it is necessary to be coupled with a PLTES system to achieve dynamic regulation of heat [63]. Ceramic-based CPCMs have excellent thermal properties and corrosion resistance and are suitable for continuous-dynamic input of high-temperature heat flow in concentrating solar heat collection systems. Xu et al. [64] prepared porous SiC with excellent thermal toughness by biomorphic loofah as a template, and its porosity could be adjusted between 64% and 87%. Then the CPCMs are further obtained by encapsulating NaCl-NaF through the SiC skeleton, which realizes broadband solar energy capture, rapid thermal transfer, and compact latent heat energy storage. 95.25% solar absorption and 20.7 W/m•K high thermal conductivity are the composites that can quickly realize solar-heat transfer and storage. Excellent solar absorption (95.25%) and thermal conductivity (20.7 W/(m•K)) enable CPCMs to quickly realize solar-heat transfer and storage. Zhang et al. [65] developed a novel microcapsule CPCM with Ti_4O_7/SiO_2 as the encapsulating shell. CPCMs have a high photo-thermal storage efficiency of 85.36%, which efficiently realizes the conversion of photo-conductive heat and preserves most of the heat. Our group initially used stainless steel spheres to encapsulate ternary nitrate for system-level charging-discharging experiments [58]. A schematic diagram of the experimental platform is shown in **Figure 10b** But this approach suffers from the defects of low thermal conductivity and easy leakage.

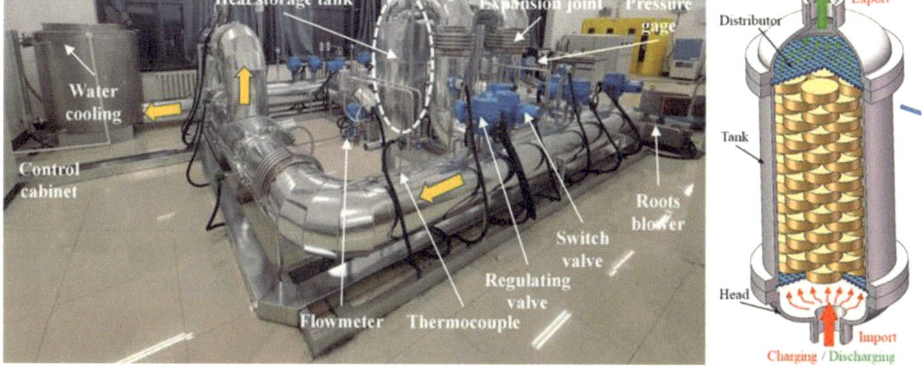

Figure 10.
Diagram of the experimental platform, the arrangement structure, and temperature evolution of SiC-based CPCMs [23, 58] (Reprinted with permission from Elsevier [OR APPLICABLE SOCIETY COPYRIGHT OWNER]).

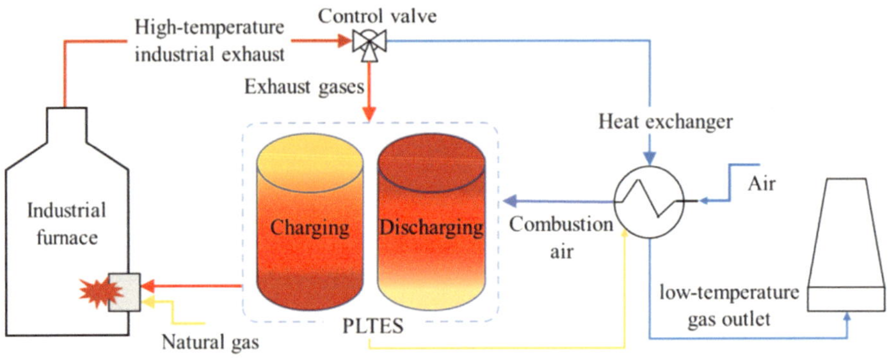

Figure 11.
The schematic diagram of the industrial waste heat recovery system based PLTES.

Therefore, we subsequently developed a simple preparation process for SiC-based CPCMs using starch as a pore-forming agent [23], which could be prepared on a large scale and operated stably for a long lifetime. The CPCMs were verified to have high heat transfer efficiency by systematic experiments.

Traditional thermal power plants and other heavy industries directly discharge a large amount of medium-high temperature waste heat during the production process, resulting in a serious waste of energy. Based on PLTES, the collection and storage of waste heat could be realized, effectively improving the efficiency of energy utilization, and reducing greenhouse gas emissions. The schematic diagram of the industrial waste heat recovery system-based PLTES is shown in **Figure 11**. Due to the continuous fluctuations, high temperatures, and high impacts of industrial waste heat emissions, ceramic-based CPCMs are a key option for the realization of waste heat recovery. Li et al. [66] developed a dynamic optimization model based on CPCMs to solve the dynamic thermal management of industrial waste heat. CPCMs were prepared by using nitrate PCM, silica carrier material, and graphite conductive addictive, and then filled in steel tubes. They developed an intelligent algorithm to study the dynamic optimization of the waste heat recovery system and built a waste heat recovery platform for steel production to verify the reliability of the shell-and-tube heat storage model.

5. Conclusions

This chapter systematically summarizes the preparation process of medium-high temperature ceramic-based CPCMs, the material types of porous ceramic skeletons and the study and application of CPCMs at the system level. So that readers can recognize the basic information of medium-high temperature ceramic-based CPCMs and screen the ideal CPCMs according to specific needs.

1. The preparation process of ceramic-based CPCM mainly includes mixing-impregnation and cold press-sintering. The former could realize the complex shape of CPCMs with high strength and thermal cycling stability but has high requirements for the skeleton's high porosity and pore size, and the process is complex and energy-consuming. The latter process is simple, economical, and suitable for mass production. However high-temperature sintering causes some PCM evaporation, and the skeleton structure is not uniform and easily damaged.

2. Skeleton materials are the key to ensuring the stability of CPCM at high temperatures for a long time. The oxide, non-oxidized ceramic materials, and clay-like mineral materials could form special porous structures to encapsulate PCMs and effectively prevent leakage. In addition, the porous ceramic skeleton could effectively increase the thermal conductivity of CPCMs and improve the absorption, storage, and transportation of heat.

3. Medium-high temperature ceramic-based CPCMs have been widely applied in the fields of solar thermal utilization, industrial waste heat recovery, etc., which makes the research of system-level CPCMs in TES an important direction in the future. The functionalization and intelligence of ceramic-based CPCMs could realize the efficient utilization/management of heat, and ensure efficient and stable energy output and targeted applications under dynamic external environmental changes.

Acknowledgements

Jun Qiu: Supervision, Conceptualization, Methodology, Investigation, Writing–review & editing, Funding acquisition.

Xibo He: Methodology, Investigation, Writing–original draft, Visualization, Data curation.

This work was supported by the National Natural Science Foundation of China (No. 52076062), National Key R&D Program of China (No. 2018YFA0702300), and Fundamental Research Funds for the Central Universities (No. 2023FRFK06007).

Conflict of interest

The authors declare that they have no known competing financial interests or personal relationships that could have appeared to influence the work reported in this chapter.

Author details

Jun Qiu* and Xibo He
School of Energy Science and Engineering, Harbin Institute of Technology, Key Laboratory of Aerospace Thermophysics, Ministry of Industry and Information Technology, Harbin, China

*Address all correspondence to: qiujun@hit.edu.cn

References

[1] Esence T, Bruch A, Molina S, Stutz B, Fourmigué J-F. A review on experience feedback and numerical modeling of packed-bed thermal energy storage systems. Solar Energy. 2017;**153**:628-654. DOI: 10.1016/j.solener.2017.03.032

[2] Alva G, Lin Y, Fang G. An overview of thermal energy storage systems. Energy. 2018;**144**:341-378. DOI: 10.1016/j.energy.2017.12.037

[3] Miró L, Gasia J, Cabeza LF. Thermal energy storage (TES) for industrial waste heat (IWH) recovery: A review. Applied Energy. 2016;**179**:284-301. DOI: 10.1016/j.apenergy.2016.06.147

[4] Ellabban O, Abu-Rub H, Blaabjerg F. Renewable energy resources: Current status, future prospects and their enabling technology. Renewable and Sustainable Energy Reviews. 2014;**39**:748-764. DOI: 10.1016/j.rser.2014.07.113

[5] Wang W, He X, Hou Y, Qiu J, Han D, Shuai Y. Thermal performance analysis of packed-bed thermal energy storage with radial gradient arrangement for phase change materials. Renewable Energy. 2021;**173**:768-780. DOI: 10.1016/j.renene.2021.04.032

[6] Tao YB, He YL. A review of phase change material and performance enhancement method for latent heat storage system. Renewable and Sustainable Energy Reviews. 2018;**93**:245-259. DOI: 10.1016/j.rser.2018.05.028

[7] Li MJ, Jin B, Ma Z, Yuan F. Experimental and numerical study on the performance of a new high-temperature packed-bed thermal energy storage system with macroencapsulation of molten salt phase change material. Applied Energy. 2018;**221**:1-15. DOI: 10.1016/j.apenergy.2018.03.156

[8] Chang Y, Yao X, Chen Y, Huang L, Zou D. Review on ceramic-based composite phase change materials: Preparation, characterization and application. Composites Part B: Engineering. 2023;**254**:110584. DOI: 10.1016/j.compositesb.2023.110584

[9] Alva G, Liu L, Huang X, Fang G. Thermal energy storage materials and systems for solar energy applications. Renewable and Sustainable Energy Reviews. 2017;**68**:693-706. DOI: 10.1016/j.rser.2016.10.021

[10] Kuravi S, Trahan J, Goswami DY, Rahman MM, Stefanakos EK. Thermal energy storage technologies and systems for concentrating solar power plants. Progress in Energy and Combustion Science. 2013;**39**:285-319. DOI: 10.1016/j.pecs.2013.02.001

[11] Liu M, Saman W, Bruno F. Review on storage materials and thermal performance enhancement techniques for high temperature phase change thermal storage systems. Renewable and Sustainable Energy Reviews. 2012;**16**:2118-2132. DOI: 10.1016/j.rser.2012.01.020

[12] Liu K, Yuan ZF, Zhao HX, Shi CH, Zhao F. Properties and applications of shape-stabilized phase change energy storage materials based on porous material support—A review. Materials Today Sustainability. 2023;**21**:100336. DOI: 10.1016/j.mtsust.2023.100336

[13] Xu B, Li P, Chan C. Application of phase change materials for thermal energy storage in concentrated solar

thermal power plants: A review to recent developments. Applied Energy. 2015;**160**:286-307. DOI: 10.1016/j.apenergy.2015.09.016

[14] Kenisarin MM. High-temperature phase change materials for thermal energy storage. Renewable and Sustainable Energy Reviews. 2010;**14**:955-970. DOI: 10.1016/j.rser.2009.11.011

[15] Ren Y, Xu C, Yuan M, Ye F, Ju X, Du X. $Ca(NO_3)_2$-$NaNO_3$/expanded graphite composite as a novel shape-stable phase change material for mid- to high-temperature thermal energy storage. Energy Conversion and Management. 2018;**163**:50-58. DOI: 10.1016/j.enconman.2018.02.057

[16] Lin Y, Alva G, Fang G. Review on thermal performances and applications of thermal energy storage systems with inorganic phase change materials. Energy. 2018;**165**:685-708. DOI: 10.1016/j.energy.2018.09.128

[17] Wei G, Wang G, Xu C, Ju X, Xing L, Du X, et al. Selection principles and thermophysical properties of high temperature phase change materials for thermal energy storage: A review. Renewable and Sustainable Energy Reviews. 2018;**81**:1771-1786. DOI: 10.1016/j.rser.2017.05.271

[18] Zhang S, Feng D, Shi L, Wang L, Jin Y, Tian L, et al. A review of phase change heat transfer in shape-stabilized phase change materials (ss-PCMs) based on porous supports for thermal energy storage. Renewable and Sustainable Energy Reviews. 2021;**135**:110127. DOI: 10.1016/j.rser.2020.110127

[19] Jiang F, Zhang L, She X, Cang D, Liu X, Xuan Y, et al. Skeleton materials for shape-stabilization of high temperature salts based phase change materials: A critical review. Renewable and Sustainable Energy Reviews. 2020;**119**:109539. DOI: 10.1016/j.rser.2019.109539

[20] Jiang F, Ge Z, Ling X, Cang D, Zhang L, Ding Y. Improved thermophysical properties of shape-stabilized $NaNO_3$ using a modified diatomite-based porous ceramic for solar thermal energy storage. Renewable Energy. 2021;**179**:327-338. DOI: 10.1016/j.renene.2021.07.023

[21] Wang H, Ren X, Zhong Y, Lu L, Lin J, He G, et al. Ternary chloride salteporous ceramic composite as a high-temperature phase change material. Energy. 2022;**238**:121838. DOI: 10.1016/j.energy.2021.121838

[22] Hou Y, Qiu J, Wang W, He X, Wang Z, Shuai Y. Controllable preparation and thermal properties of SiC spherical high temperature shape-stable composite phase change materials based on gel-casting. Journal of Alloys and Compounds. 2023;**960**:170966. DOI: 10.1016/j.jallcom.2023.170966

[23] He X, Wang W, Qiu J, Hou Y, Shuai Y. Controllable preparation method and thermal properties of composite phase change materials based on starch pore formation. Solar Energy Materials and Solar Cells. 2023;**253**:112255. DOI: 10.1016/j.solmat.2023.112255

[24] Liu X, Chen M, Xu Q, Gao K, Dang C, Li P, et al. Bamboo derived SiC ceramics-phase change composites for efficient, rapid, and compact solar thermal energy storage. Solar Energy Materials and Solar Cells. 2022;**240**:111726. DOI: 10.1016/j.solmat.2022.111726

[25] Song Y, Xu Q, Liu X, Xuan Y, Ding Y. High-performance thermal energy storage and thermal management

via starch-derived porous ceramics-based phase change devices. International Journal of Heat and Mass Transfer. 2022;**197**:123337. DOI: 10.1016/j.ijheatmasstransfer.2022.123337

[26] Zhang S, Li Z, Yao Y, Tian L, Yan Y. Heat transfer characteristics and compatibility of molten salt/ceramic porous composite phase change material. Nano Energy. 2022;**100**:107476. DOI: 10.1016/j.nanoen.2022.107476

[27] Miliozzi A, Chieruzzi M, Torre L. Experimental investigation of a cementitious heat storage medium incorporating a solar salt/diatomite composite phase change material. Applied Energy. 2019;**250**:1023-1035. DOI: 10.1016/j.apenergy.2019.05.090

[28] Fang Y, Wang K, Ding Y, Liang X, Wang S, Gao X, et al. Fabrication and thermal properties of $CaCl_2 \cdot 6H_2O$–$CO(NH_2)_2/SiO_2$ as room-temperature shape-stable composite PCM for building thermal insulation. Solar Energy Materials and Solar Cells. 2021;**232**:111355. DOI: 10.1016/j.solmat.2021.111355

[29] Jana DC, Sundararajan G, Chattopadhyay K, Colombo P. Effect of porosity on structure, Young's modulus, and thermal conductivity of SiC foams by direct foaming and gelcasting. Journal of the American Ceramic Society. 2017;**100**:312-322. DOI: 10.1111/jace.14544

[30] Nomura T, Okinaka N, Akiyama T. Impregnation of porous material with phase change material for thermal energy storage. Materials Chemistry and Physics. 2009;**115**:846-850. DOI: 10.1016/j.matchemphys.2009.02.045

[31] Liu R, Zhang F, Su W, Zhao H, Wang C-a. Impregnation of porous mullite with Na2SO4 phase change material for thermal energy storage. Solar Energy Materials and Solar Cells. 2015;**134**:268-274. DOI: 10.1016/j.solmat.2014.12.012

[32] Ren Y, Xu C, Wang T, Tian Z, Liao Z. A study of manufacturing processes of composite form-stable phase change materials based on $Ca(NO_3)_2$-$NaNO_3$ and expanded graphite. Materials (Basel). 2020;**13**:5368. DOI: 10.3390/ma13235368

[33] Ye F, Ge Z, Ding Y, Yang J. Multi-walled carbon nanotubes added to Na2CO3/MgO composites for thermal energy storage. Particuology. 2014;**15**:56-60. DOI: 10.1016/j.partic.2013.05.001

[34] Deng Y, Li J, Qian T, Guan W, Wang X. Preparation and characterization of KNO3/diatomite shape-stabilized composite phase change material for high temperature thermal energy storage. Journal of Materials Science & Technology. 2017;**33**:198-203. DOI: 10.1016/j.jmst.2016.02.011

[35] Hou Y, Qiu J, Wang W, He X, Ayyub M, Shuai Y. Preparation and performance improvement of chlorides/MgO ceramics shape-stabilized phase change materials with expanded graphite for thermal energy storage system. Applied Energy. 2022;**316**:119116. DOI: 10.1016/j.apenergy.2022.119116

[36] Qian T, Li J, Min X, Deng Y, Guan W, Ning L. Diatomite: A promising natural candidate as carrier material for low, middle and high temperature phase change material. Energy Conversion and Management. 2015;**98**:34-45. DOI: 10.1016/j.enconman.2015.03.071

[37] Wang T, Zhang T, Xu G, Xu C, Liao Z, Ye F. A new low-cost high-temperature shape-stable phase change material based on coal fly ash and K_2CO_3. Solar Energy Materials and Solar Cells. 2020;**206**:110328. DOI: 10.1016/j.solmat.2019.110328

[38] Ge Z, Ye F, Cao H, Leng G, Qin Y, Ding Y. Carbonate-salt-based composite materials for medium- and high-temperature thermal energy storage. Particuology. 2014;**15**:77-81. DOI: 10.1016/j.partic.2013.09.002

[39] Liu J, Xu J, Su Z, Zhang Y, Jiang T. Preparation and characterization of $NaNO_3$ shape-stabilized phase change materials (SS-PCMs) based on anorthite ceramic and cordierite ceramic for solar energy storage. Solar Energy Materials and Solar Cells. 2023;**251**:112114. DOI: 10.1016/j.solmat.2022.112114

[40] Jiang Y, Sun Y, Li S. Performance of novel Na_2SO_4-NaCl-ceramic composites as high temperature phase change materials for solar power plants (Part II). Solar Energy Materials and Solar Cells. 2019;**194**:285-294. DOI: 10.1016/j.solmat.2018.05.028

[41] Qin Y, Leng G, Yu X, Cao H, Qiao G, Dai Y, et al. Sodium sulfate–diatomite composite materials for high temperature thermal energy storage. Powder Technology. 2015;**282**:37-42. DOI: 10.1016/j.powtec.2014.08.075

[42] Ji M, Lv L, Liu J, Rong Y, Zhou H. $NaNO_3$-KNO_3/EG/Al_2O_3 shape-stable phase change materials for thermal energy storage over a wide temperature range: Sintering temperature study. Solar Energy. 2023;**258**:325-338. DOI: 10.1016/j.solener.2023.04.048

[43] Li C, Li Q, Cong L, Jiang F, Zhao Y, Liu C, et al. MgO based composite phase change materials for thermal energy storage: The effects of MgO particle density and size on microstructural characteristics as well as thermophysical and mechanical properties. Applied Energy. 2019;**250**:81-91. DOI: 10.1016/j.apenergy.2019.04.094

[44] Sang L, Li F, Xu Y. Form-stable ternary carbonates/MgO composite material for high temperature thermal energy storage. Solar Energy. 2019;**180**:1-7. DOI: 10.1016/j.solener.2019.01.002

[45] Jiang Y, Sun Y, Jacob RD, Bruno F, Li S. Novel Na_2SO_4-NaCl-ceramic composites as high temperature phase change materials for solar thermal power plants (Part I). Solar Energy Materials and Solar Cells. 2018;**178**:74-83. DOI: 10.1016/j.solmat.2017.12.034

[46] Luo Q, Liu X, Yao H, Wang H, Xu Q, Tian Y, et al. Fast and stable solar/thermal energy storage via gradient SiC foam-based phase change composite. International Journal of Heat and Mass Transfer. 2022;**194**:123012. DOI: 10.1016/j.ijheatmasstransfer.2022.123012

[47] Liu X, Wang H, Xu Q, Luo Q, Song Y, Tian Y, et al. High thermal conductivity and high energy density compatible latent heat thermal energy storage enabled by porous AlN ceramics composites. International Journal of Heat and Mass Transfer. 2021;**175**:121405. DOI: 10.1016/j.ijheatmasstransfer.2021.121405

[48] Wang L, Kong X, Ren J, Fan M, Li H. Novel hybrid composite phase change materials with high thermal performance based on aluminium nitride and nanocapsules. Energy. 2022;**238**:121775. DOI: 10.1016/j.energy.2021.121775

[49] Guo Q, Wang T. Study on preparation and thermal properties of sodium nitrate/silica composite as shape-stabilized phase change material. Thermochimica Acta. 2015;**613**:66-70. DOI: 10.1016/j.tca.2015.05.023

[50] Zhao S, Li J, Wu Y, Song S, Liu L. Three-dimensional interconnected porous TiO2 ceramics for high-temperature thermal storage. Renewable Energy. 2021;**178**:701-708. DOI: 10.1016/j.renene.2021.06.091

[51] Yu Q, Jiang Z, Cong L, Lu T, Suleiman B, Leng G, et al. A novel low-temperature fabrication approach of composite phase change materials for high temperature thermal energy storage. Applied Energy. 2019;**237**:367-377. DOI: 10.1016/j.apenergy.2018.12.072

[52] Rao Z, Zhang G, Xu T, Hong K. Experimental study on a novel form-stable phase change materials based on diatomite for solar energy storage. Solar Energy Materials and Solar Cells. 2018;**182**:52-60. DOI: 10.1016/j.solmat.2018.03.016

[53] Li C, Wang M, Chen Z, Chen J. Enhanced thermal conductivity and photo-to-thermal performance of diatomite-based composite phase change materials for thermal energy storage. Journal of Energy Storage. 2021;**34**:102171. DOI: 10.1016/j.est.2020.102171

[54] Li C, Wang M, Xie B, Ma H, Chen J. Enhanced properties of diatomite-based composite phase change materials for thermal energy storage. Renewable Energy. 2020;**147**:265-274. DOI: 10.1016/j.renene.2019.09.001

[55] Li R, Zhu J, Zhou W, Cheng X, Li Y. Thermal compatibility of sodium nitrate/expanded perlite composite phase change materials. Applied Thermal Engineering. 2016;**103**:452-458. DOI: 10.1016/j.applthermaleng.2016.03.108

[56] Zhao X, Tang Y, Xie W, Li D, Zuo X, Yang H. 3D hierarchical porous expanded perlite-based composite phase-change material with superior latent heat storage capability for thermal management. Construction and Building Materials. 2023;**362**:129768. DOI: 10.1016/j.conbuildmat.2022.129768

[57] Zuo X, Zhao X, Li J, Hu Y, Yang H, Chen D. Enhanced thermal conductivity of form-stable composite phase-change materials with graphite hybridizing expanded perlite/paraffin. Solar Energy. 2020;**209**:85-95. DOI: 10.1016/j.solener.2020.08.082

[58] Wang W, He X, Shuai Y, Qiu J, Hou Y, Pan Q. Experimental study on thermal performance of a novel medium-high temperature packed-bed latent heat storage system containing binary nitrate. Applied Energy. 2022;**309**:118433. DOI: 10.1016/j.apenergy.2021.118433

[59] Yao H, Liu X, Luo Q, Xu Q, Tian Y, Ren T, et al. Experimental and numerical investigations of solar charging performances of 3D porous skeleton based latent heat storage devices. Applied Energy. 2022;**320**:119297. DOI: 10.1016/j.apenergy.2022.119297

[60] Li C, Li Q, Ding Y. Investigation on the thermal performance of a high temperature packed bed thermal energy storage system containing carbonate salt based composite phase change materials. Applied Energy. 2019;**247**:374-388. DOI: 10.1016/j.apenergy.2019.04.031

[61] Li C, Li Q, Ding Y. Carbonate salt based composite phase change materials for medium and high temperature thermal energy storage: From component to device level performance through modelling. Renewable Energy. 2019;**140**:140-151. DOI: 10.1016/j.renene.2019.03.005

[62] Zhao B, Li C, Jin Y, Yang CY, Leng GH, Cao H, et al. Heat transfer performance of thermal energy storage components containing composite phase change materials. IET Renewable Power Generation. 2016;**10**:1515-1522. DOI: 10.1049/iet-rpg.2016.0026

[63] He X, Qiu J, Wang W, Hou Y, Ayyub M, Shuai Y. A review on numerical simulation, optimization

design and applications of packed-bed latent thermal energy storage system with spherical capsules. Journal of Energy Storage. 2022;**51**:104555. DOI: 10.1016/j.est.2022.104555

[64] Xu Q, Liu X, Luo Q, Tian Y, Dang C, Yao H, et al. Loofah-derived eco-friendly SiC ceramics for high-performance sunlight capture, thermal transport, and energy storage. Energy Storage Materials. 2022;**45**:786-795. DOI: 10.1016/j.ensm.2021.12.030

[65] Zhang Y, Li X, Li J, Ma C, Guo L, Meng X. Solar-driven phase change microencapsulation with efficient Ti_4O_7 nanoconverter for latent heat storage. Nano Energy. 2018;**53**:579-586. DOI: 10.1016/j.nanoen.2018.09.018

[66] Li D, Wang J, Ding Y, Yao H, Huang Y. Dynamic thermal management for industrial waste heat recovery based on phase change material thermal storage. Applied Energy. 2019;**236**:1168-1182. DOI: 10.1016/j.apenergy.2018.12.040

Chapter 10

A Critical Review on the 3D Modeling and Mitigation Strategies in the Thermal Runaway of Single-Cell and Modular Lithium-Ion Battery Architectures

Jiajun Xu, Faridreza Attarzadeh and Tanjee Afreen

Abstract

As the adaptation of lithium (Li) ion batteries (LIBs) in energy storage systems is becoming more prevalent by the day, the issue of safe and environmentally responsible design, installation, and operation of these batteries is posing a rapidly growing challenge. It is imperative to develop realistic multi-physics and multi-scale models that are useful not only for analyzing the thermal runaway (TR) events at the single-cell level but also for modular LIB designs. This needs to be accompanied by the development of easier-to-follow empirical rules and straightforward analytical models as our knowledge of TR events grows over time. The unpredictable nature of TR events and the grave fire and explosion dangers that are particularly associated with violent TR events at the modular level require employing large-scale real-time evaluation of these events as well. Although more innovative battery health indicators are being developed and employed, it is still very challenging to arrest catastrophic TR events in time. The review herein seeks to explore advanced modeling and experimental approaches holistically. The challenges and possibilities of different active and passive thermal management strategies are also critically elaborated for LIB modular designs.

Keywords: lithium-ion batteries, thermal runaway, modeling, calorimetry, thermal management

1. Introduction

In the wake of two notable thermal runaway (TR) episodes, the initial event at Boston's Logan Airport in January 2013, where a vacant Japan Airlines 787 ignited, followed by a similar incident 9 days later in Japan, the Federal Aviation Administration (FAA) took the step of grounding the entire Dreamliner fleet. The FAA sought to ensure necessary rectifications or remedies were put in place [1]. Over the past quarter-century, lithium-ion batteries (LIBs) have emerged as the predominant choice for storing energy, finding widespread use in electric vehicles, handheld electronics, and renewable energy setups [2]. Their exceptional energy density, compact size, and scalability make them a preferred choice in modern technological advancements. As the demand for higher energy density and larger storage capacities grows, the number of individual cells within LIBs increases, leading to more complex battery systems. However, these advancements increase the risks associated with electrical, mechanical, thermal, electrochemical, or mixed-mode damages or failures in these batteries, emphasizing the critical importance of battery safety.

LIBs' substantial energy density renders them prone to adverse conditions like elevated temperatures, mechanical impacts, overcharging, discharging, and internal short circuits. Moreover, the combustible nature of LIB structural components, comprising the outer casing, separator, and electrolyte, heightens their susceptibility. Currently, commercial LIB operation involves the migration of lithium ions between the cathode (consisting of lithium alloy and metal oxide) and the anode (comprising graphite), a process that generates heat. While this heat is typically dissipated effectively under low-rate operation, rapid charging, and discharging can lead to heat production rates surpassing dissipation rates, resulting in excessive heat accumulation and potential thermal runaway (TR), with the risk of combustion or explosions [3]. Furthermore, battery cycling contributes to capacity and performance degradation, influenced by factors like active material properties, electrolyte composition, and the solid electrolyte interface, directly influencing TR phenomena and their propagation [4, 5].

Gradually, this series of events unfolds, releasing heat in a manner that self-perpetuates and escalates beyond control. The initial release of heat is linked to the breakdown of the solid electrolyte interface (SEI), a phenomenon commonly observed at temperatures near 100°C [6]. The decomposition and regeneration of the SEI result in a continuous release of heat, elevating the internal temperature of the battery—a phase termed heat generation. Following this, the separator, typically composed of polypropylene or polyethylene, liquefies as the battery's temperature climbs to 120–130°C. According to Joule's law, the absence of insulation between the anode and cathode causes a short circuit, producing further heat—a phase known as heat spread. Meanwhile, the electrolyte solvent breaks down into hydrogen and hydroxide radicals, potentially initiating the electrolyte combustion process [7]. Moreover, as the cathode material deteriorates, it emits oxygen, intensifying the combustion reaction. Throughout these successive events, both temperature and internal pressure escalate due to the gas generation. Upon reaching a critical threshold, the venting cap ruptures, facilitating the release of gases and preventing case rupture and potential catastrophic events [8].

The mentioned events can vary depending on the specific chemistry of the battery. An experiment was conducted by Lei et al. [9] by using an accelerating rate calorimeter (ARC) to investigate the thermal abuse conditions of various cathode chemistries. The experiment results revealed that the onset temperature was similar across all the chemistries tested, which is around 90°C. Nonetheless, variations existed in the peak temperature, rate of temperature increase, and heat production among the batteries. Notably, the lithium-nickel-cobalt-manganese-oxide (NMC) chemistry exhibited the highest values for these parameters, followed by lithium manganese oxide (LMO) and lithium-iron-phosphate (LFP) batteries. It is essential to recognize that additional factors like state of charge (SOC), state of health (SOH), and triggers for thermal runaway (TR), such as short circuits or external heating, can also influence and alter the progression and characteristics of TR [10].

TR studies have remained oblivious to the influence of battery aging for the most part. While many studies focus on new battery chemistries or designs aiming at higher energy densities, it is also essential to understand that the safety behavior of cells can change over time. Waldmann and Mehrens conducted a study [11] on battery aging at 0°C to examine the impact of increased lithium plating and assess safety concerns using an ARC (accelerating rate calorimeter). The research outcomes highlighted that battery aging was responsible for diminished capacity and contributed to early occurrences of thermal runaway (TR), along with intensified decomposition, culminating in the ejection of the jelly roll from the cell casing. The impact of aging on TR proves intricate, as various studies have presented conflicting observations. While some researchers have documented a decline in specific safety attributes with aging, others have observed enhancements [12]. It is widely acknowledged that the processes of aging bring about changes in the characteristics of materials present within lithium-ion battery (LIB) cells [13]. Therefore, considering the influence of battery aging is crucial for a comprehensive understanding of TR behavior.

Renowned for their cylindrical design spanning 18 mm (roughly 0.71 inches) in diameter and 65 mm (approximately 2.56 inches) in length, 18,650 cells are extensively utilized across diverse sectors, including electric vehicle technology and aerospace endeavors [14]. The capacity of 18,650 cells has been progressively growing over time, with current commercially available cells achieving capacities as high as 3 Ah. However, along with increased capacity comes heightened heat and gas generation during cell failure. One particular issue revolves around the buildup of gas within the rigid casing of an 18,650 cell, essentially converting it into a pressurized container. Failure to relieve internal pressure in a controlled manner can result in violent rupture, potentially culminating in explosion. More innovative engineering solutions are necessary to address these risks and mitigate the potential for catastrophic failure [15]. Various integrated safety devices have been developed to address the potential issues related to gas buildup and pressure within 18,650 cells [16]. Among these tools are pressure relief vents, positive temperature coefficients (PTC), current-limiting switches, and current interrupt devices (CIDs). They are designed to thwart gas accumulation and, when required, effectively regulate pressure release in a secure manner.

By tradition, when evaluating the safety of LIB cells, the experimental approach is seen as the most accurate method. This means that conducting real-world tests and experiments on the batteries has been the primary way to assess their safety. It has been the go-to approach because it allows researchers to observe and measure the battery's performance under different conditions, such as temperature extremes or high loads. However, a limited number of experiments can be done as it is expensive and time-consuming. Additionally, estimating the performance of large battery packs in a statistically sound manner poses significant challenges. Moreover, the variation in size and capacity of batteries across different applications introduces complexities in scaling up the battery systems, as certain physical phenomena become more prominent at larger scales, which are less significant in single or small battery systems [17–27].

TR is a well-documented phenomenon in LIBs, which is not only equivalent to irreversible severe damage or complete failure of a LIB but also accompanied by dangerous fire risks and even explosions. It is safer to assume TR is inevitable at some point in time or under certain conditions, and all efforts must be made to ensure the battery design is not susceptible to catastrophic TR events. On top of that, TR of one cell can lead to a cascading runaway effect on neighboring cells and cause thermal or electrical damage in those cells. Even though the heat released by a single cell can be anywhere between 50 and 500 kJ, the entire energy of a pack could be one thousand times greater, and consequently, the propagation of a single point failure to the rest of the system is a daunting outcome that cannot be overlooked [28]. Capacities of modern 18,650 cells are well over 3 Ah, and it only takes ~2 s for them to generate >6 L of an essentially flammable gas mixture during a typical TR event [29]. This is also accompanied by the release of almost 70 kJ of energy. At the same time, the surface temperature can exceed 600°C. Inadequate cooling provisions or the absence of effective pressure relief mechanisms during a thermal runaway (TR) incident can escalate the likelihood of side-wall breaches. These breaches occur when thermal melting or pressure-induced splits lead to ruptures in the casings of cells. Side-wall breaches are regarded as the most severe type of breach for the sturdy casings typically utilized in standard 18,650 cells. Breaches with flares may exert enough force to surpass heat-sink materials and disrupt or collide with neighboring cells, consequently facilitating the spread of TR from cell to cell [30, 31].

Cascading failure can be due to the propagation of a premixed flame at the single-cell level. While an insignificant reactant diffusion is expected at this level, single-cell instabilities are usually linked to thermally induced diffusion. On the other hand, the interference of gaps between the cells makes the cell-to-cell propagation unsteady, and subsequent damage initiation in other cells resembles that of ignition under external heat flux. Thermal management systems should mitigate the effects of TR and simultaneously prevent cell-to-cell propagation when batteries are stacked or packed together. Specific considerations are of paramount importance when it comes to battery design and implementation of battery thermal management systems: randomness of TR events widens the range of possible outcomes. It can be arduous to determine or predict the onset, acceleration, trigger, trigger cell peak, and neighbor cell peak temperatures. Calculating the total amount of energy released through the sides and top of the battery cell is also tricky. It is also challenging to predict the exact location of such failures (e.g., top vs. side), analyze the pressure increase, identify the type of evolving gases, and the type/trajectory of ejecta material. Different active and passive cooling systems and countermeasures against damage propagation have been developed and tested. For instance, utilizing Al and Cu barriers is a simple approach for passive thermal management, which also decreases or eliminates the risk of cascading failures. Although this phenomenon has been a subject of intense scientific scrutiny, there is still plenty of room to propose novel, safer individual cells and battery pack designs. Intelligent battery management systems (BMS), battery thermal management systems (BTMS), software controls, and modular designs are becoming more focused on the safety concerns related to TR. Although LIB is not the only type of energy storage system prone to thermal runaway, its lower runaway temperature underscores the importance of thermal management or even the implementation of fire suppression systems [32]. As energy density, size, and the number of individual cells in currently used LIBs and future designs increase, so do the magnitude and probability of risks associated with electrical, mechanical, thermal, electrochemical, or mixed-mode permanent damages or failures in these batteries.

Since the inception of the Doyle, Fuller, and Newman (DFN) pseudo-2D (P2D) model in 1994, battery simulation techniques have undergone rapid advancement. MATLAB/Simulink emerges as a predominant commercially available simulation software tool in today's market, especially for vehicle system simulation. Prominent examples such as ADVISOR and Autonomie are developed on the MATLAB/Simulink platform. Additionally, Simcenter Amesim and GT-Suite are either utilized independently or provide specialized modules for co-simulation with MATLAB/Simulink, particularly for electronic vehicle (EV)-specific LIBs [33]. Battery Management Systems (BMSs) find widespread application across aerospace, EVs, and consumer electronics industries, equipped to handle intricate geometries and fluid dynamics aspects of heat transfer media like air or liquid cooling agents. Various computer-aided engineering (CAE) software packages such as Simcenter STAR-CCM+ and Fluent are employed to address computational fluid dynamics (CFD) challenges, particularly in BMS design, leveraging conjugate heat transfer (CHT). Multi-physics packages like Simcenter STAR-CCM+ enable complex geometry meshing and the application of diverse physics principles, including fluid dynamics/mechanics and electromagnetism, to generate high-fidelity simulations. A non-exhaustive list of frequently used commercial products for LIB design and simulation includes SIMENS, ANSYS, DASSAULT systems, Gama Technologies, COMSOL, MATLAB, LS-DYNA, Altair, Thermoanalytics, Materials Design, and OpenFOAM.

This short review has the following four main objectives: (1) bring some of the most challenging issues and overlooked phenomena that are faced in TR propagation simulation into light. (2) Pinpoint the most promising experimental and modeling strategies. (3) Compare the efficacy and suitability of recent successful simulations. (4) Shed light on the prospects of advanced TMSs as it is applicable to LIB modules. The exceptional capabilities of National Aeronautics and Space Administration's (NASA) state-of-the-art FTRC are explained in detail. Contrary to common belief, it is hard to understand how and why a more considerable portion of the generated heat is carried by ejecta material. Some of the most important mitigation strategies in view of structural design are also covered in this work.

2. Thermal runaway at the single-cell level

Utilizing a multi-physics and multi-scale model is essential for comprehensively understanding the intricate dynamics of battery performance. Xu and Hendricks [34] developed a model of a lithium-ion battery (LIB) to investigate both its electrochemical and thermal-mechanical processes. The model's accuracy was confirmed through comparison with experimental data obtained from large-format battery cells subjected to diverse load and boundary conditions [35]. The simulation model configuration progresses through several stages:

1. The geometry and mesh of the multi-physics models are defined.

2. The multi physics model is formulated considering the complex interactions between different physical phenomena. Simulation conditions are then identified to replicate experiments conducted on batteries. Validation of the model takes place in the third phase, where its predictions are cross-referenced with data obtained from experiments involving overcharging and oven heating [36].

3. The model incorporates the four exothermic reactions to simulate TR events accurately.

The multi-physics model is simulated by solving a series of individual physics consecutively, and the obtained results will be passed on to the next module to account for the coupling effect. This is done to ensure that the interdependencies and interactions between different physical phenomena within the battery are appropriately accounted for. This physics focuses on different areas such as fluid dynamics, heat transfer, and solid mechanics to individually address heat generation, thermal expansion, and structural deformation.

Figure 1(a) showcases the outcome of the overcharging simulation, providing insights into the temperature distribution within the lithium-ion battery (LIB). By the end of the time-dependent simulation, specifically at 3960 seconds, the highest temperature is localized at the core of the cell, gradually diminishing toward the outer circumference.

Figure 1(b) presents the simulation results under overcharging conditions, delineating the structural deformation and stress distribution within the LIB cell. At 3960 seconds, signifying the conclusion of the time-dependent simulation, stress concentration emerges along the junction between the upper and side walls of the cell, attaining its maximum intensity. Moreover, **Figure 1** highlights that the upper portion of the LIB cell experiences the most pronounced deformation, in line with empirical observations. Additionally, the simulation indicates that the entire battery cell expands due to heightened internal pressure, as depicted in **Figure 1**.

Figure 1(c) exhibits the outcome illustrating the temperature distribution within the LIB cell, with a focus on the overheating test simulation where exothermic

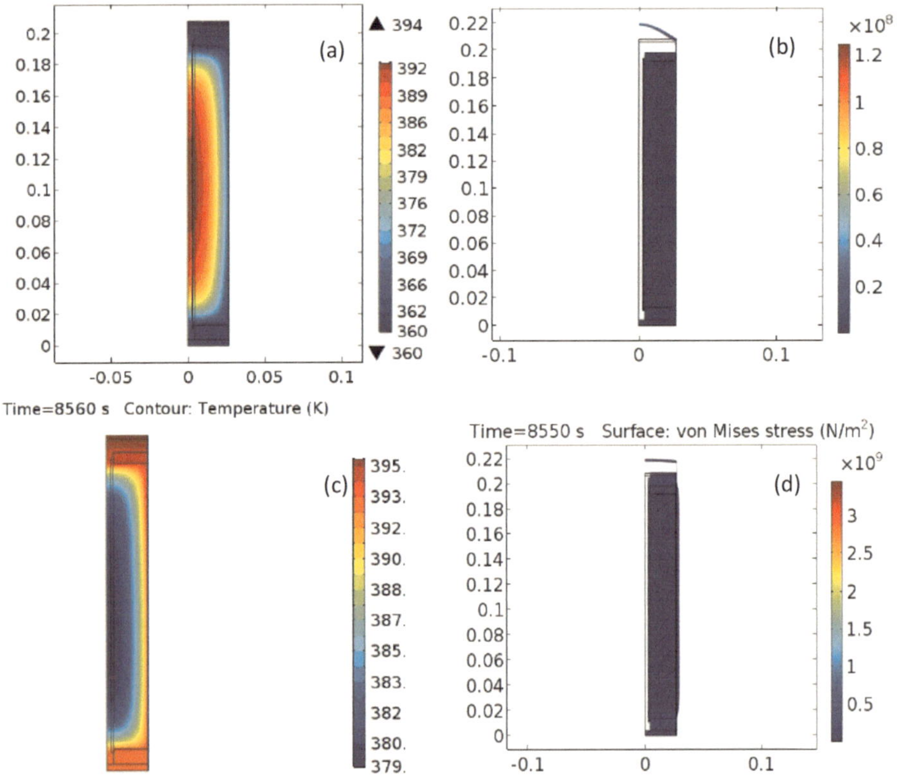

Figure 1.
(a) Temperature distribution of the LIB in the overcharging simulation at 3960 s; (b) stress distribution and structural deformation of the LIB in the overcharging simulation at 3960 s; (c) temperature distribution of the LIB in the oven heating test without exothermic reaction at 8560 s; (d) stress distribution and structural deformation of the LIB in the oven heating test without exothermic reactions at 8550 s.

reactions are disregarded. At the end of the time-dependent simulation (8560 seconds), the highest temperature is observed on the outer surface of the cell, gradually decreasing toward the center. This observation aligns with the current simulation condition, wherein the external heat input originates from a heater or heat flux.

It is pertinent to acknowledge that the maximum surface temperature achieved in this simulation is 122°C, significantly lower than the experimental temperature value of 176°C. This variance is expected due to the exclusion of internal exothermic reactions in the simulation. The absence of these reactions elucidates the lower temperature recorded in the simulation compared to the empirical values.

Figure 1(d) illustrates the implications of the simulation conducted under an overheating test scenario, focusing on the structural deformation and stress distribution within the LIB cell. As illustrated, stress concentration becomes apparent along the cell wall at the conclusion of the time-dependent simulation (8550 seconds). This stress concentration likely arises from the thermal expansion of the active battery material due to the heightened temperature. In summary, the simulation suggests that increased temperature induces thermal expansion, resulting in stress concentration along the cell wall. The observations in **Figure 1** indicate that elevated temperature conditions can significantly impact the structural integrity of the LIB cell.

As the temperature increases within a battery cell, it can trigger several exothermic chemical reactions. These reactions generate additional heat that accumulates inside the cell. If the heat generation rate surpasses the rate at which heat dissipates to the surroundings, TR may become inevitable once the cell reaches a critical temperature.

As thermal runaway (TR) ensues, heightened internal temperatures accelerate chemical reactions among cell components, giving rise to potential hazards including leakage, smoke emission, gas venting, and the initiation or propagation of flames. The unchecked progression of TR ultimately threatens the integrity of battery cells. Thus, it is imperative to comprehend and address the risks inherent in TR to uphold the secure functioning of battery systems. This study considered four specific exothermic reactions [37].

1. Solid Electrolyte Interface (SEI) decomposition reaction: The SEI layer, which forms on the surface of electrodes, is known to be metastable. At temperatures ranging from 90 to 120°C, the SEI layer can undergo an exothermic decomposition reaction.

2. Negative-Solvent reaction (NS): Involving the interaction between intercalated lithium within the negative electrode and the electrolyte, this reaction unfolds at temperatures above 120°C.

3. Positive-Solvent reaction (PS): At even higher temperatures, typically above 170°C, the positive active material in the battery can undergo a chemical reduction reaction with the electrolyte. This reaction is referred to as the positive-solvent reaction.

4. Electrolyte Decomposition reaction (ELE): The exothermic decomposition of electrolyte unfolds when the temperature goes beyond 200°C.

By considering these four exothermic reactions, the study aimed to understand their impact on the thermal behavior and potential for TR in the battery system [38] (**Table 1**).

x ε {SEI, NS, PS, ELE}, y ε {c, p, e}	Reaction Heat H_x/[J/kg]	Frequency Factor A_x/[l/s]	Activation Energy E_A/[J/mol]	Volume Content W_y/[kg/m^3]
SEI reaction	2.57*10^5	1.667*10^{15}	1.3508*10^5	1.39*10^3
NS reaction	1.714*10^6	2.5*10^{13}	1.3508*10^5	1.39*10^3
PS reaction	3.14*10^5	6.667*10^{13}	1.396*10^5	1.3*10^3
ELE reaction	1.55*10^5	5.14*10^{13}	2.74*10^5	5*10^2

Table 1.
Presents the parameters used to calculate the reactions [38].

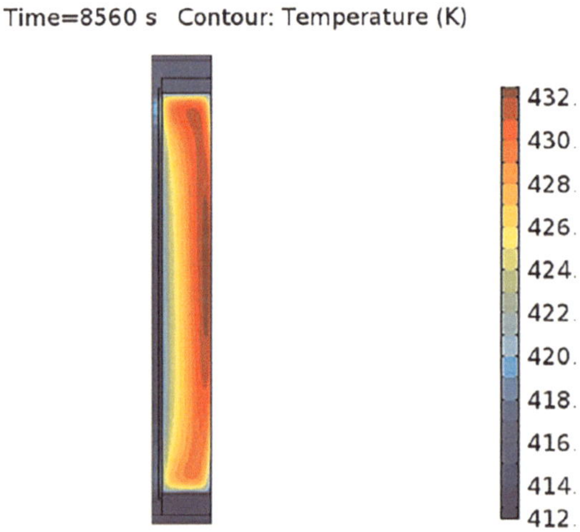

Figure 2.
Temperature distribution in the oven heating test (with exothermic reactions) of the LIB.

The temperature distribution is depicted in **Figure 2**, considering the exothermic reactions within the LIB. As a result of the increased internal temperature, these exothermic reactions are triggered. Consequently, the temperature profile displayed in **Figure 2** shows a distinct and noteworthy pattern that differs significantly from that observed in **Figure 2**. Exothermic reactions notably impact the temperature distribution within the battery cell.

Figure 3 represents the surface temperature change over time, showing significant insights. Initially, the surface temperature gradually increases due to ohmic heating during charging and discharging cycles. However, a distinct change occurs around 7500 s, when the temperature curve rises sharply, reaching approximately 90°C. This temperature rise indicates the onset of self-heating.

As the temperature exceeds 120°C, the temperature increase becomes exponential, indicating the existence of TR. Notably, the simulation results agree with the findings from destructive tests, where self-heating was observed around 110°C, and the surface temperature reached values of 170–180°C. These results also align well with experimental measurements, precisely with the recorded temperature of 176°C. Additionally, the simulation captures the time trend observed in the experimental results.

Leveraging advanced modeling tools and methodologies, their research has effectively encapsulated the electrochemical and thermal-mechanical intricacies of large-

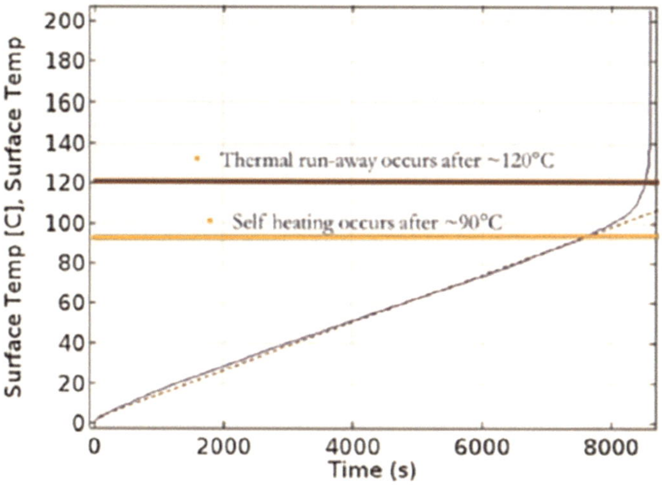

Figure 3.
Surface temperature of the LIB cell with time and TR occurrence.

scale LIBs. These numerical simulations have deepened the understanding of failure mechanisms, empowered thorough risk assessment, and facilitated the formulation of robust safety measures for LIBs. The adoption of such modeling tools stands as a cornerstone in fortifying battery safety protocols and preemptively addressing potential risks.

3. Thermal runaway at the modular level

A survey of pertinent literature shows most of the previous studies were dedicated to the TR behavior of various single-cell designs with different chemistries [39, 40]. As mentioned, TR at this level includes events such as the evolution of battery gases or fumes, ejection of cell contents, abnormally high-temperature rise, and occasionally self-ignition of the cell or ejecta material. In contrast, damage propagation in more extensive complex systems has not received the same level of attention. Although TR-related failure of a single cell may not affect the overall safety or performance of a battery module in some cases, violent thermal and electrical behavior of one cell may affect the neighboring cells, resulting in a cascading TR effect. This can potentially lead to a domino effect in the thermal failure propagation, which continues to propagate until it engulfs the entire battery pack. A rapid catastrophic release of energy within the confines of a battery module under this worst-case scenario situation may give rise to substantial damage and pose severe threats to the operator or consumer. According to Spotnitz et al. [41], the likelihood of such failures in a modular design is influenced not only by an individual cell's abuse response but also by the overall cell-to-cell insulation strategy within the battery module. They pointed out that insufficient or lack of insulation between adjacent cells increases the likelihood of thermal failure propagation. On the other hand, the electrical configuration of the battery module may also play a crucial role in how these failures propagate. Offer et al. [42] analyzed how the local temperature of individual cells is affected by the fluctuation of ohmic resistances between them. Blocks and discharge diodes are usually utilized to avoid self-discharge of the entire module through a shorted cell [43, 44]. Thermal isolation between cells is usually inadequate from the failure prevention perspective. Prismatic and pouch cells are typically stacked next to each other either with no gap or separated by cooling thin plates [45]. Electrical connections can also act as additional heat transfer pathways between cells [38]. Most existing fault detection and damage prevention strategies are based on electrical diagnosis and countermeasures. Innovative fault detection methods and battery health indicators must go beyond the traditional cell temperature and voltage monitoring. There is usually a noticeable lag between these traditional battery health indicators and TR initiation, which will not allow us to arrest an ongoing TR event soon enough. Various failure propagation testing methods have been developed to induce failure deliberately, and these will impart varying levels of energy. Among the most important ones are nail penetration testing, single-cell thermal ramp, single-cell overcharge, and TR initiation under high-intensity laser or light beams.

In their study, Xu and Hendricks constructed two distinct 3D multi-physics models using COMSOL Multi-physics, specifically tailored for an 18,650 lithium-ion battery (LIB) affixed to a G10 board [34]. One model was established as the benchmark dataset to validate experimental observations. Conversely, the second model introduced a setup with two identical batteries positioned side by side, with one cell housing a heater atop. The former model facilitated a comprehensive parametric analysis spanning various heating power inputs, aiming to discern the relationship between heating power and thermal runaway (TR). Meanwhile, the latter model was deployed to explore the repercussions of a TR event in one cell on its adjacent counterpart. They extended the second model to include thermal radiation heat transfer and investigated its effect in a battery module. They assumed the heater output changes linearly with temperature (0 to 50 W over 3000 s). Regarding boundary and ambient conditions, the entire setup was subject to natural convection with an average heat transfer coefficient of 5 W/m^2 K. Their models included four commonly observed types of exothermic reactions, namely SEI decomposition reaction (at 90–120°C), NS reaction (>120°C), PS reaction (> 170°C), and ELE reaction (> 200°C). To account for the generated heat through exothermic reactions, they relied on the following governing equations for the concentration gradient while assuming Arrhenius-law is applicable for the computation of exothermic heat sources:

$$\frac{\partial c_i}{\partial t}(x,t) = d_i \Delta c_i(x,t) + \gamma_i R_i(x,t) \tag{1}$$

$$c_i(x,0) = c_{i,0}(x,0), \forall x \in \Omega_i \tag{2}$$

$$c_i(x,t) = 0, \forall x \in \partial\Omega_i \tag{3}$$

$$Q_i(x,t) = q_i R(x,t) \tag{4}$$

$$R_i(x,t) = A_i c_i(x,t) \exp\left(-\frac{E_{a,i}}{RT(x,t)}\right) \tag{5}$$

where $i \in \{\text{SEI, NS, PS, ELE}\}$, $c_{i,0}(x)$ denotes the initial concentration, d_i for the diffusion coefficient, γ_i for the stochiometric coefficient, q_i for reaction enthalpy in J/g, A_i for the frequency factor in 1/s, $E_{a,i}$ for the activation energy in J/mol, and R for the universal gas constant. While the simulation results based on the first model accord well with the experimental results at the single-cell level, the simulation results in the second model point to an even more important conclusion. They reported that including the radiation heat transfer effect drastically changes the results at the end of

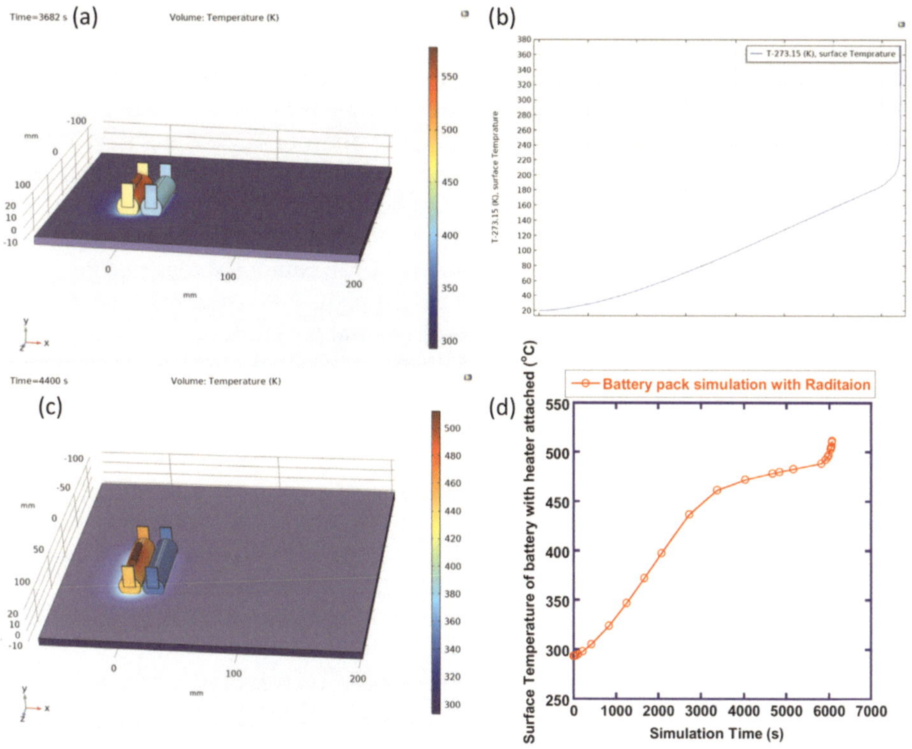

Figure 4.
(a) Temperature profile of the battery module at t = 3682 s disregarding the effect of radiation, (b) evolution of the surface temperature of the battery over time in the second model with radiation, and (c) 3D temperature profile of the battery module at t = 4400 s with [35].

the simulation. As illustrated in **Figure 4(a)**, the region with the greatest temperature is observed in close proximity to the heater and the battery it is connected to, with the neighboring cell also experiencing heightened temperatures. They noticed that surface temperature increases gradually over time when the model takes the radiation effect into account (**Figure 4(b)**), while there is an abrupt temperature rise when this effect is neglected (**Figure 4(d)**). This was attributed to the exponential increase of radiation with temperature, which is at the order of 4 according to Boltzmann's equation. While the LIB cell temperature exceeds 500°C at 3800 s without radiation (**Figure 4(a)**), it almost takes 4400 s for the LIB cell temperature to exceed the same temperature with radiation (**Figure 4(c)**). Of crucial importance is that now the G10 board and the surrounding areas also experience a more elevated temperature with the inclusion of the radiation effect (**Figure 4(c)** vs. **Figure 4(a)**). This seems to be a more realistic simulation and beneficial to risk assessment and mitigation in the safety management of modular battery designs.

The National Aeronautics and Space Administration (NASA) Johnson Space Center (JSC) has set forth Crewed Space Vehicle Battery Requirements (JSC-20793-RevD), as is relevant to human space exploration, to ensure TR severity and cell-to-cell propagation tendencies are meticulously evaluated. One of the most pressing issues is maintaining a reasonable trade-off between rigorous requirements (as in JSC-20793-RevD) and the ever-expanding need for compact, high-energy-density modular designs [46]. Researchers have been relying on copper slug calorimetry (CSC) [47–49], bomb calorimetry [50], and accelerating rate calorimetry (ARC) [51–53] methods to characterize TR events in LIBs. None of these methods can discern the portion of the total energy released through the cell casing from that of the ejected materials. In contrast, NASA's new calorimeter is meticulously designed to test many cells in a short span of time, which is on the order of 10 tests per day. The rationale behind investing in such a vast undertaking is that no two TR events are the same. Hence, it is critical to characterize enough number of like cells to ensure as many unpredictability factors are considered as possible. NASA's FTRC is equipped with the following: (1) X-ray imaging and *in situ* X-ray computed tomography, (2) gas collectors and velocity measurement features, (3) high-flux heater induced thermal runaway, (4) nail penetration testing, and (5) general portability.

Using NASA's state-of-the-art FTRC analyses, researchers could not only estimate the entire energy released during a TR event but also distinguish between the portion of the total energy released through cell casing and the part that belongs to ejecta material [54]. The following four commercially available LIBs were tested in this study: LG 18650-MJ1, LG 18650 Test Cell, Samsung 18,650-30Q, and MOLiCEL® 18,650-J. The impacts of essential factors such as separator material, bottom venting, internal short-circuiting (ISC), casing thickness, bottom rupture, and mass loss in post-TR conditions were all considered in this work. No similar research was conducted to differentiate between ejected and non-ejected materials contributions in heat generation, or at least not as detailed as a handful of FTRC-related publications from NASA. They found that between 20 and 30% of total energy is released through the casing breaches, whereas the heftier rest is released through ejecta particulates and gases. Most modular LIB designs are engineered regardless of the cells' susceptibility to eject their content violently and the amount of energy released by these violent ejections. Besides that, modeling community has also been reluctant to bring these aspects to light. From a statistical perspective, one of their significant findings was that variation of total energy release mostly follows a lognormal distribution. Moreover, they noted that high-energy cells are more likely to become unpredictable and exhibit violent ejections. As indicated in **Figure 5**, the LG 18650-MJ1 exhibited the highest median value predicted at 74.9 kJ and the highest overall value observed at 82.9 kJ. Conversely, the MOLiCEL® 18,650-J, possessing the lowest nominal capacity and electrochemical energy when fully charged, displayed the lowest median values

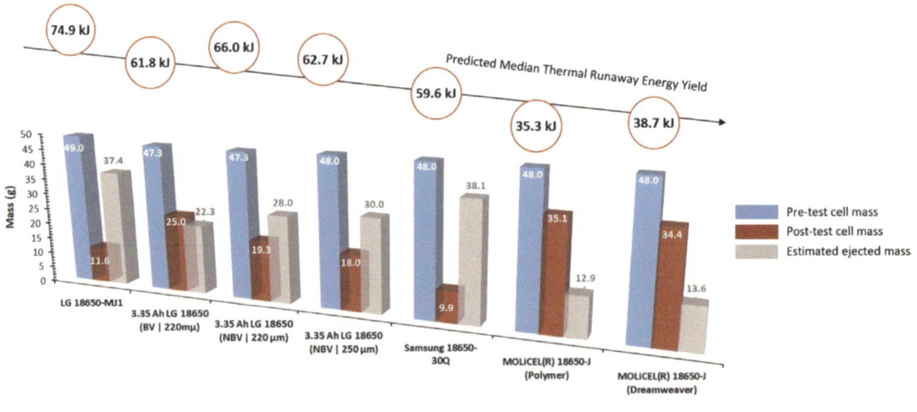

Figure 5.
Variations of pre-and post-test cell masses next to the estimated ejected mass (difference between pre-and post-mass) along with the corresponding average total TR energy yield (printed with permission from [54]).

predicted at 35.3 kJ regardless of separator material and 38.7 kJ depending on the separator material, alongside the lowest overall observed value at 31.3 kJ. Even though the total energy release was found to have a strong relationship with the stored electrochemical energy, it was hard to claim it is a perfect linear relationship. Yet another important finding was that bottom venting (BV) may alleviate the severity of TR events and let them behave more predictably. The occurrence of BV not only lowered the heat generation by ~3.9 KJ but also led to a minor mass loss in post-TR condition (6.5 g). When less material is ejected, less material is also available to burn. Besides that, bottom rupture (BR) was found to have a similar effect on the energy release (~2.6 KJ less for LG 18650-MJ1 cells with BR), but not quite as large as the BV effect either [55].

In a research conducted by Hoelle et al. on prismatic LIBs, modeling approaches have further evolved to consider more possible phenomena, such as the formation of a gas layer between roll and surrounding can, which was attributed to the electrolyte vaporization (>60 Ah) [56]. They emphasized that mass loss, can bulging, and electrolyte vaporization phenomena should be accounted for.

4. Thermal management at the modular level

Crafting an effective thermal management system (TMS) capable of upholding lithium-ion battery (LIB) modules within designated safe temperature thresholds is imperative. Modern TMS implementations offer a means to mitigate the hazards linked to catastrophic thermal runaway (TR) incidents. Despite its straightforwardness, the efficacy of air-cooling mechanisms is hindered by air's low heat capacity and thermal conductivity. This diminishes heat transfer rate from the battery to the air medium. Given air's limited heat dissipation capabilities, achieving efficient air cooling necessitates substantial airflow volume and mixing rates to counterbalance these inherent limitations. Moreover, specific applications, such as those in space exploration, lack access to air altogether. Consequently, for larger LIB modules operating under high discharge rates, active liquid cooling emerges as a more efficient alternative. Liquid coolants circulate through a thermally conductive metal plate, typically equipped with micro- or macro-channels to augment heat transfer rates. In mitigating the potential hazards associated with liquid cooling, heat pipes are considered a safer option for extracting excess heat from the battery and directing it toward a heat sink [57–59]. Heat pipes are becoming widely adopted in electronic cooling applications owing to their small form factor and exceptional thermal conductivity. Similarly, phase change material (PCM) has surfaced as a novel solution within today's thermal management systems (TMSs) [60–62]. The greatest appeal for implementing PCM is that cell temperature is curbed within a safe, narrow temperature range, which is dictated by the PCM's melting point temperature. Despite this, the PCM system encounters challenges due to its limited thermal conductivity, hindering the efficient dissipation of stored heat. However, utilizing enhanced PCMs like graphene-infused PCM variants could offer a potential solution to overcome this obstacle [63, 64].

One good example of modeling thermal control is the research led by Basu et al. in which they used the capabilities of STAR-CCM+ [65]. By leveraging lateral water coolant channels and harnessing the exceptional heat conductivity of aluminum components, they not only achieved the requisite structural reinforcement but also ensured effective heat exchange between the coolant and water. As illustrated in

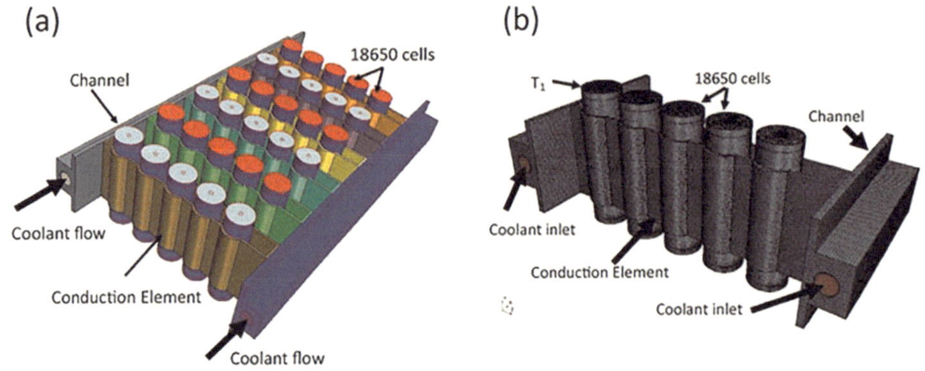

Figure 6.
(a) Battery module along with its integrated cooling system, and (b) the mesh generated by STAR-CCM+ (b). (Printed with permission from [65]. Copyright 2016, Elsevier).

Figure 6, their lithium-ion battery (LIB) module comprised 30 Li-NCA/C, 18650-sized cells (6S5P), with the initial row of parallel cells meshed using 400,000 polyhedral cells. Utilizing Battery Design Studio (BDS), they developed a model of the 18,650-sized cell, while the P2D DFN model was deployed for electrochemical aspects and current collector discretization. The CAD design of the cooling system, alongside the geometry generated by BDS, served as inputs for STAR-CCM+. This software facilitated the integration of the electrical mesh, electrochemical elements, and polyhedral mesh. Experimental validation was conducted to assess the model's accuracy in predicting cell temperature profiles, enabling an analysis of the effects of coolant flow rate, contact resistance between LIB cells and cooling elements, and discharge rate. Their compact thermal management system (TMS) successfully limited temperature elevation to within 7 K under high discharge and low coolant flow rates. This endeavor serves as a proof of concept for reliable simulation of a battery TMS, boasting an efficacy level exceeding 90%. Similar methodologies hold potential for streamlined onboard TMSs, promising reduced sensor reliance and simplified control systems.

Table 2 provides a non-exhaustive list of the existing TR propagation simulation strategies. It is common practice to take advantage of Arrhenius type of equations to at least account for the four common reactions during a TR event. Advanced studies consider additional type of reactions and other additional terms such as the ones associated with the electrochemical heat source and joule heating. The higher the number of these additional reactions or terms, the higher the level of complexity and computational cost. This can be a significant barrier to their suitability for larger modular designs. Implementation of empirical heat source terms is seen as step toward simplification, which can be in the form of time- or temperature-dependent terms or some constants.

5. Conclusion

The development of accurate models to simulate thermal runaway (TR) events with minimal computational resources, particularly in modular setups susceptible to cascading TR failures, is still evolving. Current TR simulation methodologies are highly valued for their ability to examine the electro-thermal characteristics of lithium-ion batteries (LIBs) during abusive scenarios, irrespective of TMS

DOI: http://dx.doi.org/10.5772/intechopen.114319

Author(s)	Modeling approach				Setup			
	Type	# of source terms						
		Chemical	Electrochemical	Joule heat (ISC)	Dimensions	Solver	# of cells	Design
Shen et al. [66]	Arrhenius	4	_	_	1D	LEM	6	Prismatic
Feng et al. [67]	Arrhenius	6	_	1	1D	LEM	6	Prismatic
Xu et al. [68]	Arrhenius	1 eq. fitted to ARC data			1D	ROM	4,18&3x18	Prismatic
Feng et al. [67]	Arrhenius	1 eq. fitted to ARC data + Joule heat (ISC)			3D	FEM	6	Prismatic
Citarella et al. [69]	Empirical	Temp. dependent on HRR fitted to ARC data			3D	FVM	2x12	Prismatic
Yuan et al. [70]	Arrhenius	4	_	1	3D	FVM	11	Cylindrical
Vyroubal et al. [71]	Arrhenius	4	_	_	3D	FVM	4x10	Cylindrical
Jia et al. [72]	Arrhenius	4	1	1	1D	LEM	2&9	Cylindrical
Li et al. [73]	Arrhenius	4	_	_	3D	FEM + FVM	192	Cylindrical
Jindal et al. [74]	Arrhenius	4	Full model	_	3D	FVM	10	Cylindrical
Mishra et al. [75]	Arrhenius	4	_	_	3D	FVM	25	Cylindrical
Mishra et al. [76]	Arrhenius	4	_	_	3D	FVM	25	Cylindrical
Mishra et al. [77]	Arrhenius	4	_	_	3D	FVM	25	Cylindrical
Liu et al. [78]	Arrhenius	1 eq. fitted to experimental data			3D	FEM	6	Cylindrical
Qin et al. [79]	Empirical	Temp. dependent heat release rate (ARC data)			3D	FVM	9	Cylindrical
Grimmeisen et al. [80]	Empirical	Const. heat release rate			3D	FVM	7	Cylindrical
Coman et al. [81]	Empirical	Time-dependent function for HRR			2D	FEM	48	Cylindrical
Bugryniec et al. [82]	Arrhenius	4	_	_	2D	FEM	9	Cylindrical
Coman et al. [83]	Arrhenius	3	1	_	2D & 3D	FEM	65	Cylindrical

Author(s)	Modeling approach				Setup			
	Type	# of source terms						
		Chemical	Electrochemical	Joule heat (ISC)	Dimensions	Solver	# of cells	Design
Yikai et al. [72]	Arrhenius	3	_	1	3D	FEM	2	Cylindrical
Zhang et al. [84]	Arrhenius	4	1	_	3D	FEM	2	Pouch
Chen et al. [85]	Arrhenius	6	_	_	1D	LEM	2	Pouch
Kurzawski et al. [28]	Arrhenius	3	1	_	Quasi 1D	FEM	5	Pouch
Bilyaz et al. [86]	Other	Laminar Propagation Theory with 1 reaction			1D	FDM	5&10	Pouch

**ISC denotes internal short circuit, ARC is accelerating rate calorimetry, HRR is heat release rate, LEM means lumped element method, FEM means finite element method, FVM means finite volume method, ROM means reduced order model), and FDM means finite difference method.*

Table 2.
A non-exhaustive list of empirical and modeling approaches for the heat release during TR in LIB modules.

implementation. The introduction of faults into the simulation model enables the validation of results using measurements derived from previous testing. Despite numerous efforts in this field, TR assessment is still limited to specific cell designs and chemistries, and overgeneralization of obtained results can be misleading. The exact behavior of ejecta material and the diversity of possible mechanical failure modes during TR propagation are abstruse aspects that merit special attention. It is safer to assume TR is inevitable and has specific design criteria rather than waiting for failure analysis results. Traditional battery health indicators are helpful but not fast enough to arrest catastrophic TR events in time. For larger modules, smart TMSs and even fire suppression systems could be other worthwhile investments in the long run. Nonuniformity of temperature distribution across the single cells and over the module is also another research avenue that deserves to be investigated at greater depth. Metal additive manufacturing can be used to design more efficient active or passive cooling systems with complex geometries and reduced weights. More advanced statistical analyses and predictive models are needed to propose more innovative modular designs.

Author details

Jiajun Xu*, Faridreza Attarzadeh and Tanjee Afreen
Department of Mechanical Engineering, Center for Advanced Manufacturing, University of the District of Columbia, Washington DC, DC, USA

*Address all correspondence to: jiajun.xu@udc.edu

References

[1] Cooper A. Dreamliners grounded globally due to fire risk – CNN. 2013. Available from: https://www.cnn.com/2013/01/17/travel/dreamliner-faa/index.html [Accessed: July 12, 2023].

[2] Yun S, Zhang Y, Xu Q, Liu J, Qin Y. Recent advance in new-generation integrated devices for energy harvesting and storage. Nano Energy. 2019;**60**:600-619. DOI: 10.1016/j.nanoen.2019.03.074

[3] Kumar Thakur A, Sathyamurthy R, Velraj R, Saidur R, Pandey AK, Ma Z, et al. A state-of-the art review on advancing battery thermal management systems for fast-charging. Applied Thermal Engineering. 2023;**226**:120303. DOI: 10.1016/j.applthermaleng.2023.120303

[4] Liu J, Wang Z, Bai J, Gao T, Mao N. Heat generation and thermal runaway mechanisms induced by overcharging of aged lithium-ion battery. Applied Thermal Engineering. 2022;**212**:118565. DOI: 10.1016/j.applthermaleng.2022.118565

[5] García A, Zhao P, Monsalve-Serrano J, Villalta D, Martinez-Boggio S. Optical diagnostics of the venting spray and combustion behaviour during Li-ion battery thermal runaway induced by ramp heating. Applied Thermal Engineering. 2023;**218**:119308. DOI: 10.1016/j.applthermaleng.2022.119308

[6] García A, Monsalve-Serrano J, Lago Sari R, Martinez-Boggio S. An optical investigation of thermal runway phenomenon under thermal abuse conditions. Energy Conversion and Management. 2021;**246**:114663. DOI: 10.1016/j.enconman.2021.114663

[7] Feng X, Ouyang M, Liu X, Lu L, Xia Y, He X. Thermal runaway mechanism of lithium ion battery for electric vehicles: A review. Energy Storage Materials. 2018;**10**:246-267. DOI: 10.1016/j.ensm.2017.05.013

[8] Zhang Q, Niu J, Zhao Z, Wang Q. Research on the effect of thermal runaway gas components and explosion limits of lithium-ion batteries under different charge states. Journal of Energy Storage. 2022;**45**:103759. DOI: 10.1016/j.est.2021.103759

[9] Lei B, Zhao W, Ziebert C, Uhlmann N, Rohde M, Seifert HJ. Experimental analysis of thermal runaway in 18650 cylindrical Li-ion cells using an accelerating rate calorimeter. Batteries. 2017;**3**:14. DOI: 10.3390/batteries3020014

[10] Ghiji M, Edmonds S, Moinuddin K. A review of experimental and numerical studies of lithium ion battery fires. Applied Sciences. 2021;**11**:1-29. DOI: 10.3390/app11031247

[11] Waldmann T, Wohlfahrt-Mehrens M. Effects of rest time after Li plating on safety behavior—ARC tests with commercial high-energy 18650 Li-ion cells. Electrochimica Acta. 2017;**230**:454-460. DOI: 10.1016/j.electacta.2017.02.036

[12] Ren D, Hsu H, Li R, Feng X, Guo D, Han X, et al. A comparative investigation of aging effects on thermal runaway behavior of lithium-ion batteries. ETransportation. 2019;**2**:232-245. DOI: 10.1016/j.etran.2019.100034

[13] Waldmann T, Iturrondobeitia A, Kasper M, Ghanbari N, Aguesse F, Bekaert E, et al. Review—post-mortem analysis of aged lithium-ion batteries: Disassembly methodology and physico-chemical analysis techniques. Journal of the Electrochemical Society. 2016;**163**: A2149-A2164. DOI: 10.1149/2.1211609jes

[14] Yayathi S, Walker W, Doughty D, Ardebili H. Energy distributions exhibited during thermal runaway of commercial lithium ion batteries used for human spaceflight applications. Journal of Power Sources. 2016;**329**:197-206. DOI: 10.1016/J.JPOWSOUR.2016.08.078

[15] Finegan DP, Darcy E, Keyser M, Tjaden B, Heenan TMM, Jervis R, et al. Identifying the cause of rupture of Li-ion batteries during thermal runaway. Advancement of Science. 2018;**5**: 1700369. DOI: 10.1002/advs.201700369

[16] Balakrishnan PG, Ramesh R, Prem Kumar T. Safety mechanisms in lithium-ion batteries. Journal of Power Sources. 2006;**155**:401-414. DOI: 10.1016/j.jpowsour.2005.12.002

[17] Kim M-H, Jeong Lee G, Keum C-M, Hazra G, Karak BB, Choudhuri AR, et al. Thermal modeling of the lithium/polymer battery: I. Discharge behavior of a single cell. Journal of the Electrochemical Society. 1995;**142**:3274

[18] Govar CJ, Banner JA. Safety testing of batteries for navy devices using lithium ion technology. In: Seventeenth Annual Battery Conference on Applications and Advances. Proceedings of Conference (Cat. No.02TH8576); Long Beach, CA, USA. 2002. pp. 195-198. DOI: 10.1109/BCAA.2002.986395

[19] Wang CY, Srinivasan V. Computational battery dynamics (CBD)-electrochemical/thermal coupled modeling and multi-scale modeling. Journal of Power Sources. 2002; **110**:364-376

[20] Lee KJ, Smith K, Pesaran A, Kim GH. Three dimensional thermal-, electrical-, and electrochemical-coupled model for cylindrical wound large format lithium-ion batteries. Journal of Power Sources. 2013;**241**:20-32. DOI: 10.1016/j.jpowsour.2013.03.007

[21] Cai L, White RE. Mathematical modeling of a lithium ion battery with thermal effects in COMSOL Inc. multiphysics (MP) software. Journal of Power Sources. 2011;**196**:5985-5989. DOI: 10.1016/j.jpowsour.2011.03.017

[22] Inui Y, Kobayashi Y, Watanabe Y, Watase Y, Kitamura Y. Simulation of temperature distribution in cylindrical and prismatic lithium ion secondary batteries. Energy Conversion and Management. 2007;**48**:2103-2109. DOI: 10.1016/j.enconman.2006.12.012

[23] Gu WB, Wang CY. Thermal-electrochemical modeling of battery systems. Journal of the Electrochemical Society. 2000;**147**:2910

[24] Chiu KC, Lin CH, Yeh SF, Lin YH, Chen KC. An electrochemical modeling of lithium-ion battery nail penetration. Journal of Power Sources. 2014;**251**:254-263. DOI: 10.1016/j.jpowsour.2013.11.069

[25] Guo M, Kim GH, White RE. A three-dimensional multi-physics model for a Li-ion battery. Journal of Power Sources. 2013;**240**:80-94. DOI: 10.1016/j.jpowsour.2013.03.170

[26] Zhu C, Li X, Song L, Xiang L. Development of a theoretically based thermal model for lithium ion battery pack. Journal of Power Sources. 2013; **223**:155-164. DOI: 10.1016/j.jpowsour.2012.09.035

[27] Guo G, Long B, Cheng B, Zhou S, Xu P, Cao B. Three-dimensional thermal finite element modeling of lithium-ion battery in thermal abuse application. Journal of Power Sources. 2010;**195**: 2393-2398. DOI: 10.1016/j.jpowsour.2009.10.090

[28] Kurzawski A, Torres-Castro L, Shurtz R, Lamb J, Hewson JC. Predicting cell-to-cell failure propagation and limits of propagation in lithium-ion cell stacks. Proceedings of the Combustion Institute. 2021;**38**:4737-4745. DOI: 10.1016/J.PROCI.2020.06.270

[29] Lamb J, Orendorff CJ, Roth EP, Langendorf J. Studies on the thermal breakdown of common Li-ion battery electrolyte components. Journal of the Electrochemical Society. 2015;**162**: A2131-A2135. DOI: 10.1149/2.0651510JES/XML

[30] Finegan DP, Darcy E, Keyser M, Tjaden B, Heenan TMM, Jervis R, et al. Characterising thermal runaway within lithium-ion cells by inducing and monitoring internal short circuits. Energy & Environmental Science. 2017; **10**:1377-1388. DOI: 10.1039/C7EE00385D

[31] Feng X, Ren D, He X, Ouyang M. Mitigating thermal runaway of lithium-ion batteries. Joule. 2020;**4**:743-770. DOI: 10.1016/J.JOULE.2020.02.010

[32] Bowen T, Chernyakhovskiy I, Xu K, Gadzanku S, Coney K. USAID Grid-Scale Energy Storage Technologies Primer. Golden, CO, United States: National Renewable Energy Laboratory (NREL). Availbale from: www.nrel.gov/publications. 2021. DOI: 10.2172/1808490

[33] Shen M, Gao Q. A review on battery management system from the modeling efforts to its multiapplication and integration. International Journal of Energy Research. 2019;**43**:5042-5075. DOI: 10.1002/ER.4433

[34] Xu J, Hendricks C. A multiphysics simulation of the thermal runaway in large-format lithium-ion batteries. In: 2019 18th IEEE Intersociety Conference on Thermal and Thermomechanical Phenomena in Electronic Systems (ITherm); Las Vegas, NV, USA. 2019. pp. 815-821. DOI: 10.1109/ITHERM.2019.8757424

[35] Xu J, Hendricks C. Multi-Physics Model of Lithium-Ion Battery and Battery Pack Undergoing Abuse Conditions. In: 2021 20th IEEE Intersociety Conference on Thermal and Thermomechanical Phenomena in Electronic Systems (iTherm). San Diego, CA, USA. 2021. pp. 568-574. DOI: 10.1109/ITherm51669.2021.9503201

[36] Hendricks C, Ko JK, Steffin J, Maharrey S, Stoltz C, Fuentevilla D. Gas monitoring and sampling of large-format lithium-ion cells. ECS Meeting Abstracts. 2017;**MA2017-02**:219-219. DOI: 10.1149/MA2017-02/4/219

[37] Hatchard TD, MacNeil DD, Basu A, Dahn JR. Thermal model of cylindrical and prismatic lithium-ion cells. Journal of the Electrochemical Society. 2001;**148**: A755. DOI: 10.1149/1.1377592

[38] Kim GH, Pesaran A, Spotnitz R. A three-dimensional thermal abuse model for lithium-ion cells. Journal of Power Sources. 2007;**170**:476-489. DOI: 10.1016/J.JPOWSOUR.2007.04.018

[39] Pigłowska M, Kurc B, Galiński M, Fuć P, Kamińska M, Szymlet N, et al. Challenges for safe electrolytes applied in lithium-ion cells—A review. Materials. 2021;**14**:6783. DOI: 10.3390/MA14226783

[40] Arbizzani C, Gabrielli G, Mastragostino M. Thermal stability and flammability of electrolytes for lithium-ion batteries. Journal of Power Sources. 2011;**196**:4801-4805. DOI: 10.1016/j.jpowsour.2011.01.068

[41] Spotnitz RM, Weaver J, Yeduvaka G, Doughty DH, Roth EP. Simulation of

abuse tolerance of lithium-ion battery packs. Journal of Power Sources. 2007; **163**:1080-1086. DOI: 10.1016/J. JPOWSOUR.2006.10.013

[42] Offer GJ, Yufit V, Howey DA, Wu B, Brandon NP. Module design and fault diagnosis in electric vehicle batteries. Journal of Power Sources. 2012;**206**:383-392. DOI: 10.1016/J.JPOWSOUR.2012. 01.087

[43] Hu J, Liu T, Tang Q, Wang X. Experimental investigation on thermal runaway propagation in the lithium ion battery modules under charging condition. Applied Thermal Engineering. 2022;**211**:118522. DOI: 10.1016/J. APPLTHERMALENG.2022.118522

[44] Lamb J, Orendorff CJ, Steele LAM, Spangler SW. Failure propagation in multi-cell lithium ion batteries. Journal of Power Sources. 2015;**283**:517-523. DOI: 10.1016/J.JPOWSOUR.2014.10.081

[45] Sun H, Wang X, Tossan B, Dixon R. Three-dimensional thermal modeling of a lithium-ion battery pack. Journal of Power Sources. 2012;**206**:349-356. DOI: 10.1016/J.JPOWSOUR.2012.01.081

[46] Liu H, Wei Z, He W, Zhao J. Thermal issues about Li-ion batteries and recent progress in battery thermal management systems: A review. Energy Conversion and Management. 2017;**150**: 304-330. DOI: 10.1016/J. ENCONMAN.2017.08.016

[47] Liu X, Stoliarov SI, Denlinger M, Masias A, Snyder K. Comprehensive calorimetry of the thermally-induced failure of a lithium ion battery. Journal of Power Sources. 2015;**280**:516-525. DOI: 10.1016/J.JPOWSOUR.2015.01.125

[48] Abbott KC, Buston JEH, Gill J, Goddard SL, Howard D, Howard G, et al. Comprehensive gas analysis of a 21700 Li(Ni0.8Co0.1Mn0.1O2) cell using mass spectrometry. Journal of Power Sources. 2022;**539**:231585. DOI: 10.1016/J. JPOWSOUR.2022.231585

[49] Liu X, Wu Z, Stoliarov SI, Denlinger M, Masias A, Snyder K. Heat release during thermally-induced failure of a lithium ion battery: Impact of cathode composition. Fire Safety Journal. 2016; **85**:10-22. DOI: 10.1016/J.FIRESAF.2016. 08.001

[50] Walker WQ, Cooper K, Hughes P, Doemling I, Akhnoukh M, Taylor S, et al. The effect of cell geometry and trigger method on the risks associated with thermal runaway of lithium-ion batteries. Journal of Power Sources. 2022;**524**:230645. DOI: 10.1016/J. JPOWSOUR.2021.230645

[51] Zhang L, Liu L, Yang S, Xie Z, Zhang F, Zhao P. Experimental investigation on thermal runaway suspension with battery health retention. Applied Thermal Engineering. 2023;**225**:120239. DOI: 10.1016/J.APPLTHERMALENG. 2023.120239

[52] Zhang Y, Cheng S, Mei W, Jiang L, Jia Z, Cheng Z, et al. Understanding of thermal runaway mechanism of $LiFePO_4$ battery in-depth by three-level analysis. Applied Energy. 2023;**336**:120695. DOI: 10.1016/J.APENERGY.2023.120695

[53] Liu Y, Mao G, Yang M, Jiang J, Jiang J. Slight overcharge aging behaviors and thermal runaway characteristics of Li(Ni 0.5 Co 0.2 Mn 0.3) O_2/graphite batteries at different ambient temperatures. ACS Applied Energy Materials. 2023;**6**:6760-6772. DOI: 10.1021/ACSAEM.3C00932/ ASSET/IMAGES/MEDIUM/ AE3C00932_0017.GIF

[54] Walker WQ, Darst JJ, Finegan DP, Bayles GA, Johnson KL, Darcy EC, et al. Decoupling of heat generated from

ejected and non-ejected contents of 18650-format lithium-ion cells using statistical methods. Journal of Power Sources. 2019;**415**:207-218. DOI: 10.1016/J.JPOWSOUR.2018.10.099

[55] Finegan DP, Scheel M, Robinson JB, Tjaden B, Hunt I, Mason TJ, et al. In-operando high-speed tomography of lithium-ion batteries during thermal runaway. Nature Communications. 2015; **61**(6):1-10. DOI: 10.1038/ncomms7924

[56] Hoelle S, Zimmermann S, Hinrichsen O. 3D thermal simulation of thermal runaway propagation in lithium-ion battery cell stack: Review and comparison of modeling approaches. Journal of the Electrochemical Society. 2023;**170**:060516. DOI: 10.1149/1945-7111/ACD966

[57] Wu MS, Liu KH, Wang YY, Wan CC. Heat dissipation design for lithium-ion batteries. Journal of Power Sources. 2002;**109**:160-166. DOI: 10.1016/S0378-7753(02)00048-4

[58] Greco A, Cao D, Jiang X, Yang H. A theoretical and computational study of lithium-ion battery thermal management for electric vehicles using heat pipes. Journal of Power Sources. 2014;**257**:344-355. DOI: 10.1016/J.JPOWSOUR.2014.02.004

[59] Zhao R, Gu J, Liu J. An experimental study of heat pipe thermal management system with wet cooling method for lithium ion batteries. Journal of Power Sources. 2015;**273**:1089-1097. DOI: 10.1016/J.JPOWSOUR.2014.10.007

[60] Al-Hallaj S, Selman JR. Thermal modeling of secondary lithium batteries for electric vehicle/hybrid electric vehicle applications. Journal of Power Sources. 2002;**110**:341-348. DOI: 10.1016/S0378-7753(02)00196-9

[61] Ling Z, Wang F, Fang X, Gao X, Zhang Z. A hybrid thermal management system for lithium ion batteries combining phase change materials with forced-air cooling. Applied Energy. 2015; **148**:403-409. DOI: 10.1016/j.apenergy.2015.03.080

[62] Rao Z, Wang Q, Huang C. Investigation of the thermal performance of phase change material/mini-channel coupled battery thermal management system. Applied Energy. 2016;**164**:659-669. DOI: 10.1016/J.APENERGY.2015.12.021

[63] Renteria JD, Nika DL, Balandin AA. Graphene thermal properties: Applications in thermal management and energy storage. Applied Sciences. 2014;**4**:525-547. DOI: 10.3390/APP4040525

[64] Goli P, Legedza S, Dhar A, Salgado R, Renteria J, Balandin AA. Graphene-enhanced hybrid phase change materials for thermal management of Li-ion batteries. Journal of Power Sources. 2014;**248**:37-43. DOI: 10.1016/J.JPOWSOUR.2013.08.135

[65] Basu S, Hariharan KS, Kolake SM, Song T, Sohn DK, Yeo T. Coupled electrochemical thermal modelling of a novel Li-ion battery pack thermal management system. Applied Energy. 2016;**181**:1-13. DOI: 10.1016/J.APENERGY.2016.08.049

[66] Shen K, Mao Y, Zheng Y, Yang W, Wu B. One-dimensional modeling and experimental analysis of nail penetration thermal runaway for large capacity Li-ion power battery. Journal of the Electrochemical Society. 2022;**169**: 040502. DOI: 10.1149/1945-7111/AC5CF0

[67] Feng X, He X, Ouyang M, Lu L, Wu P, Kulp C, et al. Thermal runaway

propagation model for designing a safer battery pack with 25 Ah $LiNixCoyMnzO_2$ large format lithium ion battery. Applied Energy. 2015;**154**: 74-91. DOI: 10.1016/J.APENERGY.2015.04.118

[68] Xu C, Wang H, Jiang F, Feng X, Lu L, Jin C, et al. Modelling of thermal runaway propagation in lithium-ion battery pack using reduced-order model. Energy. 2023;**268**:126646. DOI: 10.1016/J.ENERGY.2023.126646

[69] Citarella M, et al. Computational Modelling of Thermal Runaway Propagation in Lithium-Ion Battery Systems.In: 2019 IEEE Transportation Electrification Conference (ITEC-India). 2019. pp. 1-4

[70] Yuan C, Wang Q, Wang Y, Zhao Y. Inhibition effect of different interstitial materials on thermal runaway propagation in the cylindrical lithium-ion battery module. Applied Thermal Engineering. 2019;**153**:39-50. DOI: 10.1016/J.APPLTHERMALENG.2019.02.127

[71] Vyroubal P, Kazda T, Mačák M. The numerical simulation of thermal abuse and runaway in lithium ion battery pack. ECS Transactions. 2021;**105**:549-559. DOI: 10.1149/10501.0549ECST/XML

[72] Jia Y, Uddin M, Li Y, Xu J. Thermal runaway propagation behavior within 18,650 lithium-ion battery packs: A modeling study. Journal of Energy Storage. 2020;**31**:101668. DOI: 10.1016/J.EST.2020.101668

[73] Li Y, Liu G, Li Z. Numerical modeling of thermal runaway in high-energy lithium-ion battery packs induced by multipoint heating. Case Studies in Thermal Engineering. 2022;**38**: 102335. DOI: 10.1016/J.CSITE.2022.102335

[74] Jindal P, Kumar BS, Bhattacharya J. Coupled electrochemical-abuse-heat-transfer model to predict thermal runaway propagation and mitigation strategy for an EV battery module. Journal of Energy Storage. 2021;**39**: 102619. DOI: 10.1016/J.EST.2021.102619

[75] Mishra D, Shah K, Jain A. Investigation of the impact of flow of vented gas on propagation of thermal runaway in a Li-ion battery pack. Journal of the Electrochemical Society. 2021;**168**: 060555. DOI: 10.1149/1945-7111/AC0A20

[76] Mishra D, Shah K, Jain A. Investigation of the impact of radiative shielding by internal partitions walls on propagation of thermal runaway in a matrix of cylindrical Li-ion cells. Journal of the Electrochemical Society. 2021;**168**: 120507. DOI: 10.1149/1945-7111/AC3715

[77] Mishra D, Jain A. Multi-mode heat transfer simulations of the onset and propagation of thermal runaway in a pack of cylindrical Li-ion cells. Journal of the Electrochemical Society. 2021;**168**: 020504. DOI: 10.1149/1945-7111/ABDC7B

[78] Liu X, Wu Z, Stoliarov SI, Denlinger M, Masias A, Snyder K. A thermo-kinetic model of thermally-induced failure of a lithium ion battery: Development, validation and application. Journal of the Electrochemical Society. 2018;**165**: A2909-A2918. DOI: 10.1149/2.0111813JES/XML

[79] Qin J, Zhao S, Liu X, Liu Y. Simulation study on thermal runaway suppression of 18650 lithium battery. Energy Sources, Part A: Recovery, Utilization, and Environmental Effects. 2020. DOI: 10.1080/15567036.2020.1817189

[80] Grimmeisen D, Schneider MS, Numerical simulation of cell venting

within a simplifiied 18650 Li-ion battery pack, in: 13th European LS-DYNA Conference Ulm, Germany, 2021. Available from: https://www.dynalook.com/conferences/13th-european-ls-dyna-conference-2021/battery/grimmeisen_cascate.pdf [Accessed: July 21, 2023].

[81] Coman PT, Darcy EC, White RE. Simplified thermal runaway model for assisting the design of a novel safe Li-ion battery pack. Journal of the Electrochemical Society. 2022;**169**:040516. DOI: 10.1149/1945-7111/AC62BD

[82] Bugryniec PJ, Davidson JN, Brown SF. Computational modelling of thermal runaway propagation potential in lithium iron phosphate battery packs. Energy Reports. 2020;**6**:189-197. DOI: 10.1016/J.EGYR.2020.03.024

[83] Coman PT, Darcy EC, Veje CT, White RE. Numerical analysis of heat propagation in a battery pack using a novel technology for triggering thermal runaway. Applied Energy. 2017;**203**:189-200. DOI: 10.1016/J.APENERGY.2017.06.033

[84] Zhang W, Liang Z, Yin X, Ling G. Avoiding thermal runaway propagation of lithium-ion battery modules by using hybrid phase change material and liquid cooling. Applied Thermal Engineering. 2021;**184**:116380. DOI: 10.1016/J.APPLTHERMALENG.2020.116380

[85] Chen J, Ren D, Hsu H, Wang L, He X, Zhang C, et al. Investigating the thermal runaway features of lithium-ion batteries using a thermal resistance network model. Applied Energy. 2021;**295**:117038. DOI: 10.1016/J.APENERGY.2021.117038

[86] Bilyaz S, Marr KC, Ezekoye OA. Modeling of thermal runaway propagation in a pouch cell stack. Fire Technology. 2020;**56**:2441-2466. DOI: 10.1007/S10694-020-00970-6/FIGURES/11

Chapter 11

DEFEM Method and Its Application in Pebble Flows

Xu Liu, Nan Gui, Mengqi Wu, Takashi Hibiki, Xingtuan Yang, Jiyuan Tu and Shengyao Jiang

Abstract

Based on the concept of embedded discrete elements (EDEs), the discrete element-embedded finite element model (DEFEM) is extended in this work. The new method can be used to calculate the motion and stress variation of particles. This work discusses its application in granular flow simulation for particle motions with small deformations. The updated Lagrangian finite element method is used to obtain the coupling solution of the internal stress and the overall motion of particles in the DEFEM. The computation of deformation displacement is based on the concepts of displacement decomposition (translational and rotational motions and deformation displacement). The deformation displacement is the difference between particles and template particles [rigid body, using the discrete element method (DEM) to calculate translational and rotational displacements]. It is used to calculate the dynamic stress distribution of particles and the internal force of the node. Therefore, it has a wide scope of application (for example, it can be extended to non-spherical particles). The software validation proves the accuracy of this method. The application of the DEFEM in the accumulation process of particles is given. The motion characteristics and deformation of particles are discussed, and the stress distribution and force chain structure in particle accumulation are obtained.

Keywords: discrete element method, finite element method, dynamic stress distribution, pebble bed, pebble flow, coupled method

1. Introduction

A particle system is a complex system composed of many discrete materials widely existing in nature and applied to industrial production [1]. In particular, it is of great engineering significance and academic value to discuss and analyze the relationship between motion characteristics and the internal stress of particles.

The discrete element method (DEM) is widely used to study the macroscopic and microscopic physical properties of granular materials. A particle is regarded as a discrete element if its motion satisfies Newton's second law in the DEM. The DEM can quickly obtain the overall motion information of particles (position, velocity, etc.). Kačianauskas et al. [2] proposed a parallel three-dimensional DEM simulation of polydisperse materials described by normal size distribution. Fang et al. [3] developed

 IntechOpen

a CUDA-GPU parallel algorithm based on super-quadric elements, which is applicable to and reliable for the large-scale engineering applications of non-spherical granular systems. Sun et al. [4] simulate the interactions of the bionic subsoilers and an ordinary subsoiler (O-S) with the soil based on the DEM. However, particles are regarded as discrete elements, and the whole particles are considered as a whole because of ignoring the internal changes of particles. One cannot obtain the internal information of particles (internal stress distribution, deformation, etc.).

When calculating the physical information inside the particles, the finite element method (FEM) is widely used [5–7]. It is a numerical solution method for elastic mechanics problems that developed rapidly with the advancement of computer power. It is applied in continuum mechanics to obtain the deformation, stress, natural frequency, and vibration mode of the structure [8]. The analyzed objects have been extended from elastic materials to plastic, viscoelastic, viscoplastic, and composite materials and from the continuum to a discontinuity [9]. The FEM is widely used in the internal stress simulation of objects. Fang et al. [3] studied the particle–wall collision process using by the FEM, which could solve local stress and strain rate. Kabir et al. [10] used the explicit FEM to simulate the flow phenomenon of particles. The results of shear behavior, particle kinetic energy, and particle stresses within the shear cell with time were given. Wagner et al. [11] proposed a new particle flow simulation method based on the extended FEM (x-FEM), which simulates moving particles without re-meshing. Krok et al. [12] conducted a systematic finite element analysis of the thermo-mechanical behavior of pharmaceutical powders during the molding process using the finite element solver ABAQUS.

The DEM and FEM have their advantages, so the combination of DEM and FEM is widely recognized and studied [13–15]. Guo and Zhao [16] proposed a multiscale framework to simulate the mechanical behavior of granular media based on DEM and FEM. A DEM assembly with the memory of its loading history is embedded in the Gauss integral points of the finite element mesh. The DEM assembly receives the global deformation at its Gauss point from the FEM as input boundary conditions in this new multiscale framework. Onate and Rojek [17] conducted a dynamic analysis of geological mechanics problems based on the combination of DEM and FEM. The combined models can employ spherical rigid and finite elements to discretize different parts of the system. Zárate and Oñate [18] proposed a new numerical method to predict the occurrence and evolution of fractures in continuous media, which combines the FEM with DEM. Munjiza and John [19] further studied the sensitivity to the mesh size of the combined single and smeared crack model in the context of the combined finite–discrete element method. Azevedo and Lemos [20] applied the hybrid method to analyze large structures. This method uses DEM to discrete the fracture zone and a discretization based on the FEM for the surrounding areas. Argilaga et al. [21] proposed a multiscale model based on an FEM $\times$ DEM approach. The method uses discrete elements in a standard finite element framework, and it has proven to be an effective way to treat real-scale engineering problems. However, it should be emphasized that the coupling of the discrete element method and finite element method still needs further research. For example, the solution of coupled motion, internal stress, and other contents need to be further discussed and analyzed.

The discrete element-embedded finite element model (DEFEM) is proposed, and it can be used to calculate particle motion and heat transfer [22–24]. This method is applied to the calculation of temperature gradient and deformation in particles. In this paper, the DEFEM is extended based on the concept of the embedded discrete element

(EDE). The computation of deformation displacement is based on the concepts of displacement decomposition (translational and rotational motions and deformation displacement). The DEFEM mainly adopts the Lagrange finite element method to obtain the coupling solution about the stress and motion of particles. The finite element software verifies the relevant algorithms. The motion characteristics and deformation of particles are discussed, and the stress distribution and force chain structure in particle accumulation are obtained.

2. The discrete element-embedded finite element model

In this paper, the DEFEM is extended and used to calculate the translational and rotational motion and the stress of particles. In this method, the EDE is used to obtain the contact force distribution of the particle. The coupling solution of the internal stress and the overall motion of the particle are obtained by combining the advantages of the DEM and FEM.

2.1 Introduction of the EDE

There is a set of EDEs around the particles, which cover the outermost boundary of the particles (grid cells). In this way, the contact surfaces between the particles are transformed into contact between the EDEs. Among them, the concept of the EDE can refer to our team's previous articles [22]. The EDE covers boundary elements of particles, and the soft-sphere model in the DEM calculates the force distribution of particles. The obtained force distribution of particles is used as the boundary condition in the FEM [23, 24]. In this paper, the two-dimensional spherical particles are divided into triangular grid elements, and the other methods for three-dimensional cases or other grid element types are similar.

2.2 Updated Lagrangian finite element method

The numerical method is used to simulate the impact and collision process, and the discretization of the object is essential. In the FEM, the object is divided into grid elements, and the motion equation satisfied is usually established by the node of each element.

The Euler method and the Lagrange method are mainly used for solving the impact and collision problem in the FEM. The Euler method fixes the computational grid in the spatial coordinates and remains unchanged in the process of deformation and motion. This method can avoid the distortion of the grid. Still, it is difficult to track the deformation boundary of the object accurately, and it requires specific processing methods to identify the shape and position of the deformation boundary. The Lagrange method is used in the DEFEM. Its grid and object remain coincident in the whole moving process; that is, the grid and object do not move relative to each other. The Lagrangian method can track and process complex deformation boundaries. Since the small deformation problem is discussed in this paper, the Lagrange method ignores the mesh distortion. In this paper, the updated Lagrangian scheme is used in the DEFEM. The Eulerian coordinates are used for derivative and integral simulation, and the stress and strain are expressed in the form of the Eulerian metric.

2.2.1 Weak form of control equation

The material derivative of the object's momentum is equal to the sum of the external forces acting on the object. The material derivative and momentum of the object momentum are defined as follows:

$$f_i(t) = \int_V \rho b_i(x,t)dV + \int_A t_i(x,t)dA, \tag{1}$$

$$p_i(t) = \int_V \rho v_i(x,t)dV, \tag{2}$$

where f_i is the material derivative of the object momentum, b_i is the force (physical force) acting on the unit mass of the object, t_i is the force (surface force) acting on the surface of the object, p_i is the momentum of the object, ρ is the density of the object, and v_i is the velocity of the object.

According to the momentum equation, the Reynold's transport theorem, and the Gaussian theorem, the following equation can be obtained:

$$\int_V \left(\rho \frac{Dv_i}{Dt} - \rho b_i - \frac{\partial \sigma_{ji}}{\partial x_j} \right) dV = 0. \tag{3}$$

The momentum equation described by the updated Lagrangian is obtained as follows:

$$\rho \frac{Dv_i}{Dt} = \frac{\partial \sigma_{ji}}{\partial x_j} + \rho b_i \;\; or \;\; \rho \dot{\mathbf{v}} = \nabla \cdot \boldsymbol{\sigma} + \rho \mathbf{b}. \tag{4}$$

The boundary conditions are given as follows:

$$(n \cdot \sigma)|_{A_t} = \bar{t}, \quad v|_{A_v} = \bar{v}, \tag{5}$$

where A_t is the specified surface force boundary in the current configuration and A_v is the specified velocity boundary in the current configuration.

The solution region satisfies Eq. (4). However, it is difficult to directly solve the above equation in an actual situation, so it is necessary to find an approximate solution with certain accuracy. The weighted residual method is commonly used to obtain approximate solutions to differential equations. It allows the equations and boundary conditions to have quantities at each node but requires that the weighted integrals of these quantities on the region and the boundary are zero.

The following equation can be obtained by using the weighted residual method:

$$\int_V \delta v_i \left(\frac{\partial \sigma_{ji}}{\partial x_j} + -\rho b_i - \rho \ddot{u}_i \right) dV = 0 \tag{6}$$

where δv_i is the virtual velocity, which satisfies the following equation:

$$\delta v_i(\mathbf{X}) \in \mathbb{R}_0, \quad \mathbb{R}_0 = \left\{ \delta v_i^{\,0}(\mathbf{X}) | \delta v_i(\mathbf{X}) \in C^0(\mathbf{X}), \delta v_i|_{A_v} = 0 \right\} \tag{7}$$

By using partial integration and Eq. (7), the first term of Eq. (6) can be transformed into the following form:

$$\int_V \delta v_i \frac{\partial \sigma_{ji}}{\partial x_j} dV = \int_{A_t} \delta v_i \bar{t}_i dA - \int_V \frac{\partial(\delta v_i)}{\partial x_j} \sigma_{ji} dV \tag{8}$$

Substituting Eq. (8) into Eq. (6), the weak form of the momentum equation is obtained. It is also called the principle of virtual power, as follows:

$$\int_V \frac{\partial(\delta v_i)}{\partial x_j} \sigma_{ji} dV - \int_{A_t} \delta v_i \bar{t}_i dA - \int_V \delta v_i \rho b_i dV + \int_V \delta v_i \rho \ddot{u}_i dV = 0 \tag{9}$$

Eq. (9) can also be translated into the following form:

$$\delta \dot{p} = \delta \dot{p}_{int} - \delta \dot{p}_{ext} + \delta \dot{p}_{kin} = 0 \tag{10}$$

where $\delta \dot{p}_{int}$, $\delta \dot{p}_{ext}$, and $\delta \dot{p}_{kin}$ are internal force virtual power, external force virtual power, and inertia force virtual power. Their definitions are as follows:

$$\delta \dot{p}_{int} = \int_V \frac{\partial(\delta v_i)}{\partial x_j} \sigma_{ji} dV, \quad \delta \dot{p}_{ext} = \int_{A_t} \delta v_i \bar{t}_i dA + \int_V \delta v_i \rho b_i dV, \quad \delta \dot{p}_{kin} = \int_V \delta v_i \rho \ddot{u}_i dV \tag{11}$$

2.2.2 Finite element discretization

The entire solution region is divided into N nodes and several unit regions. In the initial configuration, the coordinate of the node is $X_1, X_2, \cdots X_{n_N}$; the node coordinate is $x_1(t), x_2(t), \cdots x_{n_N}(t)$ in the current configuration. In the FEM, the spatial coordinates of particle X at time t is $x_i(X,t)$, which can be obtained by the following equation:

$$x_i(X,t) = N_I(X) x_{iI}(t) \tag{12}$$

where $N_I(X)$ is the shape function of node I, and the repeated subscripts represent the summation in the range of values. Similarly, the coordinate X_i of each node in the initial configuration can also be approximately expressed by the coordinate X_{iI} of the unit node as follows:

$$X_i(t) = N_I(X) X_{iI} \tag{13}$$

The displacement of the node can approximately express the displacement of any point X in the element, and the specific form is given as follows:

$$u_i(X,t) = x_i(X,t) - X_i = N_I(X) u_{iI}(t) \tag{14}$$

Similarly, the approximate expressions of velocity and acceleration of any point X in the element are given as follows:

$$\dot{u}_i(X,t) = N_I(X) \dot{u}_{iI}(t), \quad \ddot{u}_i(X,t) = N_I(X) \ddot{u}_{iI}(t) \tag{15}$$

The approximate expression of virtual velocity is as follows:

$$\delta v_i(X) = N_I(X)\delta v_{iI} \tag{16}$$

where $\delta v_i(X)$ is virtual velocity and δv_{iI} is the virtual velocity of node I. Substituting Eqs. (16) and (14) into Eq. (9) yields the following equation:

$$\delta v_{iI}\left(\int_V \frac{\partial N_I}{\partial x_j}\sigma_{ji}dV - \int_V N_I\rho b_i dV - \int_{A_t} N_I\bar{t}_i dA + \int_V N_I\rho N_J\ddot{u}_{iJ}dV\right) = 0 \tag{17}$$

From the boundary condition of virtual velocity, Eq. (7), and due to the arbitrariness of virtual velocity δv_{iI}, the following equation can be obtained:

$$\int_V \frac{\partial N_I}{\partial x_j}\sigma_{ji}dV - \int_V N_I\rho b_i dV - \int_{A_t} N_I\bar{t}_i dA + \int_V N_I\rho N_J\ddot{u}_{iJ}dV = 0 \quad \forall I \notin A_v \tag{18}$$

Eq. (18) can also be translated into the following form:

$$M_{IJ}\ddot{u}_{iJ} + f_{iI}^{int} = f_{iI}^{ext} \quad \forall I \notin A_v \tag{19}$$

where M_{IJ} is the mass matrix of the system, f_{iI}^{int} is the internal force of the node, and f_{iI}^{ext} is the external force of the node. Their definitions are given as follows:

$$\begin{cases} M_{IJ} = \int_V N_I\rho N_J dV, \quad f_{iI}^{int} = \int_V \frac{\partial N_I}{\partial x_j}\sigma_{ji}dV \\ f_{iI}^{ext} = \int_V N_I\rho b_i dV + \int_{A_t} N_I\bar{t}_i dA \end{cases} \tag{20}$$

2.3 Introduction of the deformation displacement

The DEFEM can obtain the translational and rotational motion and the deformations of particles. The material constitutive model describes the response of the material of the object under an external force, which is mainly used to calculate the stress change of the object. The stress distribution of particles can be obtained using the deformation displacement and constitutive relation. During motion, particles have collisions and deformations. The displacement and rotation, small deformation, and the elastic material particle stress are discussed in this paper. The analysis of other constitutive models is similar to that of elastic materials.

In the DEFEM, it is necessary to decompose the overall displacement of particles when obtaining the deformation displacement, as shown in the following formula:

$$u = u_{move} + u_{rotate} + u_{deform} \tag{21}$$

where u is the absolute displacement of the node, u_{move} is the translational motion displacement, u_{rotate} is the rotational displacement, and u_{deform} is the deformation displacement. The displacement of any point can be decomposed into the above three parts, in which the deformation displacement is used to calculate the strain and stress of the element.

The translational motion displacement u_{move} and rotational displacement u_{rotate} can be solved by using the DEM. In the DEFEM, it is considered that there is a virtual template particle, which is a rigid object, and it only moves and rotates without deformation. The translational motion and rotation of the template particles satisfy Newton's equations. The translational displacement and rotation angle of the template particles at each time interval can be obtained by using the DEM and the central difference method, as shown in the following equation:

$$\chi^{t+\Delta t} = \chi^{t-\Delta t} + \dot{\chi}^{t+\Delta t/2}\Delta t, \qquad \theta^{t+\Delta t} = \theta^{t-\Delta t} + \dot{\theta}^{t+\Delta t/2}\Delta t \tag{22}$$

where $\chi^{t+\Delta t}$ and $\chi^{t-\Delta t}$ are the translational displacements of particle i at time $t+\Delta t$ and $t-\Delta t$, $\theta^{t+\Delta t}$ and $\theta^{t-\Delta t}$ are the rotation angles of particle i at time $t+\Delta t$ and $t-\Delta t$, $\dot{\chi}^{t+\Delta t/2}$ is the translational velocity of particle i at time $t+\Delta t/2$, and $\dot{\theta}^{t+\Delta t/2}$ is the angular velocity of particle i at time $t+\Delta t/2$. The formulas of translational velocity and angular velocity are as follows:

$$\dot{\chi}^{t+\Delta t/2} = \dot{\chi}^{t-\Delta t/2} + \frac{\Delta t}{m_i}\sum_{j=1}^{n} F_i^t, \qquad \dot{\theta}^{t+\Delta t/2} = \dot{\theta}^{t-\Delta t/2} + \frac{\Delta t}{J_i}\sum_{j=1}^{n} T_i^t \tag{23}$$

where m_i is the mass of particle i, J_i is the rotational inertia of particle i, n is the number of particles in contact with particle i, F_i^t is the contact force between particle i and other contact particles at time t, T_i^t is the torque between particle i and other contact particles at time t, $\sum_{j=1}^{n} F_i^t$ and $\sum_{j=1}^{n} T_i^t$ are the resultant force and resultant torque of particle i at time t, $\dot{\chi}^{t+\Delta t/2}$ and $\dot{\chi}^{t-\Delta t/2}$ are the translational velocities of particle i at time $t+\Delta t/2$ and $t-\Delta t/2$, and $\dot{\theta}^{t+\Delta t/2}$ and $\dot{\theta}^{t-\Delta t/2}$ are the angular velocities of particle i at time $t+\Delta t/2$ and $t-\Delta t/2$.

The translational velocity and angular velocity of the template particles can be calculated by Eq. (23). The translational displacement and rotation displacement of the template particles can be obtained by Eq. (22), which is also considered to be the translational displacement u_{move} and rotational displacement u_{rotate} of the particle after the decomposition of the overall displacement. Then, the deformation displacement u_{deform} of the particle can be obtained by using Eq. (21).

2.4 Numerical process of the discrete element-embedded finite element model

In this paper, the DEFEM is extended to calculate the translational, rotational, and stress variations of particles. In this method, the EDE is used to allocate the resultant contact force and torque to the boundary node of particles.

The concept of template particles is used in the whole simulation process, which only moves and rotates without deformation. The deformation displacement of the particles is obtained by calculating the difference between the coordinates of the particles and the template particles simultaneously. The related process is mainly based on displacement decomposition, as shown in Eq. (21). The deformation displacement here mainly has two functions in the DEFEM: [8] The stress distribution of particles is calculated according to the constitutive relation (2) The internal force of the node is calculated as a part of the finite element equation for the next time, as shown in Eq. (19). It participates in the whole simulation cycle, as shown in **Figure 1**.

The numerical process of the DEFEM is as follows: [8] Initialization and conditionalization: The initial time and initial conditions are assigned, and the state of the template particle is the same as that of the particle (node coordinates, node velocity,

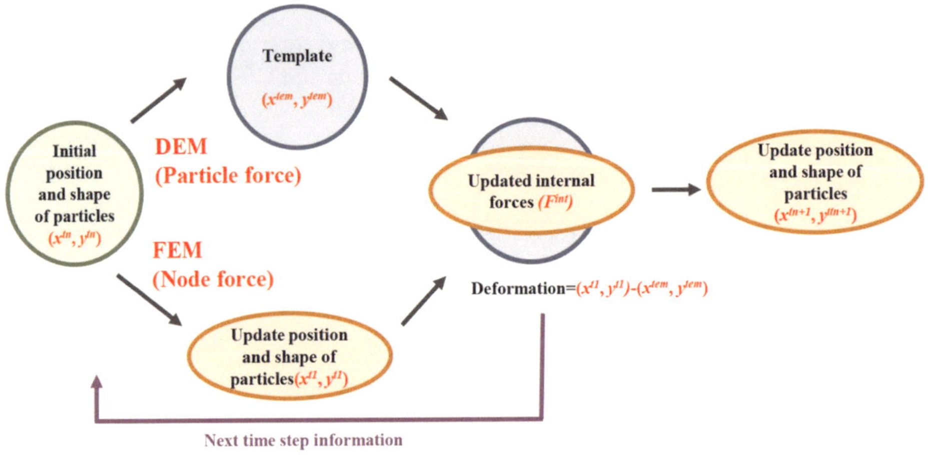

Figure 1.
Numerical process diagram of deformation displacement.

acceleration, etc.). [21] The resultant force and moment (contact force with other particles or the wall) of particles are calculated according to the soft-sphere model in the DEM. [20] The DEM is used to calculate the next time information of the template particle (node coordinates, node velocity, acceleration, etc.). [3] The resultant force of particles is transformed into the force of the EDE, and finally, the force of the particle boundary node is obtained. [5] The information about the next time step of the particle is calculated by the updated Lagrangian FEM. The coordinate difference between the particle and the template particle is denoted as the deformation displacement, and the stress distribution of the particle at this moment and the internal force of the next time step are calculated. [16] Return to step [21] until all particles are calculated. [6] Update the simulation time. [1] Output: If the program calculates the simulation deadline or maximum steps, stop the program. Otherwise, return to step [21] to calculate the particle information at the next moment (**Figure 2**).

2.5 Software validation of particle–wall collision

In this paper, the ABAQUS software is used to simulate the particle–wall collision process, and it is compared with the numerical results by using the DEFEM. The ABAQUS is a finite element software applied to engineering simulation, which can simulate the stress and strain of large structures. This paper simulates the collision process between a spherical particle and the wall using the ABAQUS. The particle with a diameter of 0.05 m vertically hit the wall at a velocity of 0.5 m/s. The elastic modulus of the particles is 0.02 GPa, and the Poisson's ratio is 0.3. The elastic modulus of the wall is 2 GPa, and the Poisson's ratio is 0.3. After the contact force of particles is counted, the simulation results of ABAQUS and DEFEM are compared, as shown in **Figure 3(a)**. It can be observed that the changing trend of the two results is the same, and the numerical value is similar. The contact force of the two numerical methods is re-zero near 0.003 seconds. In addition, the displacement of particles during collision and rebound is tracked, as shown in **Figure 3(b)**. The results obtained by the two methods are in good agreement. It can be seen from these two results that the simulation results of the DEFEM are consistent with those of the ABAQUS.

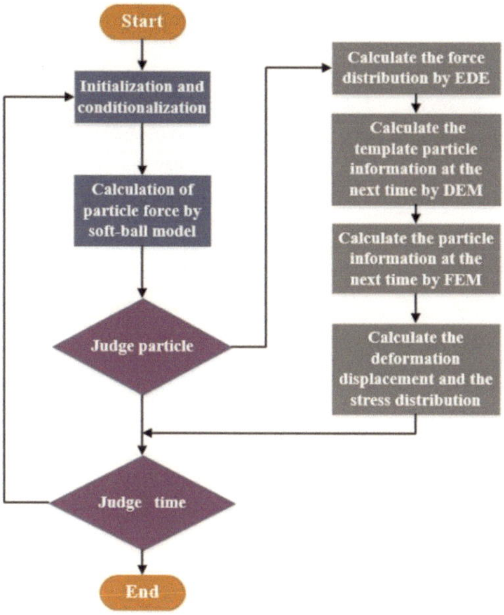

Figure 2.
Schematic diagram of the simulation process of the DEFEM.

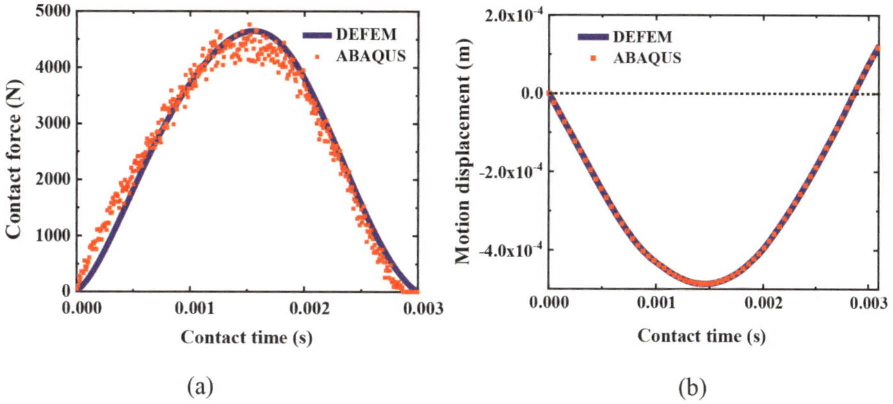

Figure 3.
(a) Comparison of the results of DEFEM and ABAQUS software for contact force of the particle; (b) comparison of DEFEM and ABAQUS software for the displacement of the particle.

Compared with the DEM, one of the advantages of the DEFEM is to calculate the dynamic deformation displacement and stress distribution of particles. It shows the comparison of deformation displacement calculated using the ABAQUS and the DEFEM, as shown in **Figure 4**. It can be observed that the distribution of the two methods is similar. The black vector in **Figure 4(b)** represents the deformation displacement at the grid node of the particle. Due to the collision between particles and the wall, the deformation near the wall is greater than that away from the wall, which also conforms to the relevant physical law.

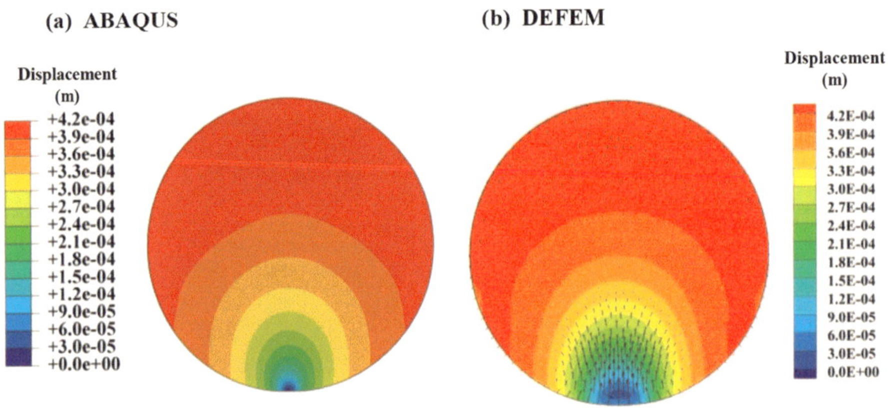

Figure 4.
(a) Static deformation displacement distribution calculated by ABAQUS software; (b) dynamic deformation displacement distribution calculated by the DEFEM (black vector represents the deformation displacement of nodes).

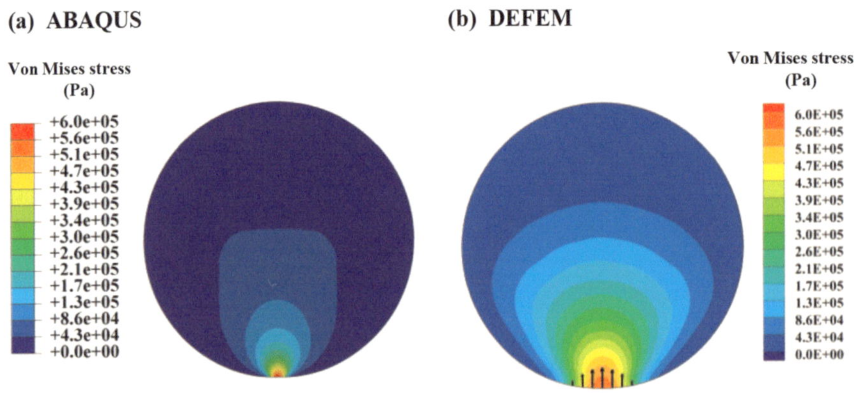

Figure 5.
(a) Static stress distribution calculated by the ABAQUS software; (b) dynamic stress distribution calculated by the DEFEM (black vector represents the contact force of the node).

As shown in **Figure 5**, the static stress distribution of the particle calculated using the ABAQUS software has the same variation trend as the dynamic stress distribution calculated by the DEFEM, but there are some differences in values. Because the FEM is used to solve the problem in the ABAQUS software, it is considered that the contact force between particles only exists at the contact point. As shown in **Figure 5(b)**, face contact is selected in the DEFEM. The multiple EDEs cover the boundary of particles. Therefore, the contact between particles can be transformed into the contact between EDEs, involving multiple nodes. The judgment and simulation process of the contact force are different between the two methods, so the stress distribution calculated by the two methods is different, but these differences are acceptable within the allowable error range.

Through the comparison between the ABAQUS software and the DEFEM, it can be observed that the DEFEM can well track the contact force, stress distribution, and movement in the process of particle collision, which also proves the accuracy of the DEFEM.

3. Simulation of particle accumulation by DEFEM

In this paper, the particle accumulation process is simulated by using the DEFEM. The motion characteristics of particles in the whole process of accumulation in pebble beds are discussed and analyzed. The stress distribution and force chain structure during particle accumulation are also obtained.

3.1 Geometric parameter settings

The particle accumulation process in a two-dimensional pebble bed is simulated. As shown in **Figure 6**, the bottom inclination angle of the pebble bed is changed in a certain range. The bottom diameter of the pebble bed is 0.2 m, and the bottom inclination angle is 15 degrees or 45 degrees, as shown in **Figure 6**. In the initial state, the particles with a regular arrangement move downward vertically at a certain velocity. The particles collide with the wall of the pebble bed or other particles and finally form a stable accumulation structure at the bottom of the pebble bed. Two regions in the pebble bed are defined: the central and the near-wall regions. The particle motion characteristics in two different regions are discussed. In the case of this paper, the near-wall region is within 0.2 m from the left and right walls, and the central region is within 0.2 m from the central axis of the pebble bed (the selection of values with careful consideration of particle diameter and pebble bed size). The velocity, stress, and deformation during particle accumulation are discussed and analyzed under the different pebble bed geometry structures. The specific parameters in the numerical simulation are shown in **Table 1**.

3.2 Motion characteristics in the near-wall and central regions

The particles initially fall vertically at a certain velocity, forming a stable accumulation structure in the pebble bed. The particle motion characteristics are different in the

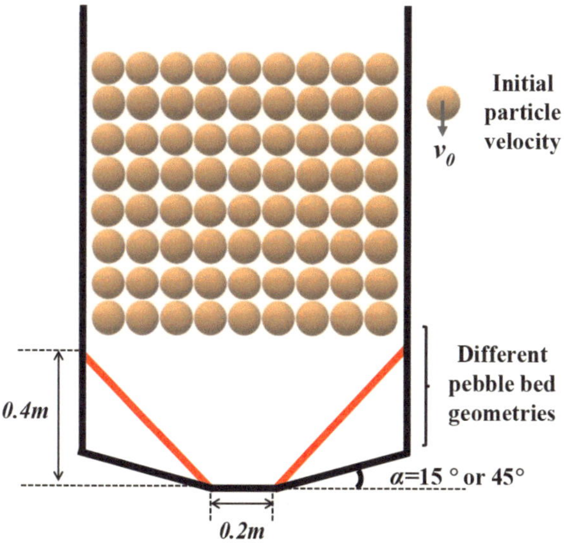

Figure 6.
Diagram of pebble bed structure.

Parameters	Values	Units
Elastic modulus	2×10^7	Pa
Poisson's ratio	0.3	
Density	1000	kg/m^3
Diameter	0.05	m
Initial velocity	0.5	m/s
Number of particles	72	
Time step	5×10^{-6}	s
Friction coefficient	0.18	

Table 1.
Parameters of the particle accumulation process.

accumulation process of the different shapes of pebble beds. Due to the effect of gravity, particles may collide with the wall or other particles. If the particles collide, the velocity of the particles changes due to the effect of contact force. The average velocity and average velocity difference of particles in different regions are defined as follows:

$$v_{ave} = \frac{1}{m}\sum_{i=0}^{m} v_i \tag{24}$$

$$v_{diff} = \frac{1}{m}\sum_{i=0}^{m}\left(v_i - \frac{1}{n}\sum_{j=0}^{n} v_j\right) \tag{25}$$

where v_{ave} is the average velocity of particles in the region, v_i is the velocity of particle i in the region, n is the number of particles in the region. v_{diff} is the average velocity difference of particles in the region, v_j is the velocity of particle j contacting particle i, and m is the total number of other particles in contact with particle i.

The difference in the average particle velocity between the central and near-wall regions at the bottom inclination of 45 degrees is shown in **Figure 7**. It can be seen that

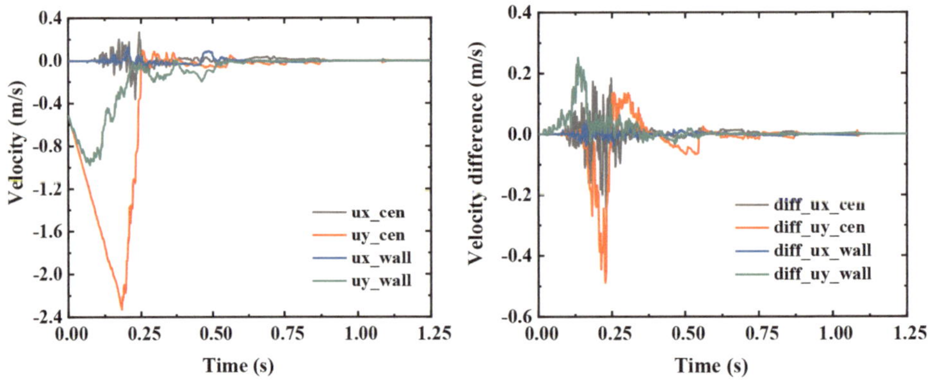

(a) (b)

Figure 7.
(a) Average velocity of particles in the central region and near-wall region with the bottom inclination of 45 degrees; (b) average velocity difference of particles in the central region and near-wall region with the bottom inclination of 45 degrees.

the horizontal velocity of particles is zero in the initial state. Moreover, the vertical velocity of particles increases steadily at the beginning of the accumulation process due to the effect of gravity. Then, it can be observed that some particles collide with the side wall, so the average velocity and average velocity difference of particles near the wall region fluctuate first.

In terms of vertical velocity, the peak velocity of particles in the central region is much larger than that in the near-wall region. This is because compared with the particles in the near-wall region, the particles in the central region can keep free falling motion for a longer time before colliding with the wall or other particles. The change in the vertical velocity of particles in the central region after the collision is more obvious; that is, the vertical velocity difference of central velocity particles is significantly greater than that of the near-wall region. In terms of the horizontal velocity, it is more about the impact of particles. Compared with the near-wall region, the probability of particle collision in the central region is greater. Therefore, the fluctuation of the average horizontal velocity and velocity difference in the central region is significantly greater than that in the near-wall region. Finally, the particles reach the state of stable accumulation; that is, the particles remain stationary (the average velocity and velocity difference of the particles are zero).

3.3 Deformation of particles in the near-wall and central regions

A major advantage of the DEFEM compared with the DEM is that it can calculate the deformation of particles. This part discusses the whole deformation of particles in the accumulation process of the pebble bed. The accumulation process of two-dimensional particles is simulated in this paper, so the deformation of particles is represented by the change of area. The average coefficient of area ratio in the defined region is given as follows:

$$S_{ave} = \frac{1}{m}\sum_{i=0}^{m}\frac{S_i}{S_0} \tag{26}$$

where S_{ave} is the average coefficient of area ratio, m is the number of particles in the region, S_i is the area of particles at the current time, and S_0 is the area of the initial moment of the particle (no deformation state). If the particle does not deform, the average coefficient of area ratio is equal to 1 ($S_{ave} = 1$); if the particle is compressed, the average coefficient of area ratio is less than 1 ($S_{ave} < 1$).

The average coefficient of area ratio in the central and near-wall regions with the bottom inclinations of 45 degrees and 15 degrees is shown in **Figure 8**. It can be seen that some particles are compressed (the average coefficient of area ratio is less than 1; $S_{ave} < 1$) during the accumulation process. When the bottom inclination is 45 degrees, the absolute value of the peak about the average coefficient of area ratio is less than that of 15 degrees, and the changing trend is the same as that of the average stress of the particles. When the particle has severe deformation, the average coefficient of the area ratio and absolute internal value of stress also increases.

3.4 Stress of particles in the near-wall and central regions

Particles may have strain and stress in the accumulation process because they may collide with other particles or walls. In this paper, the von Mises stress is used, and its formula is as follows:

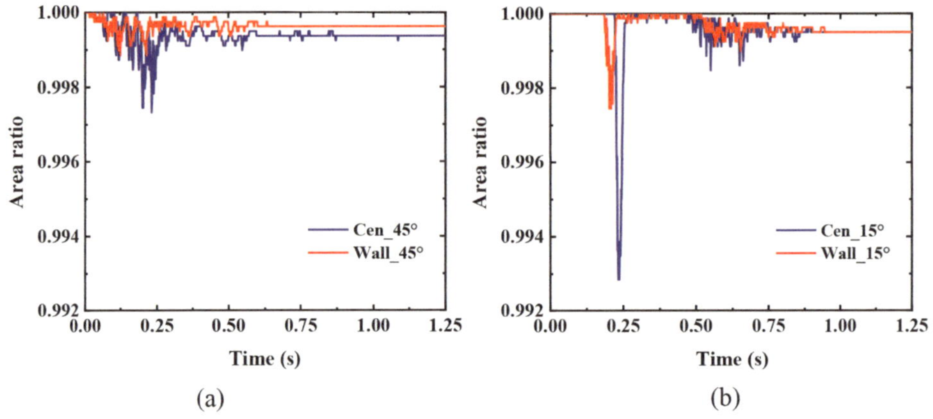

Figure 8.
(a) The average coefficient of area ratio in the central region and near-wall region with the bottom inclination of 45 degrees; (b) the average coefficient of the area ratio in the central region and near-wall region with the bottom inclination of 15 degrees.

$$\sigma = \sqrt{\frac{1}{2}\left[(\sigma_1 - \sigma_2)^2 + (\sigma_2 - \sigma_3)^2 + (\sigma_3 - \sigma_1)^2\right]} \tag{27}$$

Where σ is the von Mises stress and σ_1, σ_2, and σ_3 are the first, second, and third principal stresses, respectively.

The average stress of particles in the region is defined as:

$$\sigma_{ave} = \frac{1}{m}\sum_{i=0}^{m}\sigma_i \tag{28}$$

where σ_{ave} is the average stress in the region, σ_i is the stress of particle i in the region, and m is the number of particles in the region.

The average stress of particles in the central and near-wall regions with the bottom inclinations of 45 degrees and 15 degrees is shown in **Figure 9**.

[8] Different effects of bottom inclination.

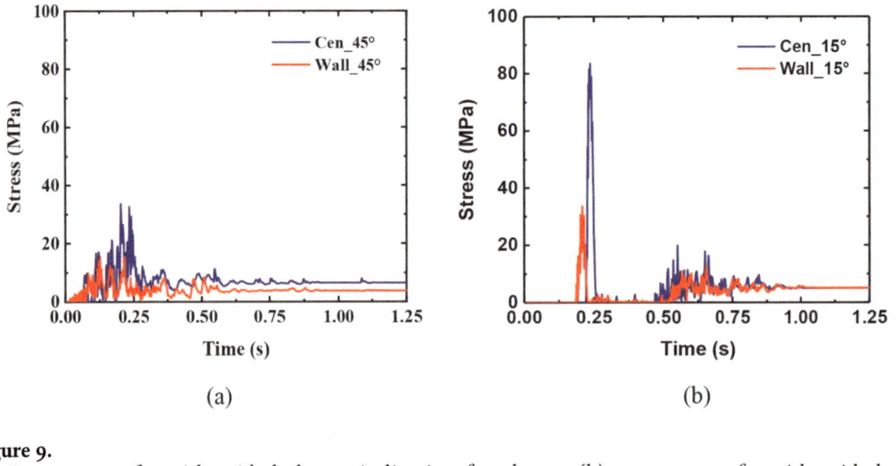

Figure 9.
(a) Average stress of particles with the bottom inclination of 45 degrees; (b) average stress of particles with the bottom inclination of 15 degrees.

DOI: http://dx.doi.org/10.5772/intechopen.109347

It can be seen from **Figure 9** that the peak stress is smaller when the bottom inclination is 45 degrees. The initial positions of particles at different bottom inclination angles are the same. When the bottom inclination is 45 degrees, the particles have contact with the wall on both sides earlier in the process of falling, which can also be verified as shown in **Figure 9(b)**. When the bottom inclination is 15 degrees, the stress remains zero at the beginning, and then the stress begins to become non-zero around 0.185 seconds gradually. If the bottom inclination angle is smaller, the acceleration time of particles is longer due to the action of gravity, and the velocity of collision with the wall is larger. Hence, the contact force and stress are larger. In the stable state formed during the accumulation process, it can be seen that there is a great difference in the average stress of particles between the near-wall region and central region when the bottom inclination is 45 degrees. When the bottom inclination is 15 degrees, there is little difference between the two. When the bottom inclination is larger, the pressure on the side wall of the pebble bed is larger, and the average stress difference between the particles in the near-wall region and central region is larger.

[21] Differences between the near-wall region and central region.

Due to the structure of the pebble bed, particles come into contact with the side wall first, so the stress value of particles in the near-wall region changes first compared with that in the central region. However, the collision of particles in the central region is more intense, so the peak stress in the central region is greater than that in the near-wall region. Moreover, the velocity of particles changes after being affected by the wall, and the velocity-changed particles are more likely to collide with the particles in the central region. Therefore, there is a certain phase difference between the average stress change of particles in the near-wall region and the central region; that is, the average stress change of particles in the central region always lags behind that in the near-wall region. The average stress of particles in the central region is greater than that in the near-wall region.

3.5 Dynamic stress distribution and force chain of particles

In the final stable stage, the change of particle position is not drastic. Particles have a stable accumulation structure and force chain due to gravity. Under different bottom inclinations, the internal stress distribution and force chain are given, as shown in **Figure 10**. It can be seen that the accumulated particles form a stable force chain structure, and the stress distribution and force chain distribution have a certain

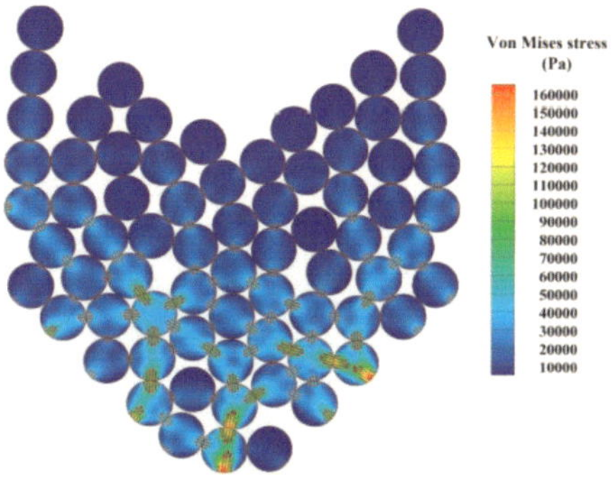

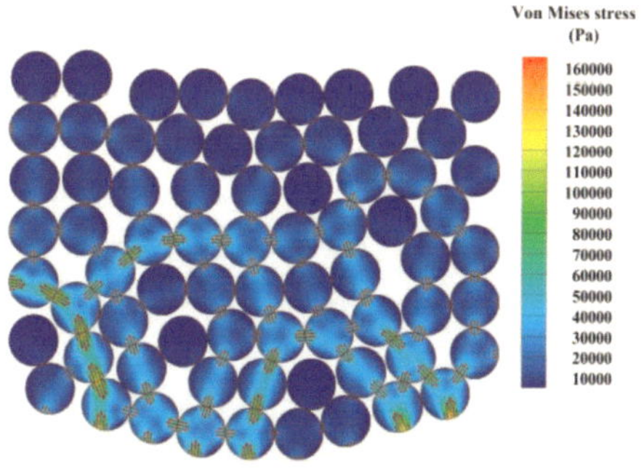

Figure 10.
(a) Dynamic stress distribution and force chain of particles with the bottom inclination of 45 degrees; (b) dynamic stress distribution and force chain of particles with the bottom inclination of 15 degrees.

corresponding relationship. The stress of particles far from the force chain is smaller, and the stress near the force chain is larger in the stable force chain structure. The red vector in **Figure 10** represents the contact force of the node.

4. Conclusions

Based on the concept of the EDE, the DEFEM is extended for simulating translational and rotational motions and small deformations of particles.

When the particle contacts other particles, it is assumed that there is a group of EDEs on the outermost boundary element of the particle. The updated Lagrangian finite element method is used to obtain the coupling solution of internal stress and the overall motion of particles in the DEFEM. It is necessary to decompose the overall displacement of particles into three parts: the translational and rotation displacement and the deformation displacement. The concept of template particles is proposed in the whole simulation process, which only moves and rotates without deformation. The translational displacement and rotation displacement of the template particles can be obtained by using the DEM. The difference between the updated coordinates of the particles and the template particles is considered the deformation displacement in the DEFEM. And it is used to calculate the stress distribution of particles and the internal force of the node. Therefore, this method has a wide range of applications, such as the simulation of non-spherical particles.

The accuracy of this method is proved by the software validation. The application of the DEFEM in the particle accumulation process is given. The motion characteristics and deformation of particles are discussed, and the stress distribution and force chain structure in particle accumulation is obtained in this paper. This paper extends the application of the DEFEM in translational and rotational motions of particles with deformations. This new method can also be used to solve multi-physical field coupling problems such as thermal and mechanical coupling in the future.

Acknowledgements

The authors are grateful for the support of this research by the National Science and Technology Major Project (2011ZX06901-003). One of the authors (T. Hibiki) appreciates the support through the Global STEM Professorship from Hong Kong government.

Author details

Xu Liu[1,2], Nan Gui[1*], Mengqi Wu[1], Takashi Hibiki[2], Xingtuan Yang[1], Jiyuan Tu[1,3] and Shengyao Jiang[1]

1 Institute of Nuclear and New Energy Technology, Collaborative Innovation Center of Advanced Nuclear Energy Technology, Key Laboratory of Advanced Reactor Engineering and Safety, Ministry of Education, Tsinghua University, Beijing, China

2 Department of Mechanical Engineering, City University of Hong Kong, Kowloon Tong, Hong Kong

3 School of Engineering, RMIT University, Melbourne, Australia

*Address all correspondence to: guinan@mail.tsinghua.edu.cn

References

[1] Jiang S, Tu J, Yang X, Gui N. A review of pebble flow study for pebble bed high temperature gas-cooled reactor. Experimental and Computational Multiphase Flow. 2019;**1**:159-176

[2] Kačianauskas R, Maknickas A, Kačeniauskas A, Markauskas D, Balevičius R. Parallel discrete element simulation of poly-dispersed granular material. Advances in Engineering Software. 2010;**41**:52-63

[3] Fang Z, Wang H, Zhang Y, Wei M, Wu X, Sun L. A finite element method (FEM) study on adhesive particle-wall normal collision. Journal of Aerosol Science. 2019;**134**:80-94

[4] Sun J, Wang Y, Ma Y, Tong J, Zhang Z. DEM simulation of bionic subsoilers (tillage depth> 40 cm) with drag reduction and lower soil disturbance characteristics. Advances in Engineering Software. 2018;**119**:30-37

[5] Fang Z, Zhang Y, Wu X, Sun L, Li S. New explicit correlations for the critical sticking velocity and restitution coefficient of small adhesive particles: A finite element study and validation. Journal of Aerosol Science. 2022;**160**: 105918

[6] Jagota V, Sethi APS, Kumar K. Finite element method: An overview. Walailak Journal of Science and Technology (WJST). 2013;**10**:1-8

[7] Sun Q, Peng W, Hai X, Yu S. Adhesion study between micron-scale graphite particles and rough walls using the finite element method. Advanced Powder Technology. 2021;**32**:1951-1962

[8] Ankem S, Margolin H. Finite element method (FEM) calculations of stress-strain behavior of alpha-beta Ti-Mn alloys: Part I. Stress-strain relations. Metallurgical Transactions A. 1982;**13**: 595-601

[9] Reddy JN. Introduction to the finite element method. McGraw-Hill Education; 2019

[10] Kabir M, Lovell MR, Higgs CF. Utilizing the explicit finite element method for studying granular flows. Tribology Letters. 2008;**29**:85-94

[11] Wagner G, Moës N, Liu W, Belytschko T. The extended finite element method for rigid particles in Stokes flow. International Journal for Numerical Methods in Engineering. 2001;**51**:293-313

[12] Krok A, Garcia-Trinanes P, Peciar M, Wu C-Y. Finite element analysis of thermomechanical behaviour of powders during tabletting. Chemical Engineering Research and Design. 2016;**110**:141-151

[13] Kh AB, Mirghasemi A, Mohammadi S. Numerical simulation of particle breakage of angular particles using combined DEM and FEM. Powder Technology. 2011;**205**:15-29

[14] Michael M, Vogel F, Peters B. DEM–FEM coupling simulations of the interactions between a tire tread and granular terrain. Computer Methods in Applied Mechanics and Engineering. 2015;**289**:227-248

[15] Rojek J, Oñate E. Multiscale analysis using a coupled discrete/finite element model. Interaction and Multiscale Mechanics. 2007;**1**:1-31

[16] Guo N, Zhao J. A coupled FEM/DEM approach for hierarchical multiscale modelling of granular media.

International Journal for Numerical Methods in Engineering. 2014;**99**: 789-818

[17] Onate E, Rojek J. Combination of discrete element and finite element methods for dynamic analysis of geomechanics problems. Computer Methods in Applied Mechanics and Engineering. 2004;**193**: 3087-3128

[18] Zárate F, Oñate E. A simple FEM–DEM technique for fracture prediction in materials and structures. Computational Particle Mechanics. 2015; **2**:301-314

[19] Munjiza A, John N. Mesh size sensitivity of the combined FEM/DEM fracture and fragmentation algorithms. Engineering Fracture Mechanics. 2002; **69**:281-295

[20] Azevedo NM, Lemos J. Hybrid discrete element/finite element method for fracture analysis. Computer Methods in Applied Mechanics and Engineering. 2006;**195**:4579-4593

[21] Argilaga A, Desrues J, Dal Pont S, Combe G, Caillerie D. FEM× DEM multiscale modeling: Model performance enhancement from Newton strategy to element loop parallelization. International Journal for Numerical Methods in Engineering. 2018;**114**:47-65

[22] Liu X, Gui N, Cui X, Yang X, Tu J, Jiang S. Discrete element-embedded finite element model for simulation of soft particle motion and deformation. Particuology. 2022;**68**:88-100

[23] Liu X, Gui N, Yang X, Tu J, Jiang S. A DEM-embedded finite element method for simulation of the transient heat conduction process in the pebble bed. Powder Technology. 2021a;**377**: 607-620

[24] Liu X, Gui N, Yang X, Tu J, Jiang S. A new discrete element-embedded finite element method for transient deformation, movement and heat transfer in packed bed. International Journal of Heat and Mass Transfer. 2021b;**165**:120714

Chapter 12

Utilizing Computational Methods to Identify Low GWP Working Fluids for ORC Systems

Zhiqiang Yang, Christophe Coquelet, Jiangtao Wu, Nian Tang and Jian Lu

Abstract

The Organic Rankine Cycle (ORC) power cycle is a well-established solution for harnessing heat sources to generate energy. Presently, ORC systems predominantly employ hydrochlorofluorocarbons (HCFCs) and hydrofluorocarbons (HFCs) as working fluids. However, these substances possess significant greenhouse gas effects and are slated for future bans. To address this, it is imperative to establish rational selection criteria and corresponding techniques for evaluating working fluids suitable for industrial ORC applications. This chapter presents the working fluid selection criteria and screening methods for environmentally friendly working fluids. The chapter is organized as follows: (1) The fundamentals of working fluids section provide a broad introduction to the core principles of working fluids; (2) the working fluid section outlines reasonable selection criteria for identifying potential alternatives; (3) the screening of ORC working fluids section discusses possible working fluid candidates, simulation approach, and thermodynamic models in detail, which is very important to access the thermodynamic performance of ORC cycle; and (4) an example of the simulation of an ORC for working fluid selection section demonstrates the strategy for the selection of a working fluid considering a defined ORC architecture.

Keywords: organic Rankine cycle, working fluid selection, process simulation, thermodynamic models, waste heat recovery

1. Introduction

According to data analysis, the energy utilization rate of developing countries is only 33%, which is 10–20% lower than that of developed countries [1]; and 60–65% of industrial energy consumption [2] is finally converted into low-grade waste heat with temperatures in the range 50–280°C, which has great potential for energy recovery. Utilizing the considerable number of undeveloped renewable waste heat is an effective means to alleviate energy shortages and achieve carbon neutrality.

The supercritical Rankine cycle, Kalina cycle, and trilateral flash cycle, and Organic Rankine Cycles (ORCs) are used for low-and medium-temperature waste heat recovery. Among them, ORCs are one of the most promising technologies for utilizing low-grade waste heat. The working fluid is the carrier for energy transfer and conversion in the ORC system. Under the same working conditions, the performance of ORC systems with different working fluids varies widely, thus, working fluid screening has always been an important direction and hotspot in ORC research. There is also the possibility, if the level of heat is low, to combine the production of electricity and heat (cogeneration); depending on the request, we can produce high quality heat using a heat exchanger (coupled with heat pump alimented by ORC for example).

In recent years, an increased scrutiny has been placed on the environmental aspect of working fluids, and regulatory pressures are driving global awareness of their impact on the environment. In 2016, the Kigali Amendment to the Montreal Protocol was adopted [3], whereby committed states must reduce the production of hydrochlorofluorocarbons (HCFCs) and hydrofluorocarbons (HFCs) according to a set schedule. The ozone depletion potential (ODP) and low global warming potential (GWP) are of particular interest, and emphasis has been placed on choosing a working fluid that demonstrates an ability to meet these climate protection initiatives. The currently used working fluids HFCs, HCFCs, and PFCs will not be evaluated because of their ODP and high GWP concerns, respectively [4]. The HVAC&R (heating, ventilation, air conditioning, and refrigeration) industry has to focus on the search for working fluids with zero (or extremely low) ODP and low GWP [5].

To find suitable low-GWP working fluids, reasonable selection criteria and corresponding knowledge for screening working fluids of ORC system should be provided for industrial applications. This chapter will introduce the working fluids selection criterion and screening methods for eco-friendly working fluids.

2. Fundamentals of the organic rankine cycle

The organic Rankine cycle is a cycle of thermodynamic processes, carried out by thermal machines with the objective to convert heat into mechanical work. It is based on four processes, and the mechanical work is extracted from the temperature difference between a heat source and a cooler, which is often water flow or ambient air.

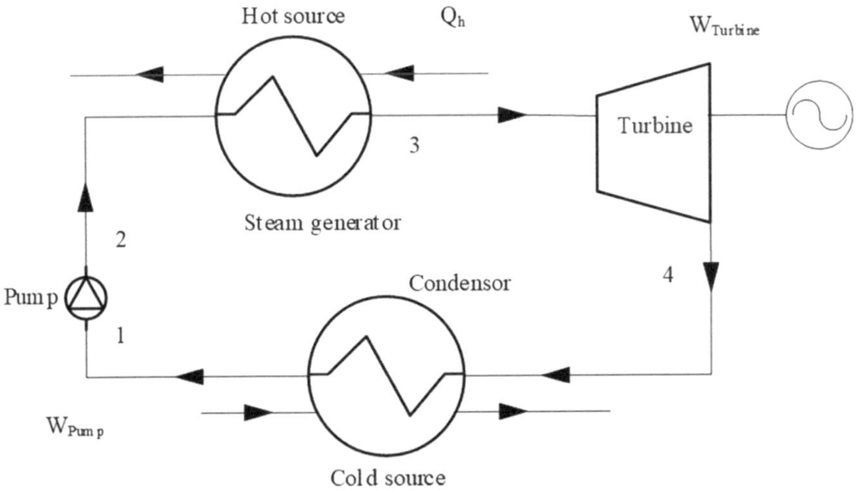

Figure 1.
Schematic of the Rankine cycle.

This type of cycle consists of a boiler/an evaporator, a condenser, a pump, and a turbine. **Figure 1** presents the cycle's elements as well as each numbered step. Then, **Figure 2** symbolizes the cycle (fluid) in a thermodynamic diagram of temperature-entropy (T-s diagram).

- Process 1–2: Adiabatic (or isentropic) pumping. The fluid has its pressure increased by the pump that, in an ideal case, should not allow any heat exchange with the environment. Therefore, the fluid has its temperature (and, consequently, its internal energy) increased.

- Process 2–3: Isobaric heat addition. The hot source transfers heat to the fluid, which in turn is vaporized, and reaches its maximum energy state in the cycle in 3. In the case of waste heat utilization, this corresponds to the energy needed to vaporize the pressurized fluid.

- Process 3–4: Isentropic expansion in a turbine. At this stage, the steam expands, and its temperature drops. Moreover, this is where the work is produced.

- Process 4–1: Isobaric heat rejection. Finally, the steam is cooled and condensed at constant pressure to the state of saturated liquid inside the condenser. The liquid then returns to the pump and the cycle starts again.

In the analysis, it is considered that the cycle operates in a steady-state, which means that all the state variables as well as the fluid flow do not change in time. Furthermore, the variation of the kinetic and the potential energy of the fluid are neglected. Thus, it is possible to calculate the efficiency linked to the ideal Rankine cycle by knowing the properties of the working fluid at different states/conditions. The ideal case occurs when the pump and turbine can guarantee isentropic conditions.

$$\eta = \frac{W_{turbine} - W_{pump}}{Q_h} \tag{1}$$

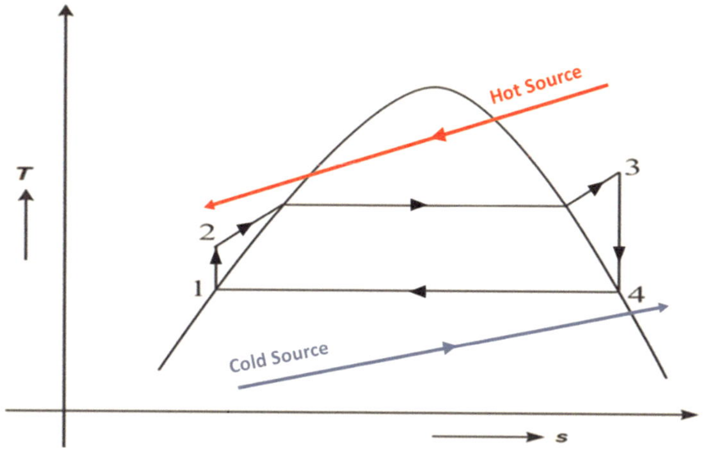

Figure 2.
Rankine cycle represented in the T-S diagram.

Eq. (1) states that the efficiency is equivalent to the difference between the work generated in the turbine and that needed to pressurize the fluid in the liquid state divided by the heat provided by the boiler. To find the value of these energies, it is necessary to apply the first law of thermodynamics in each cycle component (control volumes). In this case, the efficiency can be written in terms of enthalpy in each cycle point, as described in Eq. (2) below.

$$\eta = \frac{(h_3 - h_2) - (h_4 - h_1)}{h_3 - h_2} = 1 - \frac{h_4 - h_1}{h_3 - h_2} \tag{2}$$

There are some aspects that directly influence the efficiency of Rankine cycles. A detailed study of these aspects is carried out by Mclinden et al. [6]. Basically, they are related to the increase of the temperature difference between the heat supplied and the heat rejected. For this purpose, some modifications in the cycle architecture are proposed and developed with the objective of taking advantage of the increase in performance that occurs at higher pressures but avoiding high steam humidity levels at the end of the expansion. These solutions include the use of intermediate heat exchangers and/or modifications of the flow rate of the hot source, the system's energy source.

3. Working fluid

After establishing the parameters of the Rankine cycle, the next step is the careful study of the working fluid. The working fluid plays an essential role in the system of the organic Rankine cycle. Its properties are important to the system's efficiency, and they determine the operation conditions such as operating temperature and pressure. Moreover, given the deterioration of the environment, shortage of energy resources, and extreme climate change, its environmental impact and economic viability should be considered as well.

3.1 Working fluid classification

Above all, depending on the slope dT/ds in a T-s diagram of each fluid in the state of saturation vapor, as shown in **Figure 3** [7], the working fluids can be divided into three categories:

- Dry fluids with $\frac{dT}{ds} > 0$ (e.g., pentane)
- Isentropic fluids with $\frac{dT}{ds} = 0$ (e.g., R11)
- Wet fluids with $\frac{dT}{ds} < 0$ (e.g., water)

It is investigated that the dry and isentropic fluids are more appropriate in the application of ORCs because there would be much less formation of liquid droplets compared to the wet fluids in the two-phase region, which could reduce the equipment damage such as erosion in the turbine. Apart from this, to delete the risk of liquid droplets, a superheating process is widely recommended for the wet working

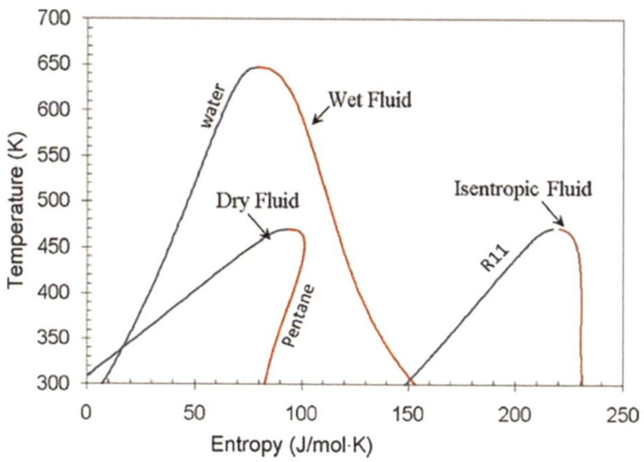

Figure 3.
Types of working fluids: Dry, isentropic, and wet.

fluids, at the expense of reduced efficiency and increased installation cost; whereas for dry fluids, superheat is unnecessary and can have an inverse impact on the ORC efficiency. In short, dry and isentropic fluids are always preferable candidates regarding operating and economic viability.

3.2 Criteria for fluid selection

When we chose the appropriate fluid candidates for the ORC application, the working fluids should be equipped with suitable chemical, thermodynamic, and transport properties and satisfy the requirements of environmental protection standards, cost, and safety.

3.2.1 Chemical property

Compared with the traditional steam Rankine cycle, organic fluids used in ORCs with high critical temperature, low triple point temperature, and low condensation temperature contribute notably to the recovery of heat from low temperature sources in terms of system performance [8].

3.2.1.1 Critical temperature

Critical temperature is considered as the primary criteria to select the working fluids by most studies. Liu et al. [8] indicated that the critical temperature should be as high as possible compared with the evaporation temperature in the application of a two-stage Rankine cycle in terms of ORC efficiency.

Some researchers [9] explored that for the pure working fluids, the hot source temperature should be 30–50°C higher than the critical temperature to gain the optimal thermodynamic performance, as for the fluids with similar critical temperature, a positive slope dT/ds of the saturation vapor line in T-s diagram is served as the second criteria.

3.2.1.2 Boiling temperature

Working fluids can be easy to handle in an ambient environment, thus, the boiling temperature is expected to be 02013–100°C.

3.2.1.3 Freezing point

The freezing point of the fluid must be lower than the lowest temperature of the cycle.

3.2.1.4 Triple point temperature

The triple point temperature should be as low as possible [7], and the condensation temperature should be below the lowest temperature of the ORC system, in case that the working fluids might freeze at any operating temperature in the cycle. Then, a condensation pressure slightly greater than the ambient pressure is preferred to prohibit air entering the ORC system [8]. In addition, the properties of organic fluids vary with operating conditions, and the thermal stability of working fluids need to be guaranteed at the range of working pressure and temperature for fear of chemical deterioration and decomposition at high temperature [10].

Material compatibility should be considered as well. Noncorrosive and non-eroding fluids are preferred for the ORC installation.

3.2.2 Thermodynamic property

The thermodynamic performance and economy of the ORC system are closely related to the physical properties of the working fluid, which will affect the efficiency, operating conditions, and equipment size of the ORC system. The thermophysical properties of working fluid mainly include the following:

3.2.2.1 Vaporization latent heat

In terms of installation size and therefore the ORC cost, fluids with high latent heat and high density are proposed to be the preferable candidates [11] because the higher the latent heat, the greater heat is absorbed in the evaporation step; but at the same time, the higher the density, the lower are the specific volume and volume flow rate. Both could mean that the system can provide more power outputs with the same equipment size. However, some researchers [12] have an inverse opinion on the latent heat of vaporization, given that better operating condition is provided at the turbine inlet for fluids with low latent heat in the saturated vapor state for the application of low-grade waste heat recovery.

3.2.2.2 Density

High vapor density is important, especially for fluids showing a very low condensing pressure (e.g., silicon oils). A low density leads to a higher volume flow rate: the pressure drops in the heat exchangers are increased, and the size of the expander must be increased. This has a non-negligible impact on the cost of the system.

3.2.2.3 Specific heat

Liquid's specific heat should be low so that it could decrease the work consumed by the pump and increase the work output indirectly. The working fluids with high liquid specific heats can absorb more heat from the hot source, which increases the output power of the ORC system.

3.2.2.4 Speed of sound

Moreover, organic fluids usually have a low speed of sound because of their large molecular weight, thus, the flowing rate is limited in the expander that largely influences the size of ORC turbines [13]; besides, it causes additional shock wave loss on account of the supersonic phenomenon appeared at the nozzle outlet of the expander.

3.2.2.5 Phase diagram

The thermodynamic property such as the phase diagram is also significant in the analysis and application of ORC.

3.2.3 Transport property

To decrease the installation cost and increase the space utilization, fluids with low dynamic viscosity and high thermal conductivity contribute to an enhancement of heat transfer and, in consequence, a reduction of heat exchanger size [13].

3.2.4 Environmental criteria

In general, there are three parameters that we need to consider in the aspect of environmental criteria: ozone depletion potential (ODP), global warming potential (GWP), and atmospheric lifetime (ALT).

ODP represents the degradation effect on the ozone layer in the atmosphere of each fluid, then GWP is a measure of the total energy that fluid gases can trap in the atmosphere relative to the heat absorbed by the same amount of carbon dioxide emission during a given time, whereas ALT represents the potential for the atmospheric accumulation of each chemical. The Montreal Protocol in 1987 proposed the phase out of ozone-depleting substances such as CFCs and HCFC, and the Kyoto Protocol proposed the reduction of greenhouse gases emissions, such as carbon dioxide, methane, nitrous oxide, hydrofluorocarbons (HFCs), perfluorocarbons (PFCs) and sulfur hexafluoride (SF_6). Generally, working fluids with zero ODP values, minimal GWP values, and low ALT values are recommended during the selection of working fluids [14].

3.2.5 Safety criteria

In choosing suitable working fluids, the potential operating dangerousness needs to be considered. According to the ASHRAE 34 safety classification standard, the fluids are classified into two toxicity groups A and B and into three flammability groups 1, 2, and 3, as shown in **Table 1** [14]. In general, fluids with nontoxic, nonflammable, and noncorrosive characteristics are preferable; however, in practice, it is hard to meet all the criteria so a relatively satisfactory fluid is sufficient during the selection.

	Class A	Class B	
Toxicity	No evidence	Evidence	
Concentration	< 400 ppm	< concentration limit	
	Class 1	**Class 2**	**Class 3**
Flammability	No flame propagation	Possible flame propagation	High flammability
Condition	In open air under normal condition	In open air under normal condition	Under any condition

Table 1.
Classification of fluids according to toxicity and flammability.

All in all, working fluids with high critical temperature, high density, high latent heat, low triple point temperature, low condensation temperature, condensation pressure slightly higher than the atmospheric pressure, zero ODP, low GWP, low ATL, and low risk of toxicity and flammability are expected in terms of system efficiency, economy, environmental protection, and safety.

4. Selection of ORC working fluids

4.1 Working fluid candidates for ORC

Several studies have been performed to find candidate molecules as alternative working fluids for refrigeration [15], heat pumping, and ORC applications [16]. Considering various criteria (particularly low toxicity, low flammability, low GWP, and high energy efficiency), only a few single-component compounds are suitable for many of the relevant applications. According to the research of Mclinden [6, 15], only a small part of elements in the periodic table would form compounds volatile enough to serve as working fluids. Molecules meeting the screening criteria for working fluids should be limited to small molecules less than or equal to 18 atoms, and the molecules comprise only the elements C, H, F, C1, Br, O, N, or S. Hundreds of substances can be considered as working fluid candidates for ORC application, including hydrofluoroethers (HFEs), hydrofluorocarbons (HFCs), ethers (HEs), unsaturated hydrofluoroolefins (HFO), hydrofluoroiodocarbons (HFIC), fluoroiodocarbons (FIC), fluorinated alcohols, and fluorinated ketones. **Table 2** exhibits pure working fluid candidates for the ORC system.

Category/Name	Product Name	ODP [c,d]	GWP [b,c]	P_c / MPa [e]	T_c /° C [e]	T_b/° C [e]	Safety classification [a]
Fluorinated ethers							
Pentafluorodimethylether	RE125	0	14,900	3.326	80.7	−34.55	—
Methyl-trifluroomethyl-ether	RE143a	0	756	3.635	104.77	−23.58	—
2-Difluoromethoxy-1,1,1-trifluoroethane	RE245fa2	0	659	3.433	171.73	29.25	—
Heptafluoropropyl-methyl-ether	RE347mcc	0	575	2.478	164.55	34.18	—

Category/Name	Product Name	ODP [c,d]	GWP [b,c]	P_c / MPa [e]	T_c /° C [e]	T_b/° C [e]	Safety classification [a]
Hydrocarbons							
Propane	R290	0	—	4.251	96.74	−42.11	A3
Isobutane	R600a	0	—	3.629	134.66	−11.74	A3
Butane	R600	0	—	3.796	151.98	−0.49	A3
pentane	R601	0	—	3.367	196.55	36.05	A3
Propene	R1270	0	—	4.555	91.06	−47.61	A3
Aromatics							
Toluene	C_7H_8	0	—	4.1263	318.6	110.6	—
Hydroflurocarbons (HFCs)							
Fluoroethane	R161	0	12	5.027	102.16	−37.7	—
Difluoromethane	R32	0	675	5.782	78.10	−51.65	A2
1,1-Difluoroethane	R152a	0	124	4.516	113.26	−24.02	A1
1,1,1,2-Tetrafluoroethane	R134a	0	1430	4.059	101.06	−26.07	A1
1,1,2,2-Tetrafluoroethane	R134	0	1100		118.59	−23	—
1,1,1,2,3,3,3-Heptafluoropropane	R227ea	0	3220	2.925	101.75	−16.34	A1
1,1,1,2,3,3,3-Heptafluoropropane	R263fa	0	9810	3.420	139.29	6.17	A1
1,1,1,3,3-Pentafluoropropane	R245fa	0	1030	3.651	153.86	15.04	B1
1,1,2,2,3-Pentafluoropropane	R245cb	0	—	3.940	174.42	25.26	—
1,1,1,3,3-Pentafluorobutane	R365mfc	0	794	3.266	186.85	40.19	—
Perfluorocarbons (PCFs)							
Hexafluoroethane	R116	0	12,200	19.88	3.048	−78.09	A1
Octafluoropropane	R218	0	8830	71.87	2.640	−36.79	A1
Octafluorocyclobutane	RC318	0	10,300	115.2	2.777	5.96	A1
Chlorofluorocarbons (CFCs)							
Trichlorofluoromethane	R11	1	4660	4.407	197.96	23.70	A1
Dichlorodifluoromethane	R12	0.82	10,200	4.136	111.97	−29.75	A1
Hydrochloroflurocarbons (HCFCs)							
Chlorodifluoromethane	R22	0.44	1760	4.990	96.14	−40.81	A1
1,1-Dichloro-1-fluoroethane	R141B	0.12	782	4.21	204	32.05	—
1-Chloro-1,1-difluoroethane	R142B	0.06	1980	4.06	137	−9.12	A2
Hydrofluoroolefins (HFOs)							
2,3,3,3-Tetrafluoropropene	R1234yf	0	4	3.382	94.7	−29.48	A2
E-1,3,3,3-tetrafluoropropene	R1234ze (E)	0	7	3.634	109.36	−18.97	A2L
Z-1,3,3,3-tetrafluoropropene	R1234ze (Z)	0	4	3.531	150.12	9.73	A2
3,3,3-trifluoropropene	R1243zf	0	1	3.518	103.78	−25.42	A2L

Category/Name	Product Name	ODP [c,d]	GWP [b,c]	P_c / MPa [e]	T_c /° C [e]	T_b/° C [e]	Safety classification [a]
E-1,2-dichloroethene	R1130(E)	0.00024	5	—	234.1	321	B2
Z-1,1,1,4,4,4-hexafluorobutene	R1336mzz (Z)	0	2.05	2.903	171.4	33.45	A1
E-1,1,1,4,4,4-hexafluorobutene	R1336mzz (E)	0	9	2.063	137.7	7.43	A1
E-1-chloro-3,3,3-trifluoropropenen	R1233zd (E)	0.00034	4.5	3.624	166.45	18.26	A1
Trifluoroemthane	R1311	0.01	1	—	123.3	7.4	A1
1,1-difluoroethene	R1132a	0	0.05	—	30.1	−20	A2
Z − 1-chloro-2,3,3,3-tetrafluoropropene	R1224yd (Z)	0.00023	0.88	—	155.5	14	A1
Alcohols							
Methanol	—	0	—	8.103	239.45	64.48	—
Ethanol	—	0	—	6.268	241.56	78.42	—
Ethers							
Dimethyl-ether	RE170	0	—	5.336	127.23	−24.78	—
Diethyl-ether	R610	0	—	3.64	193	34.55	—
Inorganics							
Ammonia	R717	0	—	1.133	132.41	−33.31	B2
Water	R718	0	—	2.206	373.95	100.00	—
Carbon dioxide	R744	0	—	7.377	3096	−78.46	A1

[a]*Reference [17].*[b]*Reference [18].*[c]*Reference [19].*[d]*Reference [20].*[e]*Reference [21].*

Table 2.
Basic properties of alternative working fluids.

What needs to be emphasized here is that the molecular screening of working fluids requires a trade-off between GWP, toxicity, flammability, and normal boiling point. Taking halogen compounds as an example (**Figure 4**), we can observe rules as follows:

- Increasing the number of Cl atoms in the molecule increases the boiling point of the molecule and increases the toxicity of the molecule.
- Increasing the number of H atoms in the molecule decreases the toxicity of the molecule, but the flammability also increases.
- Increasing the number of F atoms in the molecule increases the stability of the molecule, but the GWP value also increases.

More than 60 million single-component fluids have been screened in a comprehensive project to develop new low-GWP working fluids [22]. Many molecules meet

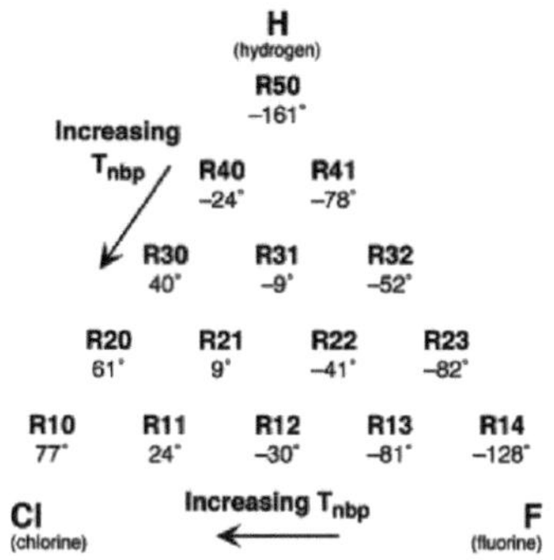

(a) Variation trend of boiling point of methane-family halogen compounds

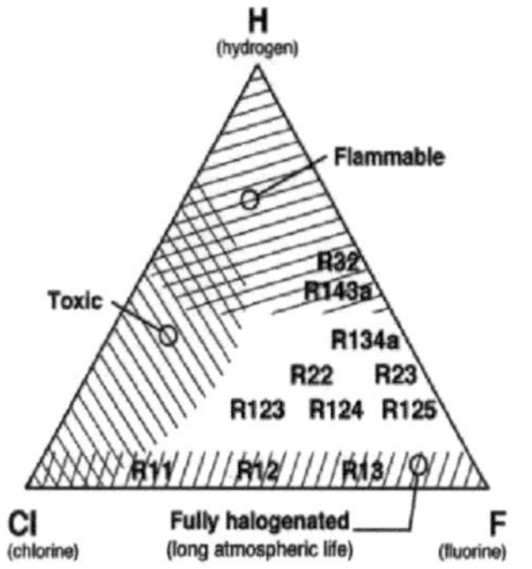

(b) Variation trend of toxicity and flammability of halogenated hydrocarbons

Figure 4.
Variation trend of boiling point, toxicity, flammability, and atmospheric lifetime for halogen compounds as a function of element [19]. (a) Variation trend of boiling point of methane-family halogen compounds. (b) Variation trend of toxicity and flammability of halogenated hydrocarbons.

the requirement of a suitable critical temperature, but the number of molecules decreases dramatically while satisfying both the requirement of low GWP. For each additional screening requirement, the number of optional molecules decreases dramatically. Only 138 molecules with an estimated critical temperature between 47 and 147°C and GWP below 1000 have been identified by Mclinden and co-workers [19]. Because the GWP became an important sector for screening suitable working fluids, some of HFOs and HCFOs have been considered as the optional fourth-generation eco-friendly working fluids for heat pumps and ORC systems [20, 23].

4.2 Assessment of cycle performance

For a given application (cold source temperature, hot source temperature, cycle structure, etc.), the screening of working fluids is basically such a process: first, build a steady-state simulation model of the ORC cycle under a given heat source and heat sink conditions, and then run this model with different working fluids.

4.2.1 Simulation approach

We propose an approach to the full modeling of an ORC system. We have only considered a system thermodynamic approach based on the thermodynamic properties of the working fluid. For doing this and as previously commented, some assumptions should be made to simplify the analysis, because geometric parameters of pump, turbine and heat exchangers are initially unknown and have to be designed. These assumptions are essential for different reasons: to ensure the functioning of the cycle according to the thermodynamics laws, avoiding, for example, at some point in the cycle, the fluid temperature exceeds the limits established by the external sources. Moreover, they can also allow a more precise selection of the best working fluid for each application, as it is possible to vary the used fluids and their evaporation and condensation pressures/temperatures. Finally, estimations such as pumping and expansion losses can be made to obtain results that are more similar to what is expected in real processes, in which the heat exchangers are not perfectly isolated from the environment; therefore, the heat exchangers that occur in the system are not considered perfectly reversible. Moreover, assumptions concerning pinch temperature can also be made to create a more realistic simulation of the ORC and the best comparison of the performance for each working fluid tested.

The proposed architecture for the analyzed cycle in this paper is shown in **Figure 5**.

It is observed that in this architecture, pre- and superheating zones are separated as well as pre- and supercooling zones to simplify the analysis of heating and cooling processes in boilers and condensers. The hypotheses that guide this cycle and make possible calculations for each state are the following:

- Perfect heat exchange in evaporator (2–3) and condenser (4–1), that is, the losses are negligible, and all heat recovered is absorbed by the working fluid.
- All heat additions between States 2 and 3 as well as all heat rejections between States 4 and 1 are isobaric.
- An isothermal compression efficiency $\eta_i p$ = 90% is assumed, and a mechanical efficiency $\eta_m p$ = 80% for the pump. Values are not standardized but are in a range commonly considered in literature.
- The working fluid leaves the state of saturated liquid in 2 and reaches the saturated vapor state in 3.
- An isothermal expansion efficiency $\eta_i t$ = 80% is assumed, and a mechanical efficiency $\eta_m t$ = 80% for the turbine. Values are not standardized but are in a range commonly considered in literature.
- The working fluid leaves the state of overheat vapor in 4 and reaches the saturated liquid state in 1.
- Pinch point temperature, which means the minimum difference between the working fluid temperature and the hot/cold source temperatures (that occurs when the fluid evaporates/condensates) $\Delta T_{evp}/\Delta T_{cnd}$ is 3 K. This value was based

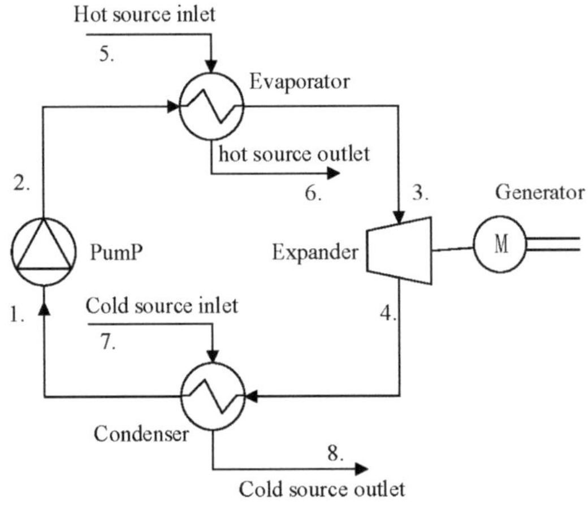

Figure 5.
Schema of the cycle used in the calculation.

on that described in [12], but the value equal to 5 K can also be considered because as the pinch point temperature decreases, the heat transfer effectiveness increases.

- The overheating temperature of the working fluid ΔT_{sup} is 3 K.

- The supercooling temperature of the working fluid ΔT_{sub} is 0.5 K [24].

In this example, we studied the application of waste heat from annealing furnaces, where the hot source is fixed as water at 363.15 K (90°C), within the temperature ranges mentioned in [25]. On the other side, the cold source is the cheapest and the most common substance: air, at 288.15 K (15°C) and 1 Atm (0.101 MPa). From this, the working fluid evaporation and condensation temperatures are adjusted to include this defined range. **Figure 6** shows a graph created to better illustrate all these cycle processes in the *T-s* diagram.

However, to calculate the power produced by the ORC and its efficiency, it is necessary to know the working fluid properties, such as enthalpy, entropy, and density in various states. These values are obtained from the NIST Reference Fluid Thermodynamic and Transport Properties Database (REFPROP) [26], a software whose library includes many substances (including organic compounds studied in this article) and which can be integrated with other calculation environments.

4.2.2 Thermodynamics models

Section 4.2.1 provides methods for calculating the heat and work quantities associated with various processes, but they are useless without knowledge of the thermodynamic and transport properties. Thermodynamic properties differ from one molecule to another, and the laws of thermodynamics themselves do not provide any description or model of material behavior.

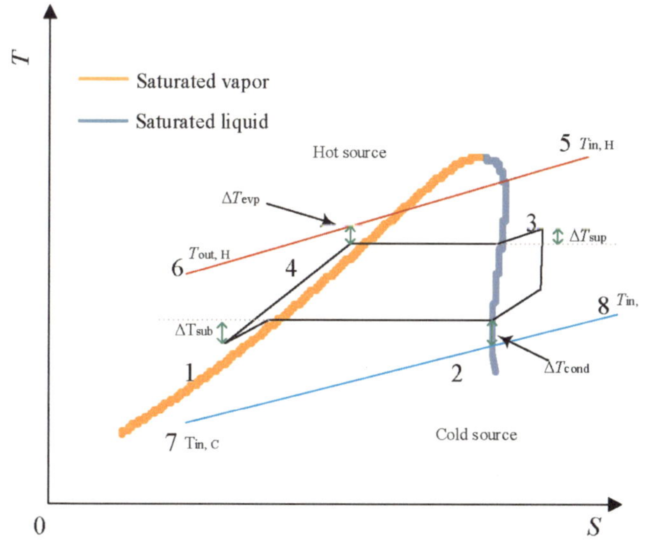

Figure 6.
T-s diagram of the modeled cycle.

An equation of state (EOS) is used to describe the thermodynamic state and the thermodynamic properties of a substance, and the functional formula is used to describe the relationship among pressure, temperature, and volume (pvT) for pure substances and mixtures. Using thermodynamic differential derivation, the thermodynamic properties and phase equilibrium of substances can be calculated from the EOS.

Equations of state can roughly be divided into four categories as follows:

1. Ideal-gas law

The ideal-gas law is the simplest EOS. It is based on two basic assumptions: (1) gas molecules do not occupy volume and are elastic molecules and (2) no interaction force between molecules. The ideal gas law can be written as Eq. (3), and the pvT relationship is obtained by molecular motion theory and statistical methods.

$$p = \frac{RT}{v} \tag{3}$$

where p, T, and v are pressure, temperature, and molar volume, respectively, and R is the gas constant.

In the low-density limit, all kinds of EOS reduce to the ideal gas law. Ideal gas law cannot be applied with liquid.

2. Virial equation of state

The virial equation of state (EOS) is a polynomial series in pressure or in inverse volume whose coefficients are functions only of T for a pure fluid.

$$Z = 1 + B\left(\frac{P}{RT}\right) + \left(C - B^2\right)\left(\frac{P}{RT}\right)^2 + \cdots \tag{4}$$

where the coefficients B and C are the second and third virial coefficients, respectively.

The expressions for the second virial coefficient can be obtained from the Tsonopoulos model [27], which is only applied to the saturated vapor because it cannot be used for liquids or at high pressures.

The Benedict-Webb-Rubin (BWR) EOS is an empirical extension of the virial EOS. The BWR expressions are based on the pioneering work of Benedict, Webb, and Rubin (1940 and 1942) who combined polynomials in temperature with power series and exponentials of density into an eight-parameter form [28]. Additional terms and parameters were later introduced by others to formulate modified Benedict-Webb-Rubin (MBWR) EOS.

The general form of BWR/MBWR correlations is

$$Z = 1 + f_1(T)/v + f_2(T)/v^2 + f_3(T)/v^n + f_4(T)\left[\left(\alpha + \gamma/v^2\right)/V^m\right]\exp(-\gamma/\mathrm{v}) \tag{5}$$

3. Cubic equation of state

The first practical cubic equation of state was proposed by van der Waals in 1873 [29]; the development of the modern cubic equation of state started from the Redlich Kwong (RK) equation published in 1949 [30], which is improvements to the van der

Waals equation. Based on the previous work, Peng and Robinson [31] proposed the Peng Robinson (PR) state equation in 1976, and its form is as follows:

$$p = \frac{RT}{v-b} - \frac{a(T)}{v(v+b)+b(v-b)} \tag{6}$$

The parameters a and b are related to the critical pressure, critical temperature, and acentric factor as shown in Eqs. (6) and (7).

$$a(T) = 0.457235\frac{R^2T_c^2\alpha(T)}{p_c} \tag{7}$$

$$b = 0.077796\frac{RT_c}{p_c} \tag{8}$$

$$\alpha(T) = \left[1 + \left(0.37464 + 1.54226\omega - 0.26992\omega^2\right)\left(1 - T_r^{0.5}\right)\right]^2 \tag{9}$$

where ω is the acentric factor, $T_r = T/T_c$, T and T_c represent reduced temperature and critical temperature.

In 1982, Patel and Teja [14] corrected the compressibility factor of the state equation to a parameter related to the substance and obtained the Patel Teja equation of state (PT EOS). The fluid properties of a certain component near the vapor-liquid critical point are difficult to obtain from cubic EOS. Modification of the Patel Teja EOS (mPT EOS) as proposed by Coquelet et al. [32] permits to better estimating the thermodynamic properties close to the critical point.

The RK, PR, PT, and mPT equations are representative of cubic EOS. This type of EOS is characterized by fewer parameters, simple form, and fast solution speed, and is widely used in industry. They differ in their accuracy to estimate the densities, particularly the liquid and supercritical fluid densities.

4. Fundamental equation of state based on Helmholtz energy

The Helmholtz energy-based equations are the most accurate EOS at a time when they were just coming into widespread use. The Helmholtz energy model was developed by NIST and included in the REFPROP software. Helmholtz-type equations of state (also called fundamental equations of state) are explained in terms of reduced molar Helmholtz free energy (see Eq. 10) wherein superscript ig concerns ideal gas contribution and superscript x concerns residual contribution. This equation contains several adjustable parameters α_i and α_k, t_k, d_k, and l_k. Experimental data are required to adjust these parameters. Therefore, numerous and accurate experimental data (phase equilibria and volumetric properties essentially) are required. In comparison to the previous cited equations, the fundamental equation of state guarantees a very accurate prediction of the thermodynamic properties at the condition of the availability and the quality of the experimental data.

$$\alpha(\delta,\tau) = \frac{A}{RT} = \alpha^{id} + \alpha^r + \sum N_k\delta^{d_k}\tau^{t_k} + \sum N_k\delta^{d_k}\tau^{t_k}\exp\left(-\delta^{l_k}\right) \tag{10}$$

Temperature and density are expressed in dimensionless variables $\delta = \frac{\rho}{\rho_C}$ and $\tau = \frac{T}{T_C}$. ρ_C and T_C are the critical density and temperature, respectively. This equation of state is parametrized for numerous working fluids that can be potentially used for ORC systems. These fluids can be pure components or mixtures. NIST REFPROP software has established a very accurate thermodynamics database for all the most commonly used working fluids based on the Helmholtz EOS. In addition, this model can be applied for the mixture. Mixing rules based on multifluid approximation are considered for the computation of the new critical properties (ρ_C and T_C) corresponding to the mixture. Binary interaction parameters have to be determined on the experimental base (phase equilibria and volumetric properties).

5. What can be one strategy for working fluid selection?

In the process of working fluid selection (**Figure 7**), the knowledge of thermodynamic properties is essential, but the step consisting of the utilization of virtual machine or process simulator software is also essential to screen some working fluids and reduce the list of potential candidates. For this essential step, the question of the thermodynamic model, predictive or not, and its accuracy is also essential. Once a list of fluids has been selected, it will be necessary to refine the thermodynamic model through data acquisition and adjustment of the parameters of the selected thermodynamic models [13].

The working fluid can be a pure fluid or a mixture. The advantage of a zeotropic mixture is its possibility to adapt the temperature glide to the variation of temperature of the heat or cold sources. With the adapted zeotropic mixture, we have the possibility to the irreversible low in heat exchanger. Zhang et al. [34] have proposed a very interesting analysis of the performance of an ORC in the context of ocean thermal energy conversion in 2022. Similar utilization of zeotropic working fluid is realized with the utilization of geothermal water as a heat source (Liu et al. [35]). It was observed that the utilization of the adapted zeotropic working fluid will lead to better matches of the temperature profiles of the working fluid and the heat source [13].

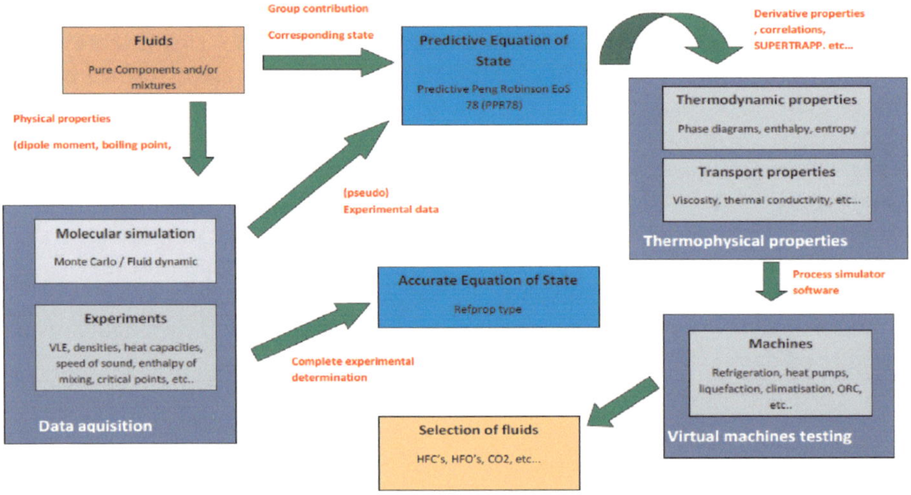

Figure 7.
A strategy for working fluid selection [33].

6. Example of simulation of an ORC for working fluid selection

This section illustrates the strategy presented on working fluid selection. We propose to discuss an example of selecting a fluid for utilizing available waste heat. In our case study, we have one heat source at the temperature of 120°C and one cold source at the temperature of 15°C.

We remind that the purpose of an ORC system is to generate power from low-temperature heat sources. Traditionally, water is used as the working fluid in Rankine cycle applications. However, because of its characteristics, water is not suitable for all applications, especially when the power of the heat source is low (unlike in coal, gas, or nuclear power stations). In a study by Dickes et al. [36] various case studies for ORC applications were presented, covering different power levels. The study discussed several working fluids, considering their impact on the environment (zero Ozone Depletion Potential [ODP] and lower Global Warming Potential [GWP]), as well as their inflammability and toxicity. To select the most suitable working fluid, a virtual machine can be used to simulate a simple ORC system. It is important to clearly define the main characteristics of the simple ORC system to conduct the simulation effectively.

1. Selection of a working fluid. In our case study, we have considered pure fluid. We will consider R1233zd(E) (*trans* 1- chloro 3,3,3-trifluoropropene), R600 (n-butane), R365mfc (1,1,1,3,3 -pentafluorobutane) and the reference fluid R245fa (1,1,1,3,3 pentafluoropropane). The properties of the selected working fluid are presented in **Table 3**.

2. The level of temperature of the heat source for the vapor generator (120°C) with a difference of temperature with the working of 5°C (fix pinch temperature).

3. The level of temperature of the cold source (condenser) is 15°C with a difference of temperature with the working of 5°C (fix pinch temperature).

4. The pump and turbine with isentropic compression of the liquid and expansion of gas, respectively. Isentropic efficiencies are 0.7 and 0.8, respectively.

Fluid	Full name	Chemical Formula	Family	T_C/ K	Pc/ bar	Acentric factor	Molar mass /g. mol^{-1}	GWP	Safety data
R245fa	1,1,1,3,3 pentafluoropropane	$C_3H_5F_3$	HFC	427	36.51	0.3776	134	1030	B1
R1233zd (E)	trans 1-chloro 3,3,3-trifluoropropene	C_3F_5Cl	HFO	440	36.24	0.3025	130.5	1	A1
R600	n-butane	C_4H_{10}	HC	425	37.96	0.2002	58.12	1	A3
R365mfc	1,1,1,3,3 - pentafluorobutane	$C_3H_5F_5$	HFC	460	32.66	0.3800	148.00	860	A2

Table 3.
Properties of the selected working fluids from NIST data base [26] and safety data issued from ASHRAE 34, 20,010 [17].

5. The mole flow is the same (1 kmol.h^{-1}).

6. The heat exchangers work at constant pressure. At the outlet of the evaporator, the working fluid is at its dew point. At the outlet of the condenser, the working fluid is at its bubble point.

7. At the outlet of the pump, we have a pre-heater connected to the regenerator located at the outlet of the turbine.

The simulation is realized using the Aspen plus software [37]. The Peng Robinson equation of state is considered for all the calculation. Four heater modules have been selected with heat flux connection between REGEN and PRE-HEAT (recuperation of heat). It can also be called Internal Heat Exchanger. GENV is the generator of vapor, and COND is the condenser. Normally, this process contributes to increased work output and so obtains a better overall efficiency. One TURBINE and one PUMP have also been considered.

The thermodynamic performance can be evaluated by considering two coefficients:

1. the energy efficiency given by the work produce over the heat absorbed by the cycle (Eq. 3)

$$\eta = \frac{W_P + W_T}{Q_{GENV}} = \frac{(h_2 - h_1) + (h_4 - h_3)}{h_1 - h_{4bis}} \tag{11}$$

2. the volumetric capacity (VC), which corresponds to the work produce per unit of working fluid. If you need an important work, energy, and if the volumetric capacity is low, you need a large flow of working fluid. This means you need to pay attention to the design and sizing of the pipe.

$$Vc = \rho_1 W_P + \rho_3 W_T = \rho_1(h_2 - h_1) + \rho_3(h_4 - h_3) \tag{12}$$

The results of each simulation are presented in **Tables 4-8**. **Table 8** presents the comparison of the performances.

Exergy efficiency can also be considered to compare the performance of the fluid (**Table 8**). This analysis is very useful for ORC: exergy can be considered as the quality of energy because it represents the useful part of energy. The definition is given by Eq. 13. T_0 is the ambient temperature and T_h is the temperature of working fluid at the outlet of the generator of vapor. For pure working fluid and for our case study, this criterion is not a very different form of energy efficiency, but it can be very

States (Figure 8)	1	2	2BIS	3	4	4BIS
Temperature/°C	110.0	61.0	20.0	20.0	20.9	49.0
Pressure /bar	15.9	1.3	1.3	1.3	15.9	15.9

Table 4.
Values of temperature and pressure for R245fa of each point of the cycle.

States (Figure 8)	1	2	2BIS	3	4	4BIS
Temperature/°C	110.0	70.3	20.0	20.0	20.4	59.5
Pressure /bar	7.0	0.5	0.5	0.5	7.0	7.0

Table 5.
Values of temperature and pressure for R365mfc of each point of the cycle.

States (Figure 8)	1	2	2BIS	3	4	4BIS
Temperature/°C	110.0	57.3	20.0	20.0	20.8	46.6
Pressure /bar	13.2	1.1	1.1	1.1	13.2	13.2

Table 6.
Values of temperature and pressure for R1233zd of each point of the cycle.

States (Figure 8)	1	2	2BIS	3	4	4BIS
Temperature/°C	110.0	59.1	20.0	20.0	21.3	49.5
Pressure /bar	18.9	2.1	2.1	2.1	18.9	18.9

Table 7.
Values of temperature and pressure for R600 of each point of the cycle.

Mass enthalpy /kJ.kg^{-1}	States (Figure 8)						η/ %	VC/ kJ. m^{-3}	Mass flow/ kg.h^{1}	η$_{EX}$/ %
	1	2	2BIS	3	4	4BIS				
R245fa	−8700.3	−8727.4	−8764.4	−8957.7	−8956.2	−8919.2	11.7	4557	134.05	49.7
R600	−2058.6	−2109.6	−2179.5	−2548.4	−2544.4	−2474.4	11.3	4902	58.12	48.0
R1233zd (E)	54.08	24.92	−5.94	−199.81	−198.48	−167.63	12.6	3754	130.50	53.4
R365mfc	−7722.39	−7755.78	−7805.78	−7989.19	−7988.47	−7938.48	15.1	2186	148.08	64.4

Table 8.
Cycle comparison between the 4 working fluids selected.

useful if a heating power is given (and not a temperature) or a zeotropic mixture is used and if the temperature of the heat or cold sources vary.

$$\eta_{EX} = \frac{\text{Exergy recovered}}{\text{Exergy consumed}} = \frac{W}{Q_h\left(1 - \frac{T_0}{T_h}\right)} \tag{13}$$

$1 - \frac{T_0}{T_h}$ is also called the Carnot's factor. Eqs. 11 and 13 are considered in our example (**Table 8**), and we can observe that the most efficient fluid is R365mfc (it will be same if exergy efficiency is calculated but the half of the energy that can be used by the steam generator is transformed into work). R365mfc has the highest critical temperature value. However, its volumetric capacity (VC) is very low, requiring a larger quantity of fluid (due to its low density compared to other fluids). Additionally, it is worth noting that the condenser pressure is lower than the atmospheric pressure,

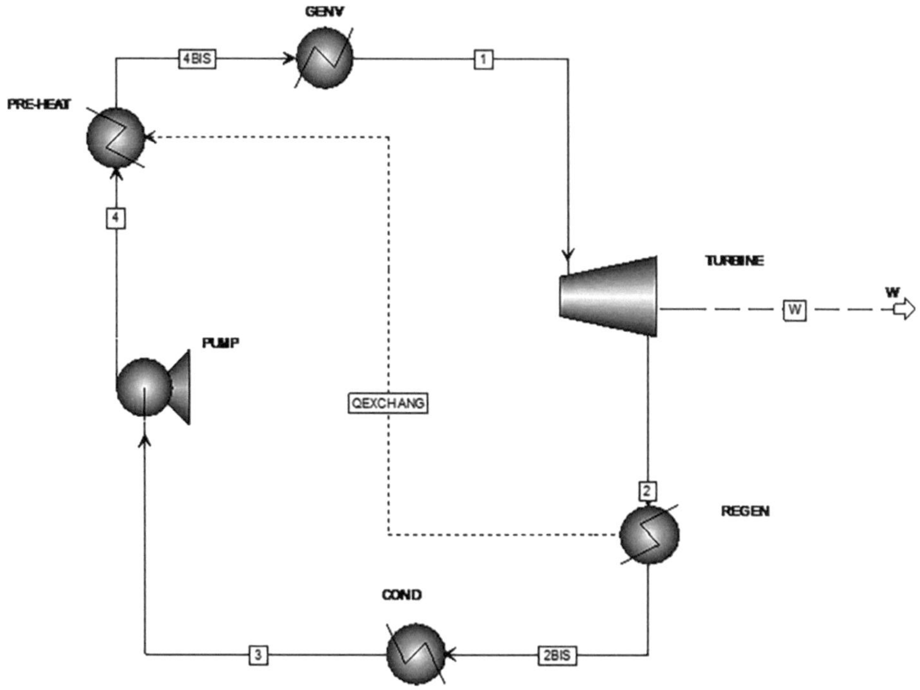

Figure 8.
Structure of the organic Rankine cycle using Aspen plus with recuperator (internal heat exchanger).

which is not recommended. Based on the ORC performance as well as environmental, hygiene, and safety criteria (EHS), it appears that the most suitable fluid for this example is R1233zd(E). The criteria related to installation volume, environmental protection, and operator safety should play an important role in future studies. For a complete selection of the fluid, it will be interesting to size the different equipment of the ORC and to evaluate if the sizing is normal in terms of the surface of heat exchangers or the speed of fluid in the turbine (subsonic flow) for example.

7. Conclusion and outlook

This book chapter proposes a virtual technique for the selection of low-GWP working fluids for the ORC system. The criteria for working fluids selection and simulation approach for assessing the cycle performance were introduced to the reads in detail. The selection criteria mainly include the chemical property (critical, boiling, and triple temperatures), thermodynamic property (Vaporization latent heat, density, heat capacity, phase equilibrium, etc.), transport property (dynamic viscosity and thermal conductivity), environmental property (low GWP and zero ODP), and safety (non-toxic, non-flammable). Depending on the application and the level of temperature of the heat source, no perfect fluid exists. It is important to proceed to a selection by considering the different characteristics of the sources and the application. In general, the differences between the cycle efficiencies of the working fluids are not very significant if the selections based on critical and boiling temperatures are correctly realized. For further investigations, it will be of interest to size the different components in the cycle. It concerns the design of heat exchangers or turbine or pump based on the estimation of transport properties of the working fluid. The thermodynamic performance of ORC systems can be improved by modifying its architecture using a regenerator as presented in our example or removing a part of the working fluid circulating in the turbine into the mixing heater. In consequence, the analysis of the performance of working fluid has to be remade. Moreover, the thermodynamic performance of ORC systems can be improved by using zeotropic mixtures as the working fluid. A zeotropic working fluid with a temperature glide of 5°C or greater can improve the efficiency and performance of the thermodynamic cycle by matching the temperature glide between the heat transfer fluid and the working fluid in a counterflow heat exchanger. This kind of matching, so-called temperature glide matching, can maintain the temperature difference between the heat exchange fluid and the working fluid in a constant, thereby reducing the irreversibility in the heat transfer process, improving performance and energy and exergy efficiency.

Conflict of interest

The authors declare no conflict of interest.

Author details

Zhiqiang Yang[1]*, Christophe Coquelet[2], Jiangtao Wu[3], Nian Tang[4] and Jian Lu[1]

1 State Key Laboratory of Fluorine and Nitrogen Chemicals, Xi'an Modern Chemistry Research Institute, Xi'an, China

2 Université de Toulouse, IMT Mines Albi, CNRS UMR 5302, France

3 Key Laboratory of Thermo-Fluid Science and Engineering, Ministry of Education, Xi'an Jiaotong University, Xi'an, China

4 Electric Power Research Institute of Guangdong Power Grid Co. Ltd, Guangzhou, China

*Address all correspondence to: zqyang@stu.xjtu.edu.cn

References

[1] Forman C, Muritala IK, Pardemann R, Meyer B. Estimating the global waste heat potential. Renewable and Sustainable Energy Reviews. 2016;**57**:1568-1579. DOI: 10.1016/j.rser.2015.12.192

[2] Cullen JM, Allwood JM. Theoretical efficiency limits for energy conversion devices. Energy. 2010;**35**:2059-2069. DOI: 10.1016/j.energy.2010.01.024

[3] UN Environment (UNEP), The Kigali Amendment to Montreal Protocol: HFC Phase-Down, OzonAction Fact Sheet, Paris, France. 2016

[4] Nair V. HFO refrigerants: A review of present status and future prospects. International Journal of Refrigeration. 2021;**122**:156-170. DOI: 10.1016/j.ijrefrig.2020.10.039

[5] Yang Z, Valtz A, Coquelet C, Wu J, Lu J. Experimental measurement and modelling of vapor-liquid equilibrium for 3,3,3- Trifluoropropene (R1243zf) and trans-1,3,3,3-Tetrafluoropropene (R1234ze(E)) binary system. International Journal of Refrigeration. 2020;**120**:137-149. DOI: 10.1016/j.ijrefrig.2020.08.016

[6] Mclinden MO, Kazakov AF, Brown JS, Domanski PA. A thermodynamic analysis of refrigerants: Possibilities and tradeoffs for low GWP refrigerants. International Journal of Refrigeration. 2014;**38**:80-92. DOI: 10.1016/j.ijrefrig.2013.09.032

[7] Chen H, Goswami DY, Stefanakos EK. A review of thermodynamic cycles and working fluids for the conversion of low-grade heat. Renewable and Sustainable Energy Reviews. 2010;**14**:3059-3067. DOI: 10.1016/j.rser.2010.07.006

[8] Liu B, David F, Riviere P, Coquelet C, Gicquel R. Air cooled two-stage Rankine cycles for large power plants operating with different working fluids: Performance, size and cost. In: American Society of Mechanical Engineers, Power Division (Publication) Power conference 2013. Vol. 56062. San Diego, California, USA; Nov 2013. p. V002T084001. DOI: 10.1115/POWER2013-98075

[9] Hærvig J, Sørensen K, Condra TJ. Guidelines for optimal selection of working fluid for an organic Rankine cycle in relation to waste heat recovery. Energy. 2016;**96**:592-602. DOI: 10.1016/j.energy.2015.12.098

[10] Quoilin S, Lemort V. Technological and economical survey of organic Rankine cycle systems. 5th European conference on Economics and management of energy in industry. 2009;**278**:12 http://orbi.ulg.ac.be/handle/2268/14609

[11] Maizza V, Maizza A. Working fluids in non-steady flows for waste energy recovery systems. Applied Thermal Engineering. 1996;**16**:579-590. DOI: 10.1016/1359-4311(95)00044-5

[12] Yamamoto T, Furuhata T, Arai N, Mori K. Design and testing of the organic rankine cycle. Energy. 2001;**26**: 239-251. DOI: 10.1016/S0360-5442(00)00063-3

[13] Coquelet C, Valtz A, Théveneau P. Experimental determination of Thermophysical properties of working fluids for ORC applications. ORC Waste Heat Recovery Applications. 2019;**32**: 137-144 https://www.intechopen.com/books/advanced-biometric-technologies/liveness-detection-in-biometrics

[14] Douvartzides S, Karmalis I. Working fluid selection for the organic Rankine

cycle (ORC) exhaust heat recovery of an internal combustion engine power plant. In: IOP Conference Series: Materials Science and Engineering. Vol. 161. 2016. p. 012087. DOI: 10.1088/1757-899X/161/1/012087

[15] Kazakov A, McLinden MO, Frenkel M. Computational design of new refrigerant fluids based on environmental, safety, and thermodynamic characteristics. Industrial and Engineering Chemistry Research. 2012;**51**:12537-12548. DOI: 10.1021/ie3016126

[16] Mateu-Royo C, Mota-Babiloni A, Navarro-Esbrí J, Peris B, Molés F, Amat-Albuixech M. Multi-objective optimization of a novel reversible high-temperature heat pump-organic Rankine cycle (HTHP-ORC) for industrial low-grade waste heat recovery. Energy Conversion and Management. 2019;**197**: 111908. DOI: 10.1016/j.enconman.2019.111908

[17] ASHRAE, American Standard ANSI/ASHRAE Standard 34. Designation and Safety Classification of Refrigerants. Atlanta, USA; 2010. ISSN: 1041-2336

[18] Available from: https://www.gov.uk/guidance/fluorinated-gases-f-gases consulted in 2023

[19] Available from: https://www.epa.gov/ozone-layer-protection/ozone-depleting-substances consulted in 2023

[20] Giménez-Prades P, Navarro-Esbrí J, Arpagaus C, Fernández-Moreno A, Mota-Babiloni A. Novel molecules as working fluids for refrigeration, heat pump and organic Rankine cycle systems. Renewable and Sustainable Energy Reviews. 2022;**167**:112549. DOI: 10.1016/j.rser.2022.112549

[21] Available from: https://webbook.nist.gov/chemistry/

[22] McLinden MO, Brown JS, Brignoli R, Kazakov AF, Domanski PA. Limited options for low-global-warming-potential refrigerants. Nature Communications. 2017;**8**:1-9. DOI: 10.1038/ncomms14476

[23] Bobbo S, Di Nicola G, Zilio C, Brown JS, Fedele L. Low GWP halocarbon refrigerants: A review of thermophysical properties. International Journal of Refrigeration. Elsevier Journal. 2018;**90**: 181-201. DOI: 10.1016/j.ijrefrig.2018.03.027

[24] Wei D, Lu X, Lu Z, Gu J. Performance analysis and optimization of organic Rankine cycle (ORC) for waste heat recovery. Energy Conversion and Management. 2007;**48**:1113-1119. DOI: 10.1016/j.enconman.2006.10.020l

[25] Johnson I, Choate ST, Davidson A. Waste Heat Recovery. Technology and Opportunities in us Industry. Laurel, MD (United States): BCS, Inc.; 2008

[26] Lemmon EW, Bell IH, Huber ML, McLinden MO. NIST Standard Reference Database 23: Reference Fluid Thermodynamic and Transport Properties-REFPROP, Version 10.0. Gaithersbu: National Institute of Standards and Technology, Stand. Ref. Data Progr; 2018. DOI: 10.18434/T4/1502528

[27] Tsonopoulos C. An empirical correlation of second virial coefficients. AICHE Journal. 1974;**20**:263-272. DOI: 10.1002/aic.690200209

[28] Reid RC, Sherwood TK, Street RE. The properties of gases and liquids. Physics Today. 1959;**12**:38-40. DOI: 10.1063/1.3060771

[29] Smith JM, Van Ness HC, Abbot MM, Swihart MT. Introduction to Chemical Engineering Thermodynamics. 8th ed. New York, USA: McGraw Hill; 2018

[30] Redlich O, Kwong JNS. On the thermodynamics of solutions. V. An equation of state. Fugacities of gaseous solutions. Chemical Reviews. 1949;**44**: 233-244. DOI: 10.1021/cr60137a013

[31] Peng DY, Robinson DB. A new two-constant equation of state. Industrial and Engineering Chemistry Fundamentals. 1976;**15**:59-64. DOI: 10.1021/i160057a011

[32] Coquelet C, El Abbadi J, Houriez C. Prediction of thermodynamic properties of refrigerant fluids with a new three-parameter cubic equation of state. International Journal of Refrigeration. 2016;**69**:418-436

[33] Coquelet C. Sur la route des propriétés thermophysiques des fluides frigorigènes. Revue générale du Froid. 2016;**1156**:46-51

[34] Zhang J, Zhang X, Zhang Z, Zhou P, Zhang Y, Yuan H. Performance improvement of ocean thermal energy conversion organic Rankine cycle under temperature glide effect. Energy. 2022; **246**:123440

[35] Liu Q, Duan Y, Yang Z. Effect of condensation temperature glide on the performance of organic Rankine cycles with zeotropic mixture working fluids. Applied Energy. 2014;**115**:394-404

[36] Dickes R, Dumont O, Guillaume L, Quoilin S, Lemort V. Charge-sensitive modelling of organic Rankine cycle power systems for off-design performance simulation. Applied Energy. 2018;**212**:1262-1281. DOI: 10.1016/j.apenergy.2018.01.004

[37] Al-Malah KIM. ASPEN PLUS: Chemical Engineering Applications. Hoboken, USA: John Wiley & Sons; 2022